Walter G. Guder, Sheshadri Narayanan (Eds.)

Pre-Examination Procedures in Laboratory Diagnostics

Pre-Examination Procedures in Laboratory Diagnostics

Preanalytical Aspects and their Impact on the Quality of Medical Laboratory Results

Edited by
Walter G. Guder and
Sheshadri Narayanan

DE GRUYTER

Editors
Prof. Dr. med. Walter G. Guder
Marianne-Plehn-Str. 4
81825 München
Germany
E-Mail: Walter.Guder@extern.lrz-muenchen.de

Prof. Dr. Sheshadri Narayanan
Weill Medical College/Cornell University
Dept. of Pathology & Laboratory Medicine
525 East 68th Street/F-715
New York NY 10021
USA
E-Mail: shn2011@med.cornell.edu

ISBN: 978-3-11-033165-3
e-ISBN (PDF): 978-3-11-033404-3
e-ISBN (EPUB): 978-3-11-039019-3

Library of Congress Cataloging-in-Publication data
A CIP catalog record for this book has been applied for at the Library of Congress.

Bibliographic information published by the Deutsche Nationalbibliothek
The Deutsche Nationalbibliothek lists this publication in the Deutsche Nationalbibliografie; detailed bibliographic data are available on the Internet at http://dnb.dnb.de.

Typesetting: Medienfabrik GmbH, Stuttgart
Cover image: Luchschen/iStock/Thinkstock
Printing and binding: CPI books GmbH, Leck
∞ Printed on acid-free paper
Printed in Germany

www.degruyter.com

Preface

Knowledge in laboratory medicine is advancing at an exponential pace.

Advances in terms of introduction of new technologies and new laboratory tests are being made in all branches of laboratory medicine.

Another dimension that has impacted on the practice of laboratory medicine in recent years is the awareness that analytical quality assurance is not sufficient to ensure safe and medically correct laboratory results since more than 70 % of the so called "laboratory errors" were found to be due to extra-analytical factors. This introduced the awareness of preanalytical errors affecting laboratory results in the laboratory diagnostic process which needed to be recognized.

These developments bring in its wake a host of analytical and preanalytical issues which need to be uncovered, understood and resolved. While analytical issues can be approached by way of quality control and standardization, the delineation of preanalytical issues requires substantially more effort. Recently this part of the diagnostic circle became part of accreditation and more recently quality assurance programs.

In conceiving this multi-authored book, our goal was to bring to our readers an up-to-date perspective on the preanalytical phase from experts in their respective disciplines. In doing this we have been selective and tailored this book to the expertise of the authors who have contributed to this volume. We thank all coauthors for their collaboration in this project. As such, and also due to space constraints we may not have covered all areas of interest to the reader, but we do hope the cited original work will help to find answers to the respective questions. After being invited by deGruyter to write this book, we would like to thank all collaborators, especially Mrs Julia Reindlmeier, Sabina Dabrowski and Katja Brockmann for their excellent collaboration and helpful suggestions.

In balance, we hope this book will be of interest not only to the practitioners of laboratory medicine, but also to the clinicians who order laboratory tests and use the results in the treatment of their patients, nurses, medical technologists, phlebotomists, medical and graduate students.

We and the contributing authors will feel gratified if this book reinforces the importance of the preanalytical phase in the practice of laboratory medicine and stimulates research in this field.

Walter G. Guder
Sheshadri Narayanan

Contents

List of Contributing Authors

Kristin Moberg Aakre
Laboratory of Clinical Biochemistry
Haukeland University Hospital
Bergen, Norway
e-mail: kristin.moberg.aakre@helse-bergen.no
Chapter 8.3

Dorothy M. Adcock-Funk
Colorado Coagulation
Englewood, CO, USA
e-mail: adcockd@labcorp.com
Chapter 4.5, 4.6

Yashpal Agrawal
Laboratory Medicine Consultant
Division of Aurora Diagnostics
Las Vegas , NV, USA
e-mail: yagrawal@lmclabs.com
Chapter 7.5

Johannes Aufenanger
Institute of Laboratory Medicine
Klinikum Ingolstadt
Ingolstadt, Germany
e-mail: johannes.aufenanger@
klinikum-ingolstadt.de
Chapter 1.2

Guiseppe Banfi
IRCCS Galeazzi
Milano, Italy
e-mail: banfi.giuseppe@unisr.it
Chapter 7.2

Raffick A.R. Bowen
Department Pathology
Stanford University
Stanford, CA, USA
e-mail: rbowen@stanfordmed.org
Chapter 4.5, 4.6

Janne Cadamuro
Universitätsinstitut für Medizinisch-Chemische
Labordiagnostik
Salzburger Landeskliniken
Salzburg, Austria
e-mail: j.cadamuro@salk.at
Chapter 1.2, 8.2

Gabriele Castelo-Rose
Greiner Bio-One GmbH
Kremsmünster, Austria
Chapter 5.1

Stephen Church
BD Diagnostics
Oxford, United Kingdom
e-mail: stephen_church@europe.bd.com
Chapter 5.3

Michael P. Cornes
Department Clinical Chemistry
New Cross Hospital
Wolverhampton, Great Britain
e-mail: michael.cornes@nhs.net
Chapter 2.3, 2.5

Rémy Couderc
Biochemistry Dept.
Armand Trousseau Hospital
Paris, France
e-mail: remy.couderc@trs.aphp.fr
Chapter 2.2

Joris Delanghe
Dept Clinical Chemistry
Gent University Hospital
Gent, Belgium
e-mail: joris.delanghe@ugent.be
Chapter 2.7

Carl-Erik Dempfle
Gerinnungspraxis
Mannheim, Germany
e-mail: mail@gerinnungspraxis-mannheim.de
Chapter 7.1

Arnold von Eckardstein
Institute of Clinical Chemistry
Universitäts-Spital Zürich
Zürich, Switzerland
e-mail: arnold.voneckardstein@usz.ch
Chapter 1.2

Thomas Ehrenfellner
Greiner Bio-One GmbH
Kremsmünster, Austria
Chapter 5.1

Friedrich da Fonseca-Wollheim
Berlin, Germany
e-mail: fonseca-wollheim@arcor.ce
Chapter 4.1, 4.3

Peter Findeisen
Institute of Clinical Chemistry
Universitätsmedizin Mannheim
Mannheim, Germany
e-mail: peter.findeisen@umm.de
Chapter 7.7

Manuela Födinger
Institute of Laboratory Diagnostics
Wiener Krankenanstaltenverbund
Wien, Austria
e-mail: manuela.foedinger@wienkav.at
Chapter 1.2

Walter G. Guder
München, Germany
e-mail: walter.guder@extern.lrz-muenchen.de
Chapter 1.1, 1.3, 1.4, 2.1, 2.3–2.8, 3.1, 3.2, 3.3, 4.1–4.3, 6.1, 7.2, 7.4, Annex.

Peter Hagemann
Zürich, Switzerland
e-mail: peter.hagemann@bluewin.ch
Chapter 2.1, 2.4

Georg Hoffmann
Verlag Trillium GmbH
Grafrath, Germany
e-mail: ghoffmann@trillium.de
Chapter 1.2

Walter Hofmann
Institute of Clinical Chemistry
Medizet des städtischen Klinikums
München Schwabing
München, Germany
e-mail: walter.hofmann@klinikum-muenchen.de
Chapter 1.2

Helene Ivanov
Greiner Bio-One GmbH
Global Product Mangement Preanalytics
Kremsmünster, Austria
e-mail: helene.ivanov@gbo.com
Chapter 5.1

Martha Kaeslin-Meyer
Laboratory Medicine
Kantonsspital Aarau
Aarau, Switzerland
e-mail: martha.kaeslinmeyer@ksa.ch
Chapter 1.2

Gunn B. B. Kristensen
The Norwegian EQA Program
NKK
Bergen, Norway
e-mail: gunn.kristensen@noklus.no
Chapter 8.3

Ann Helen Kristoffersen
Laboratory of Clinical Biochemistry
Haukeland University Hospital
Bergen, Norway
e-mail: ann.helen.kristoffersen@helse-bergen.no
Chapter 8.3

Alexander B. Leichtle
University Institute of Clinical Chemistry
Center of Laboratory Medicine Inselspital
Bern, Switzerland
e-mail: alexander.leichtle@insel.ch
Chapter 7.7

Alexander von Meyer
Laborkliniken Nordoberpfalz AG
Weiden, Germany
e-mail: alexander.vonmeyer@kliniken-nordoberpfalz.ag
Chapter 6.2

Sheshadri Narayanan
Department Pathology and Laboratory Medicine
Weill-Medical College of Cornell University:F–715
New York, NY, USA
e-mail: shn2011@med.cornell.edu
Chapter 2.6, 3.1–3.4, 6.1, 7.2, 7.4–7.6

Matthias Nauck
Institute of Clinical Chemistry and
Laboratory Medicine
Universitätsmedizin Greifswald
Greifswald, Germany
e-mail: matthias.nauck@uni-greifswald.de
Chapter 8.1

Nora Nikolac
University Department of Chemistry
University Hospital "Sestre Milosrdnice"
Zagreb, Croatia
e-mail: nora.nikolac@gmail.com
Chapter 4.2

Astrid Petersmann
Institute of Clinical Chemistry and
Laboratory Medicine
Universitätsmedizin Greifswald
Greifswald, Germany
e-mail: astrid.petersmann@uni-greifswald.de
Chapter 8.1

Jaqueline Präuer
Greiner Bio-One GmbH
Kremsmünster, Austria
e-mail: jaqueline.praeuer@gbo.com
Chapter 5.1

Sverre Sandberg
The Norwegian Quality Improvement of
Primary Care Laboratories (NOKLUS)
Bergen, Norway
e-mail: sverre.sandberg@Noklus.no
Chapter 8.3

Melanie Schimpl
Greiner Bio-One GmbH
Kremsmünster, Austria
e-mail: melanie.schimpl@gbo.com
Chapter 5.1

Kathrin Schlüter
Scientific Affairs
BD Diagnostics
Heidelberg, Germany
e-mail: kathrin_schlueter@europe.bd.com
Chapter 5.3, 8.1

York Michael Schmitt
Praxis für Laboratoriumsmedizin
Marienhospital Darmstadt
Darmstadt, Germany
e-mail: york.schmitt@laborpraxis-darmstadt.de
Chapter 4.1–4.3

Audrey N. Schuetz
Department Pathology and Laboratory Medicine
NY Presbyterian Hospital
New York, NY, USA
e-mail: ans9112@med.cornell.edu
Chapter 7.6

Christa Seipelt
Sarstedt
Nümbrecht, Germany
e-mail: marketing@sarstedt.com
Chapter 5.2

Ana-Maria Simundic
Institute of Clinical Chemistry
University Hospital Center "Sestre Milosrdnice",
Zagreb, Croatia
e-mail: am.simundic@gmail.com
Chapter 2.9, 7.3

Oswald Sonntag
Eichenau, Germany
e-mail: osonntag@sonntaggmbh.de
Chapter 4.4

Thomas Streichert
Institute of Clinical Chemistry
Universitäts-Klinikum Köln
Köln, Germany
e-mail: thomas.streichert@uk-koeln.de
Chapter 6.2

Gottfried Töpfer
Schoepstal, Germany
e-mail: dietoepfers@t-online.de
Chapter 4.1–4.3, 7.1

Nils Tryding
Ahus, Sweden
e-mail: nils@tryding.se
Chapter 4.4

Anne Vassault
Service de Biochimie Metabolique
Hopital Universitaire Necker Enfant Malades
Paris, France
e-mail: anneva@free.fr
Chapter 2.2

Claire Wiesner
Greiner Bio-One GmbH
Kremsmünster, Austria
Chapter 5.1

Hermann Wisser (†)
Korntal-Münsingen, Germany
Chapter 1.4

Donald S. Young
University of Pennsylvania Medical Center
Philadelphia, PA, USA
e-mail: donaldyo@mail.med.upenn.edu
Chapter 3.4

1. General Part

Walter G. Guder

1.1 Introduction and History of the Preanalytical Phase

1.1.1 Introduction

It is extremely important to obtain adequate sample, ensure its appropriate transport and maintain proper storage before medical laboratories can perform diagnostic analysis [1]. Thirtyfive years after the introduction of the term preanalytical phase [2], it is now appropriate to summarize the existing knowledge, terms and recommendations about all biological and technical aspects between the patient and laboratory, which need to be controlled in order to obtain accurate laboratory result.

Most important of the whole diagnostic process, but often not considered as part of the quality criteria, is the extralaboratory part of the total turnaround time (TTT or total TAT) from the medical question to the sample [3]. This has been shown to be the source of a majority of "laboratory errors" [4] and is included in the ISO/EN/DIN 15189 requirements for quality and competence (ISO 2007). It was only recently that quality indicators of the preanalytical phase were harmonized [5] (see Chapters 8.1–8.3).

Guidelines and diagnostic pathways helped to improve the selection of type of sample, time and site of sampling and also the definition of reference intervals for clinical interpretation of results (see Chapter 1.2). Various biological and technical aspects can either influence the laboratory result in vivo or interfere with the analytical procedure, thus changing the result and causing misinterpretations (see Chapter 1.3). These influences include age and gender differences, circadian rhythms as well as food, herbs and drinking habits (see Chapters 3.1–3.4). In addition, drugs can either change analyte concentrations by their biological actions or interfere with the analytical method used (see Chapter 4.4). Interferences can also be caused by secondary changes in sample color and turbidity by lipaemia, hemolysis or icterus (see Chapters 4.1–4.3). The site of sampling determines the result by biological variables (see Chapters 2.1–2.8). Biological influences can be the cause of differences between serum and plasma (see Chapter 2.6). Besides, anticoagulants, additives or contamination can also cause interferences.

The sampling is performed by varied professionals worldwide. The technical procedure including patient and sample identification, sampling techniques and storage and transport conditions are described in separate chapters (Chapters 5.1–5.3 and Chapter 6.1). An additional list provides information on the present state of knowledge about analyte stability in samples and their temperature dependence as well as stabilizing additives (see 10, Annex).

When a sample reaches the laboratory, several processes are mandatory before the requested analysis is performed (see Chapter 8.2). Individual aspects for the various analytes are discussed separately (on hemostaseology, hematology, acid–base state, biochemical and immunological analytes as well as therapeutic drug monitoring, microbiology and biobanking: see Chapters 7.1–7.7). Auditing and internal and external quality assurance covering all the facets of preanalytical phase are explained in Chapters 8.1–8.3, opening the discussion on unresolved questions [6].

Hopefully, this booklet contributes to the increasing awareness of the multiple aspects of laboratorial diagnostic function. Also, the editors wish to thank all the authors for their competent contributions; they have made this volume an updated component of the ongoing progress in medical treatment.

1.1.2 The long history prior to the appearance of the term preanalytical phase

Definition and present state of concepts of preanalytical phase that emerged would not have been possible without more than hundred years of history of the preanalytical phase preceding it. Since chemical and microscopic analysis of body fluids improved medical diagnosis, the type of sample, time of sampling as well as the preparation of patients have been part of the method description. In addition, many inventions and technical improvements contributed to the standards that we use today. Table 1.1.1 summarizes some of the preanalytical techniques and inventions, which although existing since many decenniums are still considered standardized prerequisites for preanalytical procedures.

Table 1.1.1: History of technical products and procedures involved in preanalytical phase

Procedure	Material	Time of Invention	Reference
Urine sampling	Urine collection vessel (matula)	Middle Ages	1, 7
Blood sampling	Blood collection tubes	Post- 1800	1
	Evacuated blood collection tubes	1949	8, 9
	Aspiration tube system	1975	10
	Clot activators	1975	
	Safety needles	1980s	
Anticoagulation	Citrate	19th	
	Oxalate	century	
	EDTA	1930s	11
	Heparinates	1930s	12
	Hirudin	1995	13
Plasma/serum separation	Serum or plasma separator	1998	14
Centrifugation	Centrifuge for urine sediment,	1892	15,1
	centrifuge for blood samples,	1890–1900	
	hand driven and electrically driven		
	Ultracentrifugation	1924	16

Fig. 1.1.1: Uroscopy showing the preanalytical (man carrying the vessel protecting the matula filled with urine), analytical (uroscopist looking through the urine) and postanalytical phases (hand pointing to the brain of the uroscopist and rolling papers waiting for the prophetic (!) interpretation. (From a wood cut of Steffen Arndes from Lübeck 1510–20; from [17]).

Among the first preanalytical materials documented is the special urine sample container, the "matula" (Fig. 1.1.1), which based on Galen's humoral theory allowed to separate hypostases (sediment) from sublimia (floating matter) and nubes (cloud).

As seen in Fig. 1.1.1 (circa 1500 AD), this matula was carried by the patient to the uroscopist in a matching cylindrical basket, which could be closed with a lid and carried with a handle. The basket protected urine from sunlight that would change the color or turbidity of urine. It was well known that without this protective basket urine was too unstable to get true informations about the color, consistency and contents.

Three hundred years later (post 1800 AD), only when newly developed chemical methods made the detection and quantitation of the constituents of body fluids possible, was blood used for chemical analysis. Initially, whole blood and/or serum, contained in blood tubes standing upright at room temperature for hours, were used as samples. The first use of a centrifuge was for separating urine sediment (Fig. 1.1.2). A hand-driven centrifuge rotor was used to spin two tubes with urine to examine sediment under microscope [15]. This remained a standardized procedure until electrically driven centrifuges with higher speed were available for blood centrifugation from 1920 onwards (although mentioned by Stenbeck in 1892 [15]). Ultimately, the invention of ultracentrifuge facilitated the separation of molecules according to molecular weight and specific weight [16].

Fig. 1.1.2: Hand driven centrifuge used to form urine sediment.

Anticoagulants to preserve samples were used in the second half of the 19th century (citrate and oxalate, later EDTA and heparinates and recently hirudin), while standardized EDTA plasma was only possible after the 1940s [11]. These anticoagulants were added to a vessel before letting the blood flow into it and could only be roughly quantitated. Only with the invention of vacuum tubes and aspiration tubes [9, 10] standard volumes and anticoagulant concentrations were achieved. Years later, serum or plasma separators and additives stimulating coagulation in plastic tubes were introduced.

It is well known among clinical scientists that preparation of the patient for sampling (diet, prolonged fasting prior to sampling, posture before and during sampling, physical activity, etc.), time and site of sampling (early morning versus afternoon, venous or capillary sample, etc.), choice of anticoagulant (serum or citrated-, EDTA- or heparinized blood, etc.), transport and storage (whole blood or serum, room temperature or cold, etc.) and centrifugation time and temperature exert their influence on laboratory results [18, 19]. However, the influences of these factors were not quantitated. Because their impact on the analytical results could not be separated from the analytical variability, their contribution to the results remained largely unknown. Trying to define the possible causes of unexpected results stimulated discussions on the possible mechanisms involved.

In 1977, based on the observations of Bürgi from Switzerland [20], we defined the influence factors as biological variables changing the concentration of the analyte in the matrix analyzed. Interferences on the other hand, were defined as factors which are different from the analyte intended to be measured altering the result [21, 22]. The definition of each factor seemed important for reasons beyond its theoretical importance. Interferences, however, are method dependent and can in many cases be reduced or even eliminated by changes in the analytical procedure [22]. Thus, drug interferences were reduced by specific reagents and analytical procedures [23, 24], while the effect of serum color changes as appearing in hemolytic, turbid and icteric samples could be reduced by changing reference wavelengths or time or mechanism (kinetic) of colorimetric analysis [25, 26]. The effect of influence factors on the other hand can be reduced only by standardization of the preanalytical processes. These were part of the old recommendations on the preparation of patient (fasting prior to sampling, posture before and during sampling, etc.) [18].

After statistical quality assurance of the analytical procedures was introduced in the late 60s and early 70s [27, 28], it became apparent that in addition to the analytical errors, other major variables need to be controlled in order to obtain accurate laboratory results [22]. Only in 2002, Bonini et al quantitated the contribution of the extraanalytical variables on total laboratory error [4].

1.1.3 The term preanalytical phase was coined

In 1977 the term "preanalytical factors" was used by Statland and Winkel for variables influencing the result before sampling [2]. In the 1980s the terms "influence and interference factors" were included into the educational and professional programs [29–31]. For the first time Einer and Zawta published a book exclusively on preanalytical variables [32]. The terms influence and interference factors became part of the terminology of laboratory sciences [33] and national as well as international standards [e. g. 6, 34]. The NCCLS (Now CLSI) preanalytical standards introduced in 1981 were partially followed by respective European Standards (ECCLS). After the analytical process was redefined as examination procedure, the term "preanalytical procedures" was changed into "preexamination procedures" [6].

Subsequent to these measures, the term preanalytical phase was included in textbooks [35] and other teaching manuals of laboratory medicine [36, 37]. In 2002, Bonini et al published that preanalytical errors make up more than 60 % of errors in laboratory medicine [4]. Ricos et al [38] helped to define biological variation for each analyte as a basis for defining preanalytical goals. Around this time, many national quality assurance programs were initiated. In six meetings on preanalytical variables, organized as satellites of European or international meetings [39–44], various activities and results related to the preanalytical phase were exchanged and intensively discussed. In addition, WHO published the recommendations on sample type and stability in 2002 [45], which have appeared in printed version in several languages in several editions [46] (e. g. 7th edition in German 2009, 3rd edition in English 2010, App–Version in 2013).

This increasing importance and awareness of preanalytical variables led to the first and now second European Conference on preanalytical phase [47–49].

It is to be hoped that the awareness about the importance of preanalytical phase on the quality of medical laboratory results leads to a broader implementation of preanalytical aspects into national and international quality assurance programs that may help to significantly reduce the proportion of the often underestimated portion of errors in laboratory results.

1.1.4 Conclusions and future aspects

Although the term preanalytical phase seems rather novel, the procedures and variables of the preanalytical phase remain a crucial component of the diagnostic laboratory process. The introduction of quality assurance in the analytical process in the 1960s together with a perceived barrier in linking the collection and sampling of specimens with testing in the laboratory, have created a state of unawareness about the many variables that influence the final laboratory result. Prior to the emergence of the concept of the preanalytical phase, results not in agreement with the patient's condition were sometimes not accepted by the clinician, and resultantly defined as laboratory errors. Laboratorians and clinicians have realized that preanalytical variables may be the cause of "unsuitable" results. Thus it is important to define the evidence of each ordered test [50], as was recently documented in diagnostic pathways recommendations [51]. In future, we can look forward to the introduction of quantitative quality programs that include various preanalytical variables in the routine estimation of trueness of medical laboratory results. This can be accomplished on the basis of international and national quality assurance programs [6]. These approaches may include criteria for the rejection of tests ordered as either being unnecessary or preanalytically inappropriate [52].

References

[1] Senses, Sensors and Systems. A journey through the history of laboratory diagnosis. Editiones Roche 2004.
[2] Statland BE, Winkel P. Effects of preanalytical factors on the intraindividual variation of analytes in the blood of healthy subjects: consideration of preparation of the subject and time of venipuncture. Crit Rev Clin Lab Sci 1977; 8:105–44.
[3] Lundberg GD. Critical (panic) value notification: an established laboratory practice policy (parameter). J Am Med Ass 1990; 263:709.
[4] Bonini PA, Plebani M, Ceriotti F, Rubboli F. Errors in laboratory medicine. Clin Chem 2002; 48:691–8.
[5] Plebani M, Sciacovelli L, Aita A, Chiozza ML. Harmonization of pre-analytcal quality indicators. Biochem Med 2014; 24:105–13.
[6] ISO (International Organization for Standardization), European Commiteee for Standardization (EN), Deutsches Institut für Normung (DIN) 15189. Medical Laboratories – Particular Requirements for Quality and Competence. Geneva ISO 2007.

[7] Fogazzi GB, Ponticelli C, Ritz E. The Urinary Sediment, an integrated view. 2nd edn Oxford: Oxford University Press 1999.

[8] Kleiner J. Blood collection apparatus. US patent No 2,460,641; august 1945.

[9] Bush V, Cohen R. The evolution of evacuted blood collection tubes. Lab med 2003; 34:304–10.

[10] Sarstedt international. 50 years of Sarstedt 1961–2011. 2011 Nümbrecht Sarstedt AG.

[11] Banfi G, m Salvagno GL, Lippi G. The role of ethylene-diamine tetraacetic acid (EDTA) as in vitro anticoagulant for diagnostic purposes. Clin Chem Lab Med 2007; 45:565–76.

[12] Barrowcliffe TW. History of Heparin in Lever R, Mulloy B, Page CP. Heparin - A Century of Progress. Handbook of Experimental Pharmacology 2012; 207:3–22.

[13] Mennssen HD, Melber K, Brandt N, Thiel E. The use of hirudin as universal anticoagulant in haematology, clinical chemistry and blood grouping. Clin Chem Lab Med 2001; 39:1267–77.

[14] Coleman CM. Separation float for blood collection tubes with water swellable material. April 7, 1998; US Patent 5,736,033.

[15] Stenbeck T. Eine neue Methode für die mikroskopische Untersuchung der geformten Bestandteile des harns und einiger anderen Secrete und Excrete. Z Klin Med 1892; 20:457–75.

[16] Svedberg T. Sedimentation constants, molecular weights, and isoelectric points oft he respiratory proteins. J Biol Chem 1933; 103:311–25.

[17] Guder WG, Kutter D, Bonini P. From uroscopy to molecular analysis – improving diagnostic information from urine analysis. Clin Chim Acta 2000; 297:1–3.

[18] Peters JP, Van Slyke DD. Quantitative Clinical Chemistry. Interpretations Vol 1. Baltimore: Williams and Wilkins 1946.

[19] Richterich R. Klinische Chemie. Theorie und Praxis. Basel: S Karger 1964.

[20] Bürgi W, Richterich R, Mittelholzer ML. Der Einfluss der Enteiweißung auf die Resultate von Serum- und Plasma-Analysen. Klin Wschr 1967; 45:83–6.

[21] Guder WG. Einflussgrößen und Störfaktoren bei klinisch chemischen Untersuchungen. Internist 1980; 21:533–42.

[22] Keller H, Guder WG, Hansert E, Stamm D. Biological influence factors and interference factors in clinical chemistry: general considerations. Clin Chem Clin Biochem 1985; 23:3–6.

[23] Young D. Effects of Drugs on Clinical Laboratory Tests, Washington: AACC Press, 1990.

[24] Tryding N, Tufvesson C, Sonntag O. Drug Effects in Clinical Chemistry. Analytical Interferences and Biological Effects of Drugs on Biochemical and Haematological Laboratory Investigations. 7th edn Stockholm: AB Realtyck 1996.

[25] Guder WG, da Fonseca -Wollheim F, Heil W, Schmitt YM, Töpfer G, Wisser H, Zawta B. The haemolytic, icteric and lipemic sample. Recommendations regarding their recognition and prevention of clinically relevant interferences. J Lab Med 2000; 24:357–64.

[26] Sonntag O. Haemolysis as an interference factor in clinical chemistry. J Clin Chem Clin Biochem 1986; 24:127–39.

[27] Büttner J. Statistische Qualitätskontrolle in der Klinischen Chemie. Z Klin Chem 1967; 5:41–8.

[28] Fraser CG, Hyltoft-Petersen P. Desirable standards for laboratory tests if they are to fulfill medical needs. Clin Chem 1993; 39:1447–55.

[29] Guder WG, Narayanan S, Wisser H, Zawta B. Samples: From the Patient to the Laboratory. The Impact of Preanalytical Variables on the Quality of Laboratory Results. Darmstadt, GIT-Verlag 1996. Diagnostic Samples: From the Patient to the Laboratory. 4th updated edn Weinheim: Wiley-Blackwell 2009.

[30] Young DS Effects of preanalytical variables on clinical laboratory tests. Washington, AACC Press 1993.

[31] Guder WG, Hagemann P, Wisser H, Zawta B. Fokus Patientenprobe; Kompendium Präanalytik. CD-Rom; Heidelberg, Schwechat, Basel BD 2007.

[32] Einer G, Zawta B. Präanalytikfibel. Kooperation von Arzt und Labor. Leipzig Johann Ambrosius, Barth 1987.

[33] Dybkaer R. Vocabulary for use in measurement procedures and description of reference material in laboratory medicine. Eur J Clin Chem Clin Biochem 1997; 40:1416–20.
[34] Clinical Laboratory Standards Institute (CLSI), Collection, Transport and Processing of Blood Specimen for Testing Plasma-based Coagulation Assays. Approved standard 6th edn 2004. Wayne, PA:CLSI, Document H4-A5.
[35] Guder WG, Wahlefeld A-W. Specimens and samples in clinical laboratory; the preanalytical phase. In Bergmeyer HU. Methods of Enzymatic Analysis 3rd edn 1983. Weinheim Verlag Chemie Vol II, pp. 2–20.
[36] Wisser H. Präanalytik in Greiling, Gressner Lehrbuch der Klinischen Chemie und Pathobiochemie. 1987. Stuttgart New York, Schattauer, pp. 37–49.
[37] Hagemann P. Preanalytical Phase, in Thomas L. Laboratory and Diagnosis. 2005. Frankfurt TH-Books, pp. 1965–74.
[38] Ricos C, Alvarez V, Cava F, Garcia-Lario JV, Hernandez A, Jiminez CV et al. Current databases on biological variation: pros et cons and progress. Scand J Clin Lab Invest 1999; 59:491–500.
[39] The Impact of the Preanalytical Phase on the Quality of Laboratory Results. 1. European Meeting on the Pre-Analytical Phase in Laboratory Sciences. Tampere Finland July 1–2. 1995. Meylan: France; Becton Dickinson Vacutainer Systems.
[40] The Impact of the Pre-Analytical Phase on the Quality of Laboratory Results. 2. Symposium Oxford July 5-7 1996. Proceedings, Meylan: Becton Dickinson Vacutainer Systems.
[41] The Impact of the Pre-Analytical Phase on the Quality of Laboratory Results in Haemostasis.1st Symposium. Montpellier October 14. 1996. Proceedings, Meylan: Becton-Dickinson Vacutainer Systems.
[42] The Impact of the Preanalytical Phase on the Quality of Laboratory Results. Proceedings, 3rd Symposium; European expert meeting; FESCC Educational Program. Basel August 15–16. 1997. Meylan: Becton Dickinson Vacutainer Systems.
[43] Biological and Technical Aspects of the Preanalytical Phase in Clinical Chemistry and Laboratory Medicine. Fourth Expert Symposium. Karlovy Vary, Sept. 24–25. 1998. Proceedings, Meylan: Becton-Dickinson Vacutainer Systems Europe.
[44] Pre-Analytical Phase on the Quality of Laboratory Results, a satellite symposium of the IFCC 2002, Kyoto; Tokyo, Leuwen: Terumo Corp Europe.
[45] World Health Organization. Use of Anticoagulants in Diagnostic Laboratory Investigations and Stability of blood, plasma and serum samples. 2002 Geneva, Svitzerland: WHO and WHO/DIL/LAB/99.1 Rev.2.
[46] Guder WG, Ehret W, da Fonseca-Wollheim F, Heil W, Müller Plathe O, Töpfer G, Wisser H, Zawta B. Die Qualität diagnostischer Proben. 1. Auflage Deutsche Gesellschaft für Klinische Chemie und Deutsche Gesellschaft für Laboratoriumsmedizin 1999 Heidelberg Becton Dickinson GmbH.
[47] European Federation of Clinical Chemistry and Laboratory Medicine (EFLM). 1st EFLM-BD European conference on preanalytical phase. Padua 2012.
[48] European Federation of Clinical Chemistry and Laboratory Medicine (EFLM). 2nd EFLM-BD European conference on preanalytical phase. Preanalytical quality improvement – in quality we trust. Biochem Med 2013; 23: A1-A55.
[49] Quality of the Preanalytical Phase in Europe. Satellite meeting of the Euromedlab Prague, Czech Republic, May 26th 2001. Leuwen: Terumo Europe.
[50] Price CP, Christenson RH. Evidence based Laboaratory Medicine. 2003 Washington AACC Press.
[51] Hofmann W, Ehrich JHH, Guder WG, Keller F, Scherberich J. Diagnostic pathways for exclusion and diagnosis of kidney diseases. Clin Lab 2012; 58:871–89.
[52] Guder WG, Müller OA. Unnecessary laboratory tests (Unnötige Laboruntersuchungen). D Med Wchschr. 2009; 134:575–84.

Georg Hoffmann, Johannes Aufenanger, Manuela Födinger, Janne Cadamuro, Arnold von Eckardstein, Martha Kaeslin-Meyer, Walter Hofmann

1.2 Requesting Laboratory Tests: Benefits and Limitations of Laboratory Diagnostic Pathways

1.2.1 Introduction

Around the year 2000, several hospitals in Germany and other European countries started using industrial management strategies, in order to improve their competitiveness in times of increasing cost pressure. Originally, these models and methods were developed in Anglo-Saxon countries, where "clinical pathways" for optimization and control of clinical business processes have been in use since the 1990s [1, 2]. The fact that Germany lagged behind in the introduction of these strategies was at least in part due to the delay in adapting diagnosis-related groups (DRGs) by decades as compared to the USA, Canada, UK and Australia [3].

Clinical pathways have been established in practically all countries with DRG-based reimbursement systems. They represent a crucial ingredient of clinical workflow management systems (WMS) [4] and are a more or less logical consequence of the fact that fixed prices require standardized processes; otherwise costs tend to get out of control. This is especially true for diagnostic processes, since they represent the very beginning of most hospital stays. Errors made at the start may entail many consecutive mistakes, and wrong diagnoses may lead to insufficient therapies. Although these remarks may seem self-evident, "diagnostic pathways" attracted much less attention in the past than clinical "care pathways", probably because treatment is more expensive than diagnostics.

In order to put more emphasis on diagnostic issues in the context of clinical pathways, the German Association for Clinical Chemistry and Laboratory Medicine [DGKL, 5] started an initiative in 2006, aiming to define the specific rules for the implementation of laboratory diagnostic pathways as a subset of clinical pathways. This activity led to the publication of a handbook for laboratory diagnostic pathways in 2012, which recently appeared in its second edition [6]. It includes more than 80 graphs, presenting pathway examples for various medical fields, such as metabolism, hematology and immunology, to name a few. In 2012, laboratory representatives of the German-speaking countries Austria, Switzerland and Liechtenstein joined this initiative. To our knowledge, this is the first international group dedicated especially to laboratory diagnostics pathways. A comparable initiative exists in Australia for diagnostic imaging pathways [7].

1.2.2 Definitions

A laboratory diagnostic pathway describes the entire process from the initial medical question to the final result [6]. It encompasses the right tests for defined questions (WHAT) with a professional reasoning (WHY) and – if necessary – with a time label (WHEN). In contrast to clinical pathways, where the correct timing of all actions needs to be guaranteed by humans, laboratory tests (biomarkers) can often be sequentially analyzed without human interaction, just using information technology and analytical automation systems. Exceptions are follow-up tests (e. g. drug monitoring) or certain cases, where an unexpected laboratory result needs a confirmation test or additional sample material.

Diagnostic pathways combine the well-known principle of stepwise reflex and reflective testing [8] with a management concept that helps to fulfill medical needs with organizational and economic efficacy [9]. In essence, diagnostic pathways are "smart" test profiles, which – in contrast to conventional inflexible laboratory profiles – are not necessarily worked off completely, but just to a point, where the diagnostic decision can be made. Formally, such pathways may be represented either in a rule-format (if...then...else...) or by graphical decision trees (so-called "algorithms"). Diagnostic pathways play an essential role in the beginning of diagnostics, notably to rule out frequent causes of acute and chronic syndromes such as diabetes mellitus (Fig. 1.2.1). They usually follow international guidelines, for example those of the American Diabetes Association [10], and are implemented as standard operating procedures (SOPs) in the clinical setting.

Another beneficial field of application is the differential diagnosis of specific syndromes like porphyrias (Fig. 1.2.2), which by their rarity and/or complexity are prone to either under- or over-diagnosing.

Although early definitions of diagnostic pathways date back to the 1970s [11], there is still some confusion with regard to overlapping terms such as "recommendations", "guidelines" or "directives" [12]. All of them are closely related, but describe different aspects of a complex topic with many facets. A frequent saying is that directives "must", guidelines "should" and recommendations "can" be obligatory [6]. For more definitions see addendum.

Diagnostic pathways are not identical with any of these terms, but must of course respect regulatory demands (directives) and evidence-based facts (guidelines) whenever possible. In contrast to guidelines, they are not universally valid but need to be adapted to local requirements and resources. For example, a laboratory diagnostic pathway for NSTEMI (myocardial infarction without ST-elevation in the ECG) may start with a highly sensitive troponin assay [13]; this test should be repeated after 3 h [14], if the result is normal. However, if no sensitive troponin is available (e. g. in a point-of-care situation), the repeat measurement needs to be performed after 6–9 h due to the higher decision limits of conventional troponin assays [13, 14].

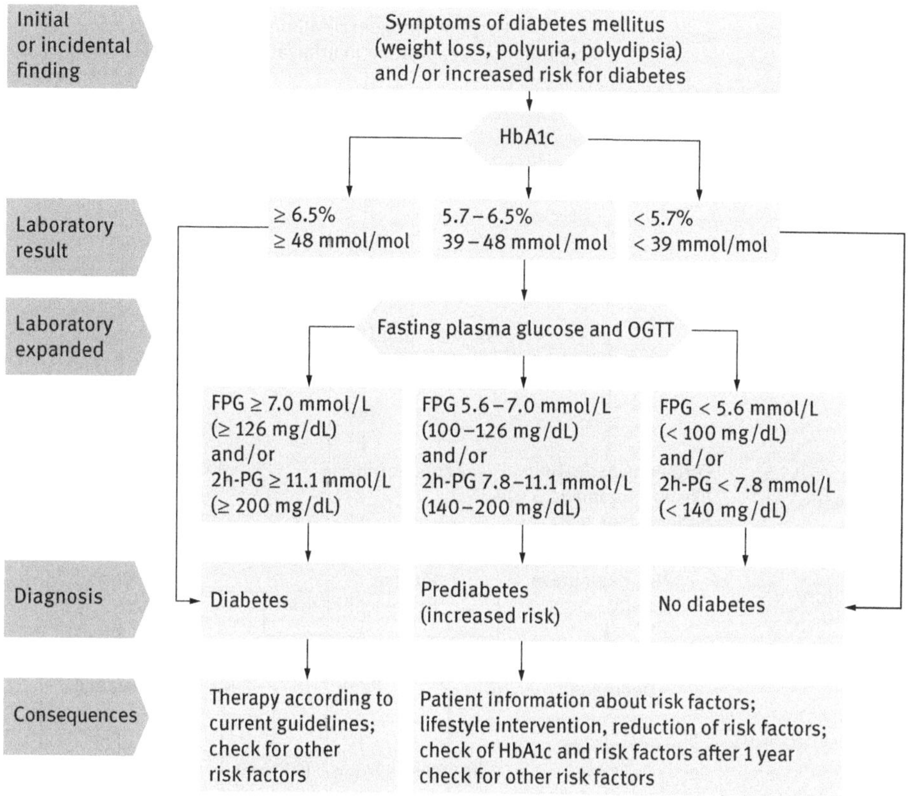

Fig. 1.2.1: Example of a relatively simple diagnostic pathway for the diagnosis of diabetes mellitus (translated from ref. (von Eckardstein and Minder in [6]). The pathway follows the recommendations of the American Diabetes Association (ADA) as modified by Deutsche Diabetes Gesellschaft (DDG). FPG = fasting plasma glucose; OGTT = oral glucose tolerance test; 2h-PG = Plasma glucose level two hours after OGTT. For risk factors of diabetes see ADA recommendations [10].

Figuratively speaking, diagnostic pathways are comparable to car navigation systems, which are based on street maps ("guidelines") and respect traffic laws ("directives"), without being one or the other; although it is usually wise to follow the electronic recommendations, in certain situations it may be life saving to deviate. And finally another parallel: diagnostic pathways can be plotted on paper as depicted in Figs 1.2.1 and 1.2.2 ("street map"), but unless they have been implemented in an electronic system (laboratory or hospital information system), they will not be helpful in daily routine. This is especially true for iterative and comprehensive pathways that try to cover all frequent and rare causes of a syndrome. For example, the stepwise differential diagnostics of icterus covering frequent causes such as infectious hepatitis and then rarer causes such as autoimmune hepatitis, metabolic disorders, intoxications, etc. will lead to a very complex pathway diagram which is difficult to read and to interpret by visual inspection but can be used for navigation if implemented within the laboratory or hospital information system.

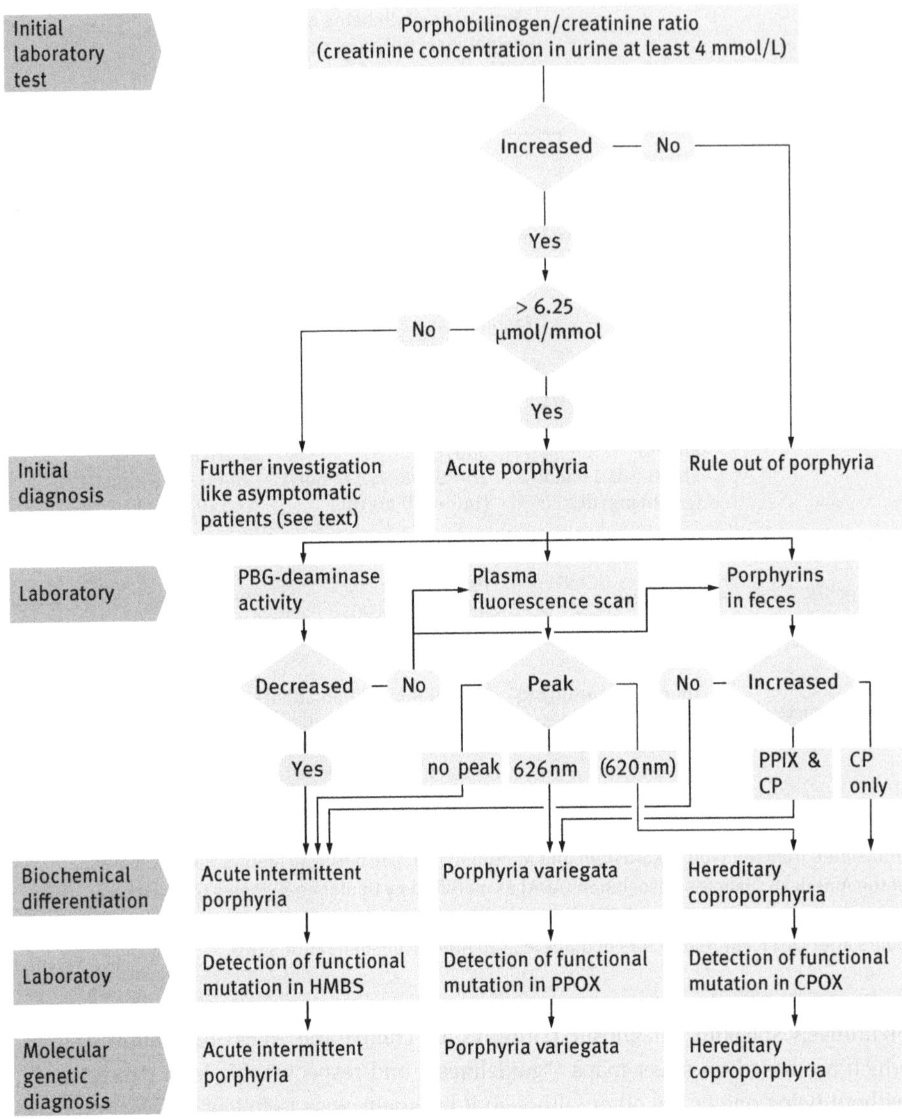

Fig. 1.2.2: Example of a relatively complex diagnostic pathway for the differential diagnosis of acute porphyrias causing an abdominal crisis (translated from ref. [6]). Typically, porphobilinogen/creatinine ratios in urine (μmol/mmol) are increased above 5 and may persist for at least one week after onset of symptoms. A normal quantitative porphobilinogen level excludes porphyria as the cause of an acute abdominal crisis (right branch of the decision tree). Three major inherited diseases can underly acute porphyrias (middle branch): acute-intermittent porphyria due to mutations in porphobilinogen (PBG)- deaminase (= hydroxymethylbilansynthase [HMBS]), porphyria variegata due to mutations in protoporphyrinogen-oxidase (PPOX) and hereditary coproporphyria due to mutations in coproporphyrinogen-oxidase (CPOX). The lower part of the diagram describes genetic (HMBS, PPOX, CPOX) and biochemical tests (e. g. PBG desaminase activity and plasma fluorescence), details of which can be found in ref. [6].

Clinical care pathways are usually more complex than their diagnostic counterparts, because they represent frameworks for the inter-professional description and steering of all medical services within the hospital. Formally, they rely mostly on flow charts or swim lanes with time points, responsibilities and decision points [4, 6]. Quite often, clinical pathways as well as medical guidelines and directives do not sufficiently take into account current diagnostic developments and issues. In these cases, diagnostic pathways can fill the gap on a local basis, by seeking consensus among laboratory and clinical experts while respecting evidence, personnel, technical as well as economic resources. For each laboratory-diagnostic pathway in the above-mentioned book [6], an interdisciplinary expert group was formed, consisting of clinicians and laboratory experts.

1.2.3 Paradigm Shift

If established broadly, laboratory diagnostic pathways may provoke a paradigm shift in that the end users (physicians, nurses) are no longer faced with the time-consuming selection of appropriate laboratory tests. In simple words, ordering a medical condition in the electronic order entry system is the only step required. The system then provides the user with the standard tests that will be analyzed, the respective patient materials to be taken, and the preanalytical specifications to be considered. Additional tests that are required at specific decision points are under the responsibility of the laboratory, acting along the previously consented diagnostic pathways. A recent study confirmed that physicians often feel uncertain about which tests to order (both for clinical and economic reasons), and that they endorse the help of information technology for test selection and even for interpretation of test results [15].

1.2.3.1 Benefits

It has been shown in rigorous evaluations that clinical pathways have the potential to improve the quality of patient care [1, 4, 16] and outcomes [17]. No such systematic studies have been published so far for diagnostic pathways, but projects are underway in the authors' institutions. Table 1.2.1 presents a list of potential benefits, modified according to Bruni [18]. In summary, establishing diagnostic pathways means standardizing diagnostic decision processes according to the best available knowledge. Such an initiative can serve as an integral part of total quality management and internal quality assurance.

The concept of rule-based systems and decision trees came up more than 20 years ago, when large test profiles became available on automated analyzers and computers were increasingly used to regulate the flood of data [19]. The term "reflex testing" [8] was introduced at that time as an antipode to profile testing. It indicated

Table 1.2.1: Ten major benefits of diagnostic pathways.

1.	Clinicians rapidly obtain the right result with the appropriate test(s) for the clinical question based on the best available evidence.
2.	Based on diagnostic pathways, frequent routine processes follow a fixed and uniform standard of quality across all medical disciplines and professional groups.
3.	Weak points in the process are readily recognized, necessary changes are triggered.
4.	Dispensable and redundant tests are avoided.
5.	Crucial tests, leading to the correct diagnosis, are requested automatically and cannot be missed.
6.	Human errors are minimized and legal certainty is improved.
7.	Diagnostic decisions become transparent and accessible to all those involved in patient care.
8.	The nature and extent of resource consumption can clearly be identified. Departments receive valid data for process and finance control.
9.	Diagnostic pathways provide a platform for cross-sectional consensus and improve the interaction between the clinicians and the laboratory.
10.	Diagnostic pathways are a valuable educational tool for clinicians, nurses and students.

"algorithmic" (i. e. computer based) decision making as opposed to "reflective" decisions made by physicians [20].

The essential benefit of the selective paradigm is statistical in nature: while a shot-gun strategy produces many false positive results, the stepwise approach increases the probability of a suspected diagnosis with each new result, thus reducing the risk of false positives. Diagnostic pathways ("smart profiles") as described above have an additional advantage over the traditional stepwise ordering process – since all relevant patient materials are initially available in the laboratory, the whole sequence of appropriate tests can be performed without drawing new specimens and ordering consecutive tests. This makes workflows much smoother and reduces the number of phone calls and other time-consuming activities.

Diagnostic pathways can also help to integrate the laboratory more closely in the medical and administrative process of the hospital, which will boost the effectiveness of medical services with respect to DRG requirements. Most importantly, well-defined pathways ensure that state-of-the-art laboratory tests are ordered, outdated and unnecessary tests are avoided, and that for identical diagnostic issues identical strategies are applied throughout the hospital.

The laboratory can thus contribute to patient outcome and value creation, improve case management by faster and more reliable laboratory results, and finally support DRG coding by delivering the appropriate ICD codes, whenever a defined diagnosis can be attributed to an endpoint in the decision tree. Since (primary and secondary) diagnoses are the backbone of any DRG reimbursement system, this service can also improve the profitability of the hospital [21, 22]. Potential non-monetary benefits of increased transparency are greater acceptance of the laboratory services by the hospital staff and a better work climate on both sides.

Finally, the presentation of diagnostic pathways in the format of decision trees seems to be a value in itself, since it makes the underlying *if...then...else* rules easy to read and to understand, especially for medical professionals, who are not experienced in computer science. It is not surprising that many medical guidelines include such "algorithms" to reduce the complexity of diagnostic reasoning and to make the course of all possible decisions transparent at a glance.

1.2.3.2 Limitations and pitfalls

The advantage of simplicity is, however, also one of the biggest problems of decision trees. Each branch stands for a Yes or No with no gradation in between, for example:

IF troponin is elevated
THEN diagnosis is myocardial infarction.

This rule looks very straightforward, but in reality, the IF part ("troponin is elevated") is not that clear-cut; rather it indicates just a probability for the presence of myocardial infarction. Assuming sensitivities and specificities of around 90 % for troponin [14, 23], one out of ten such decisions will be wrong – provided that the prevalence of myocardial infarction (MI) in the examined patient population is 50 %. If the prevalence is lower, the error rate of positive results will be even higher. In the typical case of a chest pain unit with about 15 % MI patients, four out of ten positive decisions will be wrong (positive predictive value PPV = 61 %).

Thus, once a wrong decision has been made by the computer, the whole subsequent decision chain will run in the wrong direction. This is the reason, why computer-based algorithm can never take "responsibility" for medical decisions and therefore will never replace the human professional in the laboratory. It remains a human task to ensure correct decisions by considering the full clinical context rather than just a computer flag.

Table 1.2.2 gives an overview of the potential drawbacks from an organizational and an economic point of view. An important limitation of laboratory pathways is that they are not suited for patients suffering from several diseases at a time and/or showing ambiguous clinical symptoms, because each pathway needs a clear starting

Table 1.2.2: Five major limitations of laboratory diagnostic pathways.

1.	Not applicable among patients with complex diseases
2.	Lack of flexibility and individuality
3.	Only suitable for patients with a clear suspicion/symptom
4.	Other diagnostic techniques like imaging not included
5.	Lack of suitable IT tools embedded in routine laboratory information systems

point – a well-defined suspicion or a predefined symptom. Another relevant limitation is the lack of evidence for many of the "good old" laboratory tests, which have been ordered for decades without really evaluating their cost/efficiency ratio. This opens the door for pathways that have been generated under the pressure of other interests (economy, society, insurances, etc.) with potentially negative health outcomes.

Generating and maintaining laboratory diagnostic pathways requires time, money, personnel – and not to forget endurance. Laboratory decision trees must be developed in close and prudent collaboration between laboratory and clinical physicians, in order to achieve the highest possible quality of evidence in the respective clinical setting. Algorithms plotted solely by the laboratory staff will not find sufficient adherence in clinical practice. In many cases, the biggest bottleneck is lack of IT support. Although many providers of information systems claim that their products provide rule generators and graphics software for pathway implementation, it often turns out that these tools are not powerful enough and also too complicated to handle. Specialized staff with knowledge both in medical diagnostics and information technology is needed, but rarely available.

Finally it should be mentioned that the seeming clarity of well-defined diagnostic pathways provokes the abuse of primarily medical intentions for mere economic purposes. Shortly after the aforementioned book [6] was published, we observed that some hospital and insurance managers tended to overrate the relevance of published decision trees and to prohibit tests, which were not contained. Potential abuse of diagnostic pathways is certainly not a reason for resignation, but it must be considered when starting a pathway project.

1.2.4 Conclusion and Outlook

Diagnostic pathways, if adequately generated and implemented, undoubtedly ensure that patients receive the laboratory test results at the best level of evidence. They represent an essential part of an innovative way to optimize processes in medicine, and contribute to cost, results and performance transparency in health care. Medical decisions are made visible through diagnostic pathways, especially if the technical requirements are given. Particularly under the conditions of DRGs, they allow for a clear determination of which laboratory test should be used for which question. Moreover, laboratory diagnostic pathways represent an important contribution to the integration of modern disease management in a complex health policy development with finite resources, and thus build a bridge between medical and economic needs in the health care system.

Health care quality policy emphasizes that any diagnostic process should be based on validated outcomes and consequently the best available evidence. However, to date, there is a lack of robust evidence regarding diagnostic tests – especially those, which have been in use for decades – because most studies dedicated to this concept do not meet methodologically acceptable standards. The key question therefore is how to evaluate the test results in terms of global clinical outcome, to determine utility and effectiveness of diagnostic rules, and to derive high quality recommendations.

Using tumor markers as an example, McShane et al [24] stated that fundamental discussion and collaboration between clinical physicians, statisticians, and laboratory scientists are essential, in order to select the best markers out of a large number of potential candidates for beneficial guidelines. In their study, they described a specific review process called REMARK (REporting recommendations for tumor MARKer prognostic studies). In essence, only studies that are built upon a certain framework of quality should be mentioned and included in a review. Diagnostic pathways should be generated only, if suitable prerequisites such as study design, statistical significance and appropriate patient collective are met.

Another potent tool to evaluate the potential of a study [25] is QUADAS (QUality Assessment of studies of Diagnostic Accuracy included in Systematic reviews). It provides a list of requirements for a study to be fulfilled. The respective 14 questions can be answered with Yes, No or Maybe, thus leading to very comprehensible and illustrative recommendations. With this kind of instrument, it should be possible to bring more light into the process of developing new diagnostic pathways.

Finally, a case study based on GRADE (Grading of Recommendations Applicability, Development and Evaluation), tried to establish clinically feasible guidelines for cow milk allergy in cooperation with the World Allergy Organization [26]. This case showed how important but also how time consuming and complex it is to develop a constructive and consistent guideline. Aiming to evidence-based medicine, it must be our ambition despite all barriers and pitfalls to construct reasonable, practicable, and innovative guidelines for clinical physicians, out of a flood of studies, to generate high standard diagnostic pathways.

Applying such tools is not an end in itself. The ultimate goal must be to reduce residual risks of any diagnostic decision for patients and physicians, all the more with regard to some negative experience made after the introduction of DRGs in other countries (e. g. increase in the discharge of unstable patients and increasing release to nursing facilities) [27]. Fears that are based on catchwords such as limiting the freedom of diagnostic ordering must be taken seriously and considered in the light that many hospital patients suffer from multifactorial complex diseases – a condition that cannot be dichotomized in the simple diagram of a diagnostic pathway. To improve this situation, better multivariate algorithms and more sophisticated software packages [28, 29] are needed, which are currently under development especially in the bioinformatics community [10].

Acknowledgments: The authors wish to thank Prof. Dr. Karl Lackner, University of Mainz, Germany, Dr. Daniela Buhl, Kantonsspital Luzern, Switzerland, Prof Dr. Wolfgang Korte, Kantonsspital St. Gallen and Dr. Martin Risch, Labormedizinisches Zentrum Schaan, Liechtenstein for fruitful discussion.

Figs 1.2.1 and 1.2.2 are the English reproduction of the German original published in von Eckardstein and Minder's chapter on Diabetes and Metabolism in ref. [6]. We thank Prof. Elisabeth Minder, Stadtspital Zürich, Switzerland for her kind permission to use this artwork.

Addendum: Definition of Complementary Terms

Directives are compulsory rules for acts of commission or omission, released by a legitimate institution. The infringement entails defined sanctions, in particular in accordance with social and professional legislation. The most famous directive for laboratory diagnostics in Germany has been published by the Bundesärztekammer [30] and contains instructions for quality management.

Guidelines are systematically developed aids for decision making for an appropriate medical approach in each single case. They are not compulsory, but rather practice orientated corridors for decisions and actions. A German consortium of scientific medical expert panel (AWMF) provides more than 1,000 guidelines on the internet (www.awmf.org/leitlinien/leitlinien.html) at three levels of evidence:
- S1: Recommendations by experts
- S2k: Guidelines based on consensus
- S2e: Guidelines based on evidence
- S3: Guidelines based on consensus and evidence

Recommendations, in this definition, are the weakest form of guidelines.

References

[1] Gerardi T. A regional hospital association's approach to clinical pathway development. J Healthc Qual 1994; 16:10–14.
[2] Peters M, Broughton PMG. The role of expert systems in improving the test requesting pattern of clinicians. Ann Clin Biochem 1993; 30:52–9.
[3] Hoffmann G, Schenker M, Kammann M Meyer-Lüerssen D, Wilke M.The significance of laboratory testing for the German diagnosis-related group system. Clin Lab 2004; 50:599–607.
[4] Eckardt J, Sens B: Praxishandbuch Integrierte Behandlungspfade. Economica Verlag, Heidelberg, 2006, ISBN 978-3-87081-430-4.
[5] www.dgkl.de
[6] Hofmann W, Aufenanger J, Hoffmann G. Klinikhandbuch labordiagnostische Pfade. 2. Auflage. Walter de Gruyter, Berlin, 2014; ISBN 978-3-11-031400-7.

[7] www.healthdirect.gov.au/partners/diagnostic-imaging-pathways
[8] Srivastava R, Bartlett W, Kennedy I, Hiney A, Fletcher C, Murphy M. Reflex and reflective testing: efficiency and effectiveness of adding on laboratory tests. Ann Clin Biochem 2010; 47:223–7.
[9] Vanhaecht K, de Witte K, Panella M, Sermeus W. Do pathways lead to better organized care processes? J Eval Clin Pract 2009; 15:782–8.
[10] American Diabetes Association. Standards of medical care in diabetes-2012. Diabetes Care. 2012; 35(Suppl) 1:11–63.38.
[11] Essex B. Diagnostic pathways in clinical medicine. An epidemiological approach to clinical problems. Cabdirect, www.cabi.org 1976. ISBN 0-443-01514-7.
[12] Bundesärztekammer: Verbindlichkeit von Richtlinien, Leitlinien, Empfehlungen und Stellungnahmen, 2006 (www.Bundesaerztekammer.de).
[13] European Society of Cardiology: Guidelines for the management of acute coronary syndromes in patients presenting without persistent ST-segment elevation. 2011. www.escardio.org/guidelines
[14] Keller C, Zeller T, Ojeda F. Serial changes in highly sensitive Troponin I assay and early diagnosis of myocardial infarction. J Am Med Assoc 2011; 28:2684–93.
[15] Hickner J, Thompson P, Wilkinson T, Epner P, Sheehan M, Pollock A et al. Primary care physicians challenges in ordering clinical laboratory tests and interpreting results. J Am Board Fam Med 2014; 27:268–74.
[16] Pearson D, Goulart-Fisher D, Lee T. Critical pathways as a strategy for improving care: problems and potential. Ann Int Med 1995; 123:941–8.
[17] Lawson D, Revelino K, Owen D. Clinical pathways to improve patient outcomes. Maney Online 2006; 11:269–72.
[18] Bruni K. Welche Bedeutung gewinnt die Labormedizin mit der Einführung der Swiss DRG? 2007; Diploma thesis University Hospital Zürich, Switzerland.
[19] Hoffmann G, Stephans E. Barriers and chances for computer-based decision support in the medical laboratory – results of an expert survey. J Lab Med 1997; 21:231–7.
[20] Murphy M. Reflections on reflex thresholds. Ann Clin Biochem 2012; 49:551–17.
[21] Aufenanger J. Do diagnostic pathways contribute to economic improvement of a hopsital? Considerations from the perspective of laboratory medicine. J Lab Med 2011; 35:285–9.
[22] Hoffmann G. DRG Watchdog Highlights New Financial Dimensions of Laboratory Diagnostics. Clin Lab 2003; 49:507–10.
[23] Ross G, Bever F, Uddin Z, Hockman E. Troponin I sensitivity and specificity fort he diagnosis of acute myocardial infarction. J Am Osteopath Assoc 2000; 100:29–32.
[24] McShane L , Altman D, Sauerbrei W, Taube S, Gion M, Clark G. Reporting recommendations for tumor MARKer prognostic studies (REMARK), J Clin Oncol 2005; 23:9067–72.
[25] Withing P, Rutjes A, Reitsma J, Bossuyt P, Kleijnen J. The development of QUADAS: a tool for the quality assessment of studies of diagnostic accuracy included in systematic reviews, BMC Medical Research Methodology, 2003; 3:57–78.
[26] Hsu J, Brozek J, Terracciano L,,Kreis J, Compalati E, Stein T et al. Application of GRADE: Making evidence-based recommendations about diagnostic tests in clinical practice guideline. Implementaion Sci 2011; 6:62, doi: 10.1186/1748-5908-6-62.
[27] Aiken L, Sermeus W, Kutney-Lee A. Patient safety, satisfaction, and quality of hospital care: cross sectional surveys of nurses and patients in 12 countries in Europe and the United States. Brit Med J 2012; 344:e1717.
[28] Lee Y, Lin Y, Wahba G. Multicategory support vector machines. J Am Stat Assoc 2004; 99(465):67 doi: 10 1198.
[29] http://www.dgkl.de (AG Bioinformatik).
[30] Bundesärztekammer: Richtlinien Der Bundesärztekammer zur Qualitätssicherung laboratoriums-medizinischer Untersuchungen. D Ärztebl 2010; 107:C301–15.

Walter G. Guder

1.3 Definition of the Influence and Interference Factors in the Preanalytical Phase

1.3.1 How the Terms Developed

Prior to the introduction of the quality assurance programs in the nineteen sixties, laboratory results which did not fit into the clinical picture of the patient were categorized as laboratory errors. With the improved analytical precision and accuracy, we became aware of the numerous variables unrelated to the condition of the patient that could affect the analytical result. These variables could not be standardized or controlled by the analytical quality assurance programs. Based on my observations while consulting in a 2000-bed acute care hospital from 1970 onwards, a number of cases revealed that influences before, during and after the sampling altered the analytical result to a degree exceeding the maximal allowable error in the analytical phase [1]. Also in the nineteen seventies, Statland and Winkel defined the phase prior to analysis as the "preinstrumental phase" [2] which later was termed "preanalytical phase" [3].

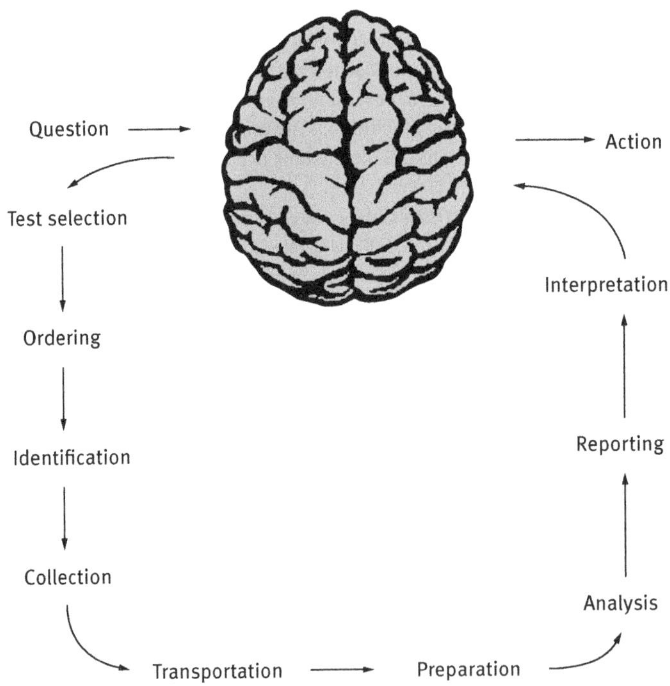

Fig. 1.3.1: The laboratory diagnostic procedure [4].

The total laboratory procedure consists of several stages and courses of action that begins with the physician requesting the performance of a laboratory investigation (Question) in a patient [4] (Fig. 1.3.1). Hence the patient is the starting point and aim of the laboratory examination.

The total process of laboratory investigation can be divided into three consecutive phases:

- The **preanalytical phase** encompasses all procedures and the time frame starting from the formulation of the medical question to the end of the sample preparation.
- The **analytical phase** covers the analytical process including the analytical validation of the results.
- The **postanalytical phase** is defined by the time interval between the reporting of the laboratory finding and the action that follows the physician's interpretation of the result.

The total time required for this process was defined as the **total turnaround time (TTT)**. The different stages exhibit apparent differences regarding the time needed, cost and frequency of errors. Fig 1.3.2 shows that the preanalytical phase takes more than 50 % of the turnaround time (TAT).

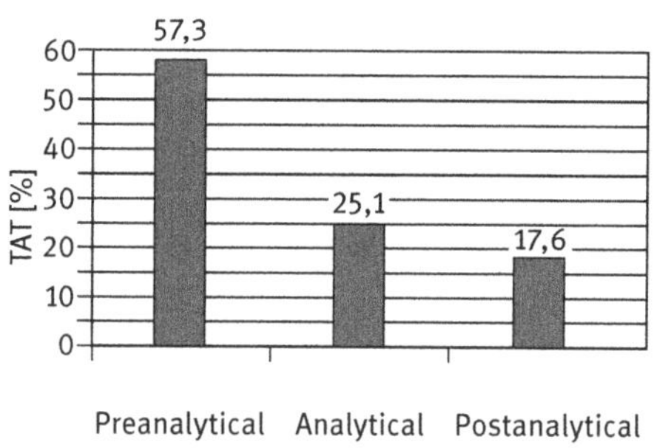

Fig. 1.3.2: Time needed for the different phases of laboratory diagnostic process [5].

When electrolytes were determined, 57 % of TAT was required for the preanalytical, 25 % was needed for the analytical and 18 % for the postanalytical phases [5]. These data raised queries regarding the preanalytical processes contributing to this long duration of time. Technically, the transportation process was improved by refining transport systems inside and outside the hospitals. Since half of the time was spent once the samples arrived at the laboratory, an attempt was made to improve these procedures by mechanizing sample distribution, centrifugation and storage activities (see Chapter 8.2) [6].

There are numerous variables influencing the laboratory results. While variables like fasting before sampling, circadian rhythms and time of blood transport affected the laboratory results, the awareness about these impacts was often underestimated. After years of discussions in the national and international expert groups in the nineteen sixties and seventies [2, 7], the term “biological influence factor” was coined and was compared to interference factors. This led to its definition in 1980 [8, 9, 10], which is valid even today.

1.3.2 Preanalytical Influence Factors

Biological influence factors lead to variations in the quantity of the analyte to be measured in a defined matrix. They are by definition independent of the analytical method used. These factors are either present in the uninjured individual to be studied prior to the specimen collection like circadian rhythms [7, 11] or appear as side effects of a disease and its treatment changing other analyte concentrations due to drug effects. In fact biological influences are the basis of laboratory diagnostic criteria. Here the influences not required for the diagnostic function only are important. They can be changeable like climate or unchangeable like sex or other genetic aspects. Some can be influenced by previous actions like diet, and some do not (like age). Of special importance here are the influence factors whose side effects can be reduced by standardizing preanalytical conditions. Table 1.3.1 provides some examples of how recommendations can help to reduce the effect of biological influence factors.

Table 1.3.1: Variable factors influencing human blood composition and recommendations to reduce their influences (modified from [9]).

Influence Factor	Examples of blood constituents affected	Recommendation to standardize specimen collection.	References
Food	Urea, uric acid, triglycerides, glucose, urine constituents	Sampling after 12 h fasting	Steinmetz et al 1973 [12]
Prolonged fasting	Coagulation factors, cholinesterase, oral glucose tolerance	2 days feeding before sampling	Lamers et al 1985 [13]
Circadian Rhythms	Potassium, iron, cortisol, eosinophile leucocytes	Sampling between 7 and 9 a.m.	Stamm 1967 [7], Wisser and Breuer 1979 [11]
Muscular activity	Creatine kinase, hemoglobin, LDH, glucose, coagulation factors, leucocytes, troponin I	No strenous muscular activity three days before sampling	Stansbie and Begley 1991 [14]
Body position before and during sampling	Proteins, lipids, enzymes	15 min supine position before sampling	Fielding et al 1980 [15]
Ethanol and drugs	Glucose, uric acid, lactate, amylase, γGT	Control of alcohol consumption and drug effects	Young et al 1975 [16]

Table 1.3.1: (continued)

Influence Factor	Examples of blood constituents affected	Recommendation to standardize specimen collection.	References
Smoking	Lipase, amylase, cholesterol, glucose	Control and consideration of smoking	Balldin et al 1980 [17], Statland and Winkel 1977 [3]
Coffee	Glycerol (triglycerides), cortisol	Sampling after fasting	Young 1979 [18]
Extended time of tourniquet	Enzymes, proteins and cells	Tourniquet should be released after the vein is entered and not exceed 1–2 min. Repeat puncture should be performed on the other arm.	Lippi et al 2006 [19]

All these influences occurred in vivo. However, analyte concentration can also be altered in vitro. Thus potassium will increase in plasma/serum when blood samples are stored at lower temperatures without visible hemolysis. Hence Büttner suggested that influences can also biologically affect after sampling in vitro [20]. Chapters 3.1–3.4 provide details on various influences.

1.3.3 Interference Factors

Interference factors alter the result of a sample constituent after the sample has been collected. They are different from the measured analyte and interfere with the analytical procedure. Therefore their effect is method dependent and may thus be reduced or eliminated by selecting a more specific method.

The interferent can either be a biological constituent of the sample (like acetoacetate interfering with the Jaffe procedure for measurement of creatinine), an exogeneous molecule in the sample (like a drug interfering with the analytical procedure) or an interferent entering the sample in vitro (like an anticoagulant [EDTA] interfering with the procedure of the intended analyte [alkaline phosphatase]). Therefore, interferences can be reduced or eliminated by selecting a more specific analytical procedure. Typical interferences will be covered in Chapters 4.1–4.6.

References

[1] Guder, WG. Einfluss von Probennahme, Probentransport und Probenverwahrung auf klinisch chemische Untersuchungen. Ärztl Lab 1976; 22:69–75.

[2] Statland BE, Winkel P. Physiological variation of the concentrations values of selected analytes determined in healthy young adults. In Proceedings of the 1976 Aspen conference on Analytical Goals in Clinical Chemistry (Elevitch FR ed.) College of American Pathologists Chicago 1977.
[3] Statland BE, Winkel P. Effects of preanalytical factors on the intraindividual variation of analytes in the blood of healthy subjects. Consideration of preparation of the subject and time of venipuncture. Crit Rev Clin Lab Sci 1977; 8:105–44.
[4] Lundberg GD. Critical (panic) value notification: an established laboratory practice policy (parameter). J Am Med Ass 1990; 263:709.
[5] Godolphin W, Bodtker K, Uyeno D, Goh L.-Q. Automated blood sampling handling in the clinical laboratory. Clin Chem 1990; 36:1551–5.
[6] Hoffmann GE. The third generation of laboratory systems. Clin Chim Acta 1998;278: 203–16. and "Development in Pre-analytical Automation" Video Grafrath:Trillium 1998.
[7] Stamm D. Tagesschwankungen der Normalbereiche diagnostisch wichtiger Blutbestandteile. Verh Dtsch Ges Inn Med 1967; 73:982–9.
[8] Guder WG. Einflussgrößen und Störfaktoren bei klinisch chemischen Untersuchungen. Internist 1980; 21:533–42.
[9] Guder WG, Wahlefeld A-W. Specimens and samples in clinical chemistry. In Bergmeyer. Methods of Enzymatic Analysis.1983 Weinheim, Verlag Chemie; 3rd ed Vol II, pp 2–20.
[10] Keller H, Guder WG, Hansert E, Stamm D. Biological influence factors and interference factors in clinical chemistry: general considerations. Editorial. J Clin Chem Clin Biochem 1985; 23:3–6.
[11] Wisser H, Breuer H. Circadian changes of clinical chemical and endocrinological parameters. J Clin Chem Clin Biochem 1979; 19:323–37.
[12] Steinmetz J, Panek E, Sourieau F, Siest G. Influence of food intake on biological parameters. In: Siest G. (ed), Reference Values in Human Chemistry. Basel, Karger 1973, pp 195–200.
[13] Lamers KJB, Doesburg WH, Gabreels FJM, Lemmens WAJG, Romson AC, Wevers RA et al. The concentration of blood components related to fuel metabolism during prolonged fasting in children. Clin Chim Acta 1985; 152:155–63.
[14] Stansbie D, Begley JP. Biochemical consequence of exercise. IFCC J 1991; 3:87–91.
[15] Fielding P, Tryding N, Hyltoft Petersen P, Hørder M. Effect of posture on concentrations of blood constituents in healthy adults: practical application of blood specimen collection procedure recommended by the Scandinavian Committee of reference values. Scand J Clin Lab Invest 1980; 40:615–21.
[16] Young DS, Pestander LC, Gibbermann V. Effect of drugs on clinical laboratory tests Clin Chem 1975; 21:1D–423D.
[17] Balldin G, Borgström A, Eddeland A, Genell S, Hagberg L, Ohlson K. Elevated serum levels of pancreatic secretory proteins in cigarette smokers after secretin stimulation. J Clin Invest 1980; 66:159–62.
[18] Young DS Biological variability, in: Chemical Diagnosis of Disease, Amsterdam, Elsevier/North Holland Biomedical Press 1979, pp 1–113.
[19] Lippi G, Salvagno GL, Montagnana M, Franchini M, Guidi GC. Venous stasis and routine haematologic testing. Clin Lab Haematol 2006; 28:332–7.
[20] Büttner J. Unspecificy and interference in analytical systems: concepts and theoretical aspects. Klin Chem Mitt 1991; 22:3–12.

Walter G. Guder, Hermann Wisser (†)*[1]

1.4 Extraanalytical Procedures and their Management in Total Turnaround Time

1.4.1 Introduction

The total turnaround time (TAT) of a diagnostic process using laboratory investigations as defined in Chapter 1.3 includes the following preanalytical and postanalytical functions (Table 1.4.1), which can be defined, covering 45–80 % of the total turnaround time. The broad variation of these data is partially due to the continuous changes in technical improvement and fuctional speed in analytical as well as pre- and postanalytical phases.

Table 1.4.1: Extraanalytical functions and persons involved (partly from [1]).

Function	Persons involved	% Time of TAT	Technical aids
Preanalytical Functions		39–60	
Question and selection of test	Physician	2–3	Diagnostic pathways
Request of Laboratory test	Physician, nurses,	2–3	Request form, order entry system
Patient identification and sampling	Physician, nurses, phlebotomists	5–15	Patient and sample identity labels
Sample transport	Any person, taxi driver, posting persons	10–20	Transport systems, containers, cooling package
Sample preparation	Lab technician	20	Centrifuges, sample distributer
Sample storage	Lab personnel	0	Storage container
Analytical phase including analytical assessment	Lab personnel	20–30	Analyzers with sample storage and
Postanalytical functions	Lab personnel, Physicians, nurses	5–25	
Reporting result	Lab personnel	2–5 (10–20 if not electronically)	Hospital information system, printers
Interpretation of result in relation to patient's state	Physician	1–3	Interpretative recommendations online
Decisions on medical action (treatment, further diagnostic actions, etc.)	Medical personnel	2–3	Diagnostic pathways

1 Partially translated from Guder et al 2007

1.4.2 Effective Time Management

The aim of laboratory examination of patients is to improve the outcome of their medical treatment. Thus evidence-based laboratory medicine always has to follow patient outcome to define its aim. Since time is a critical aspect in all diagnostic assessments, the definition of medical needs regarding time seems essential.

Set priorities accordingly: Define the medical need and importance in the treatment of the patient and, directly related to this, its urgency. In general, laboratory analyses can be classified according to criteria of importance and urgency (Fig. 1.4.1).

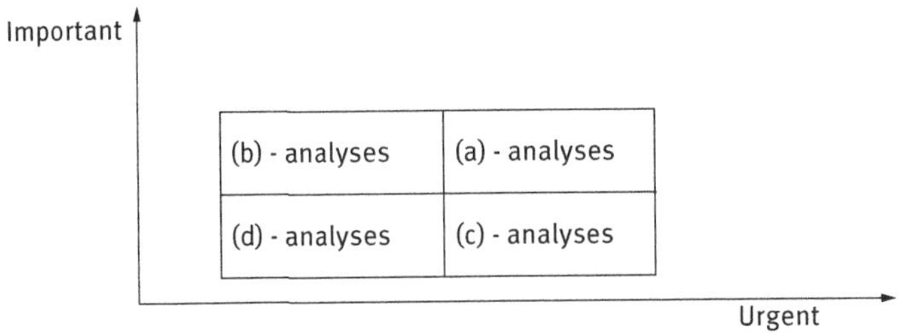

(a) - analyses:	important and urgent:	work up instantly
(b) - analyses:	important, not urgent	work up continuously
(c) - analyses:	not important, but urgent	may disturb workflow
(d) - analyses:	not important, not urgent	less disturbing potential, but cost relevant

Fig. 1.4.1: Priority scheme of laboratory investigations.

Problem: Usually, the urgency is determined outside the lab, because the informations needed to decide are not available.

Conclusions: According to the intended aim to optimize timing, C- and D- investigations should be avoided as much as possible. A- and B- analyses need different organizations of work flow.

1.4.2.1 Timing of laboratory investigations

The indicator of an effective organization of timing a laboratory finding is the "turnaround time"(TAT).
TAT can be defined differently [2, 3]:

Test TAT: Time span between sampling and report of laboratory result.
Laboratory TAT: Time span between sample arrival in the laboratory and report of the laboratory result.

Preanalytical TAT: Time between sampling and receipt of sample in the laboratory.

Most authors recommended to define as the starting time of TAT the question (indication) of laboratory tests and the clinical interpretation of the laboratory result by the physician treating the patient as the end point [4, 5, 6]. Both time points are however difficult to document.

Realized TATs

Laboratory TAT of 94 % of all laboratory emergent requests (hematological, coagulation and clinical chemistry tests including acid base status) were examined and analytically confirmed after 30 min. Figure 1.4.2 shows two maxima, one after 10 min and the second 10 min later. The first group consisted of blood tests done in whole blood, and the second group was for examinations performed in serum or plasma. The time difference between both the groups is exclusively due to the centrifugation time needed [7].

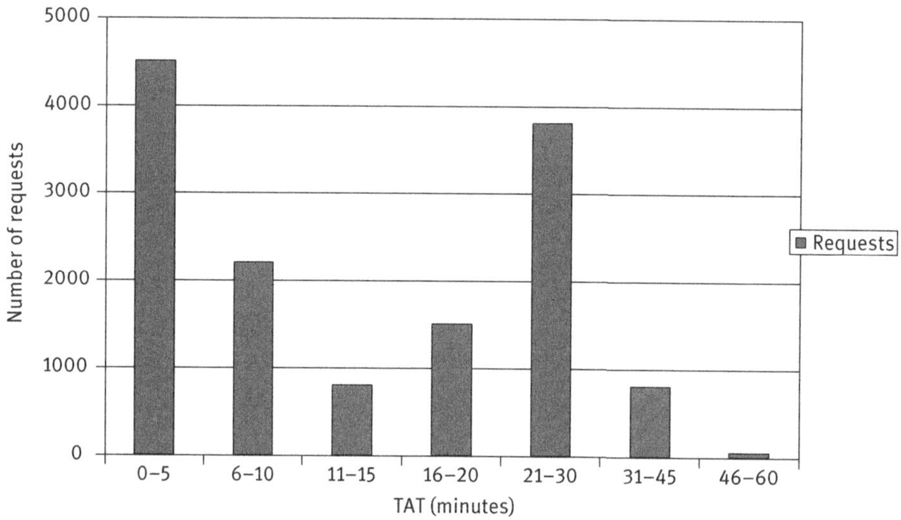

Fig. 1.4.2: Turnaround time of emergency requests (means of 28 days) [6, 7].

TAT of emergency stat tests

The most frequently requested stat test as urgent are hemoglobin and potassium, TAT for which was determined in a study at the College of American Pathologists with more than 700 participants [2]. Both analytes were selected assuming that their TAT is representative for all emergency tests requested simultaneously (Table 1.4.2). In addition, the results were segregated according to the size of hospitals. It can be seen, that the TATs depend not only on the personnel performing the blood sampling procedure, but likewise on the type of sample transport to the laboratory.

With the increasing size of the hospital, TAT also increases. Mechanical transport systems seem more rapid than courier services and personal transport by clinic personnel. If pneumatic tubes were used, short distance between ward and pneumatic system was important to reduce TAT. In these studies, shortest times were obtained when sampling was performed by phlebotomists carrying the samples directly into the laboratory.

Table 1.4.2: Real test-TATs for performing measurement of hemoglobin- and potassium concentration [2].

Clinic size (number of beds)	< 250	251–500	> 500
TAT of hemoglobin Median (10–90 percentile) in min	24 (13–39)	25	30 (27–56)
TAT of potassium	33 (24–53)	36	44 (23–60)
Sampling done by	phlebotomists	intensive care nurses	Intensive care physicians
Transport done by	Mechanical or pneumatic transport system	courier	clinic personnel

TAT test for microbiological investigations

Table 1.4.3 summarizes the data obtained over 11 weeks [8].

Table 1.4.3: TAT of microbiological examinations [8].

Sample	n	TAT in hours (min–max)
Urine samples	88	28.7–67.8
Other bacteriological samples	52	40.0–89.4
Total samples	140	34.2–67.8

Which TATs are expected by physicians?

Based on the results of a questionnaire answered by 2763 clinicians in 722 institutions summarized in Table 1.4.4, it can be seen which turnaround times are expected for different analytes needed [3].

The result is surprising insofar as in all analytes in question, clinicians expect shorter TATs than laboratory personnel. Thus 92 % of surgeons and 94 % of emergency physicians expect hemoglobin results to be available within 30 min, but only 28 % of laboratory physicians confirm this, whereas 90 % expected the result in 60 min.

Similar results were obtained regarding potassium and glucose determinations. While 77 % of clinicians expect results in emergency cases for potassium and glucose within 30 min, this is expected by only 19 % of laboratory physicians. For pO_2-determination 58 % of clinicians expected results in 10 min, but only 22 % of laboratory physicians had the same opinion. Surgeons and emergency physicians expect a TAT of <30 min at a higher percentage compared to internal medicine physicians. Again, the time limit of most laboratory personnel was considerably higher compared to that of clinically active colleagues.

Table 1.4.4: Turnaround times expected by clinicians [3]. The numbers give % of clinicians expecting results reported below the time given. Based on a questionnaire answered by 2763 physicians in 722 institutions.

Category of physician	Hb	potassium	glucose	pO_2
Surgeons	92 % ≤ 30 min	85 % ≤ 30 min	88 % ≤ 30 min	50 % ≤ 10 min
Internal medicine	76 % ≤ 30 min	70 % ≤ 30 min	74 % ≤ 30 min	46 % ≤ 10 min
Emergency	94 % ≤ 30 min	72 % ≤ 30 min	76 % ≤ 30 min	62 % ≤ 10 min
Laboratory physician	28 % ≤ 30 min	20 % ≤ 30 min	19 % ≤ 30 min	22 % ≤ 10 min
	90 % (≤ 60 min)	88 % ≤ 60 min	92 % ≤ 30 min	96 % ≤ 60 min
All clinicians	82 % ≤ 30 min	73 % ≤ 30 min	77 % ≤ 30 min	58 % ≤ 10 min

These results were one of the reasons for the rapid development of preanalytical technologies [9]. In addition, point of care examinations increasingly replaced central laboratory functions in the field of acid–base status and glucose measurement also because of their rapidity [10].

1.4.2.2 Measures to reduce TAT

Table 1.4.5 illustrates the measures that help to reduce turnaround times in practice [7].

Table 1.4.5: Measures to reduce turnaround times (TAT) (modified according to Wisser et al 2002 [7]).

Phase	Measures
Preanalytical	Insert order entry for requesting laboratory tests, reduce marking labels Lower number of tubes needed Use pneumatic or mechanical tube transport replacing personal courier services Use conveyer belts for intralaboratory distribution and storage shorten distances between sample entering place and analyzers More smaller centrifuges may replace few larger ones Less sample splitting

Table 1.4.5: (continued)

Phase	Measures
Analytical	Analyzers with a broad test menu and high performance Adequate personnel distributed to the workload Replace serum- by whole blood or plasma analysis Speed up the release of lab results
Postanalytical	Establish ward terminals for electronic data transport via hospital information system Provide results before all findings are finished
Total procedure	Continuous TAT monitoring

Several aspects of this kind have recently been documented. Providing a short distance between the sampling cabin and the laboratory [11] reduced TAT considerably.

How to control turnaround times

The control of TAT is possible using the following statistical criteria as illustrated in Table 1.4.6.

Table 1.4.6: TAT-control [12].

1.	**Descriptive statistics: calculate**
	1.1 mean 1.2 median 1.3 90th percentile 1.4 The percentage of laboratory results accepted
2.	**For summarizing statistics create the following:**
	2.1 Shewhart quality control charts 2.2 p-charts (control charts for percentile parts)

Summary

- Effective time management presumes clearly defined aims and priorities in relation to their importance and urgency.
- A clear quality criterion for the timely organization of reporting laboratory findings is the test-turnaround time (time span between sampling and provision of test result).
- The medical needs for TAT are not the same for all measured analytes, but depends on the importance and urgency of the respective finding for diagnosis and treatment of the patient.

- For continous control of the preanalytical phase, the determination of the median of time between sampling and arrival of sample in the laboratory is a suitable determinant.
- As a measure of time span between sampling and report of finding can the median of TAT, containing 90 % of all laboratory tests requested be accepted. This may eventually be separately determined for hematology, hemostasiology, microbiology and clinical chemistry.

References

[1] Guder WG, Hagemann P, Wisser H, Zawta B. Fokus Patientenprobe; Kompendium Präanalytik. CD-Rom Heidelberg BD 2007.
[2] Howanitz PJ, Steindel St J, Cembrowski GS, Long Th A. Emergency department stat test turnaround time. Arch Pathol Lab Med 1992; 166:122–8.
[3] Howanitz PJ, Cembrowski GS, Steindel St J, Long Th A. Physician goals and laboratory test turnaround time. Arch Pathol Lab Med 1993; 117:22–8.
[4] Manor PG. Turnaround time in the laboratory: A view of the literature. Clin Lab Sci 1999; 12:85–9.
[5] Steindel St J, Howanitz PJ. Changes in emergency department turnaround time performance from 1990–1993. Arch Pathol Lab Med 1997; 121:1031–41.
[6] Valenstein PN. Preanalytic delays as a component of test turnaround time. Lab Med 1990; 21:448–51.
[7] Wisser H, Bertsch Th, Wisser D. Preanalytical prerequisites for the quality of samples. J Lab Med 2002; 26:284–90.
[8] Roger S, Bywater MJ, Reeves DS. Audit of turn-around times in a microbiology laboratory. J Clin Pathol 1991; 44:257–8.
[9] Hoffmann GE. Development in Pre-analytical Automation (Video) Grafrath, Trillium 1998.
[10] Price CP, John AS. Point of Care Testing: Making Innovation Work for Patient-Centered Care. 2013 Washington AACCPress.
[11] Stenders S, Mandir IA. Transportation of blood samples through adjacent phlebotomists cabins to the laboratory by a simple conveyer belt reduces tunaround time for the samples and waiting time for the patients and may make recommended hand inversions of samples with additive superfluous. Biochem Med 2013:23:A 31 (abstract).
[12] Valenstein P. Laboratory turnaround time. Am J Clin Pathol 1996; 105:676–88.

2. Types of Samples and Anatomic Site of Origin

Walter G. Guder and Peter Hagemann*

2.1 Arterial, Venous or Capillary Blood?

2.1.1 Introduction

While studying human blood circulation (Fig. 2.1.1), one has to distinguish arteries, which carry blood from the (left) heart to the different tissues, from veins, carrying venous blood back from tissues to the (right) heart. A wide net of capillaries connect both types of vessels. These are the major sites of metabolic and gas exchange between tissue cells and vessels. Since oxygen is one of the major molecules tranported and extracted by tissues, thus making arterial blood light red, and venous blood darker red, the arteries are usually depicted in red and veins in blue. Accordingly, different concentrations of metabolites are to be expected depending on their metabolism. For example, glucose usually decreases during the capillary blood flow, whereas lactate is a product of anoxic glucose metabolism and increases from the arterial to the venous blood flow.

Regarding blood sampling for diagnostic purposes, it has to be decided which type of sample is more appropriate for the medical question intended to be answered. Besides medical knowledge about the physiology of the intended analytes to be measured, criteria like practicability of sampling and risk assessment may influence the decision. For these reasons venous blood is the most often used type of sample, followed by capillary sample, which became the preferred type of sample in newborns and young children because of the lower amount of blood needed. Arterial blood on the other hand is mainly used for analysing blood gases and acid base state.

2.1.2 Criteria involved in decision

The decision about the site of collection of blood not only determines the procedure, but also is of substantial influence on the laboratory result. Thus, the concentration of glucose and therefore the diagnostic criteria for diabetes mellitus are different, because glucose is higher in capillary than in venous blood, whereas carbon dioxide (CO_2) increases from capillary to venous blood. Which type of sample is most appropriate depends therefore on the analyte intended to be measured.

In adults venous blood is the standard material used for most analytes.

* translated from Guder et al 2007 [1]

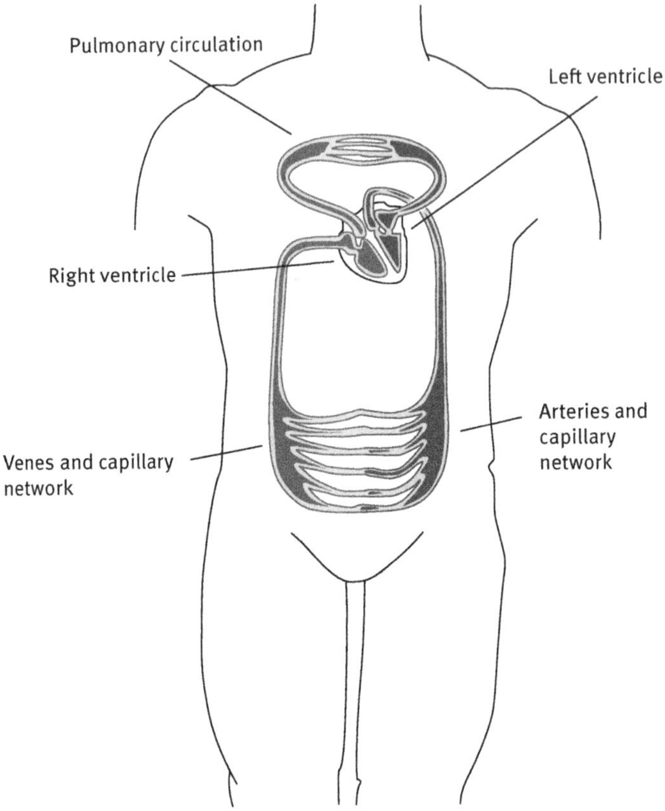

Human circulation

Fig. 2.1.1: Human circulation containing arteries (red) and venes (blue) with the capillary network in between.

Table 2.1.1: Pros and cons for different kinds of blood sampling.

	Arterial	Capillary	Venous
Disadvantages	Higher risk Special materials needed	Contamination with tissue fluid or skin material	Results not always representative because of local conditions and changes (like glucose)
Advantages	Results diagnostically more representative (e. g. blood gases)	Lower blood loss, mixed arterial/ venous, less risky	phlebotomy easier to perform than arterial sampling, sufficient amount of sample, low risk

Table 2.1.2: Comparison of concentrations of different analytes in different kinds of blood sample.

	Arterial	Capillary	Venous
glucose	⇑	▸	⇓
lactate	⇓	◂	⇑
pO2	⇑	▸	⇓
pCO2	⇓	◂	⇑
pH	⇑	▸	⇓
calcium	⇔	⇔	⇔
ionized calcium	⇓	◂	⇑
potassium	⇓	◂	⇑
sodium	⇔	⇔	⇔
cholesterol	⇔	⇔	⇔

⇔ no clinically relevant difference between different samples
⇑ higher compared to veneous/arterial or capillary sample
⇓ lower compared to veneous/arterial or capillary sample
▸ decreasing from arterial to veneous blood
◂ increasing from arterial to veneous blood

Reference

[1] Guder WG, Hagemann P, Wisser H, Zawta B. Fokus Patientenprobe; Kompendium Präanalytik. CD; Heidelberg:BD 2007.

Anne Vassault, Rémy Couderc

2.2 Special Pre-Examination Conditions in Newborns and Pediatric Patients

2.2.1 Introduction

The main requirements for the management of pre-examination process, essential part for the reliability of the results provided by the medical laboratory are given by the standard EN ISO 15189 [1].

Pediatric blood collection has to deal with specific issues, taking into account the pain during blood collection, the difficulties in localizing the vein for phlebotomy, the opportunity to use capillary blood, the handling of small tubes and blood volume, the identification of the samples and labelling, an increased frequency of hemolysis, hyperbilirubinaemia, microclots and bubbles and the need for communicating the results rapidly. The impact of these factors and their effects on the quality management and the choice of method and instruments are discussed in this chapter.

2.2.2 The impact of pain during blood collection

The quality of the sample for analysis depends on the collection step during blood collection. Pain associated with puncture is an undesirable effect especially for pediatric patients. To avoid pain, Eutectic Mixture of Local Anesthetics (EMLA) cream is used routinely as a local treatment prior to venipuncture in children. There are patches (EMLA PATCH®, 1g by patch) available for application 1 h before venipuncture, which are easier to use and more efficient than EMLA cream (Fig. 2.2.1).

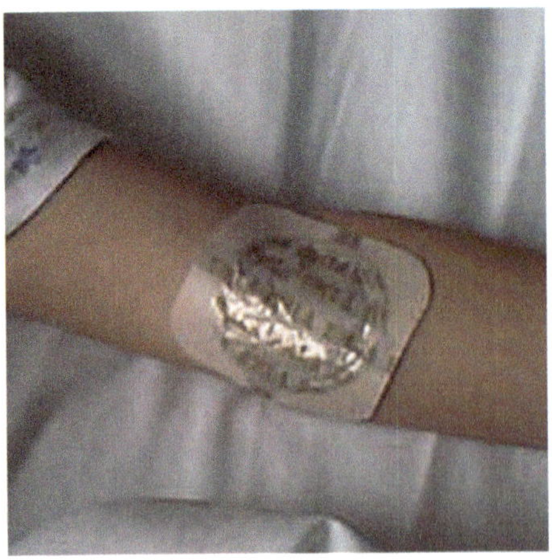

Fig. 2.2.1: Eutectic mixture of local anesthetics (*EMLA* PATCH®).

The time of deposit can be written on the patch. The use of EMLA PATCH® is possible for newborns (> 37 weeks), 0.5 to 1 g, during 1 h, with 12 h as a minimum interval between 2 applications.

It is better to use venous blood collection with EMLA (without pain) instead of skin puncture, which is painful.

Nevertheless, a significant proportion of children will still be distressed and some studies support the proposal that in some children receiving distraction therapy for venipuncture; EMLA may not be required [2].

2.2.3 Difficulty in finding the veins and blood collection

A novel vein imaging system using near-infrared technology (VeinViewer®) has been described as a tool for the identification of superficial veins, thus reducing the number of skin punctures. Visibility of the peripheral veins was improved with this device [3]. The equipment is beneficial to patients in whom venous access is particularly difficult, for instance, in obese children or those affected with poor perfusion, and which enhanced first-attempt success rates in such cases [4]. In a future development, this technology can be used to create robotic systems that allow fully automated phlebotomy [5].

A device (Buzzy®) combining cold, vibration and distraction containing many of the optimal components to overcome barriers to pain treatment proved to be an effective pediatric pain management intervention. The technology of vibration and cold is inexpensive and reusable, requires little extra time, can be implemented by the patient rather than medical staff and may be beneficial for patients across a range of ages [6]. The Buzzy® can be used during diagnostic blood specimen collection by venipuncture for the majority of the routine biochemistry tests except where protein, albumin and transferrin determinations should be performed [7]. However, cold-induced hemoconcentration promotes the efflux of water, diffusible ions and low molecular weight molecules from the blood vessel, thus increasing the concentration of other blood analytes at the puncture site. These variations may influence test results, especially for erythrocytes, hemoglobin and hematocrit. The Buzzy® device should, therefore, be used with caution when collecting blood for conventional hematological testing [8].

In term neonates, venipuncture is less painful as assessed by validated pain evaluation and parental rating and is associated with less maternal anxiety than heel lance [9].

Blood sampling through peripheral venous catheters in children was shown to be pain reducing and, except for glucose measurements, was reliable for determining selected basic analytes. The findings demonstrate the importance of considering technical problems that may occur during blood sampling.

2.2.4 Collection of small sample volumes

The decreased total blood volume that is needed to collect, especially in neonates whose total blood volume lie between 80 and 100 mL/kg body weight, can be accomplished with less than 0.2 mL, thus requiring only one tube. For CSF, only 8 drops could be collected instead of 40 drops that are needed for adults.

Urine specimens are very difficult to collect for infants.

The laboratory shall periodically review the sample volume required and requirements for phlebotomy to ensure that neither insufficient nor excessive amounts of sample are collected. Provisions to reduce the sample volume to collect according to the body weight are to be implemented by the laboratory (e. g. 2 weeks old, 3.5 kg: 350 mL, 15 months old, 10 kg: 800 mL).

53 % of the infants required transfusion during their neonatal intensive care admission [10].

The blood volume collection according to the weight of the child is given in Table 2.2.1.

Table 2.2.1: Maximum amount of blood to be drawn at one time (mL) according to the weight of the child [10].

Weight (Kg)	Maximum amount to be drawn at one time (mL)	Maximum amount to be drawn during stay 1 month or less (mL)
3–4	2.5	23
4–5	3.5	30
5–7.5	5	40
8–10	10	60
10–12.5	10	70
13–15	10	80
15–17.5	10	100
18–20	10	130

Newborns and premature infants (Fig. 2.2.2)

- The neonatal icterus associated with a very high level of bilirubin can lead to interferences with a lot of the methods used. Hence, the choice of methods according to their ability to overcome such interferences is preferable.
- The importance of the influence of hemolysis is also a criterion for the choice of methods. At times, the severity of hemolysis precludes obtaining reliable result and it becomes necessary to re-collect the blood.
- The hematocrit of new borns is greater than 50 %, sometimes even 60 %, so that the plasma or serum yield is very low (maximum 0.2 mL). The laboratory then would be left with the choice to prioritize tests with the collaboration of the clinicians.

Fig. 2.2.2: Premature infant.

2.2.5 Capillary blood

Most of the blood collections are from venous blood in hospitalized children. It is easy to collect capillary blood using heel lance, and is supposedly not traumatic for the infant (Fig. 2.2.3). However, frequent hemolysis can be encountered with such collection.

The use of this type of collection depends on the test – usually, it is used for blood gas analysis, glycemia, HBA_{1C}, hematocrit and blood cell count. However, microclots preclude their use for coagulation tests, and hemolysis invalidates the results obtained for potassium.

Capillary collection by using heel lance is used for newborn screening, with a drop of blood collected on specialized paper. However, improper deposition of the drop of blood on to the spot of paper can be problematic as illustrated in Fig. 2.2.4.

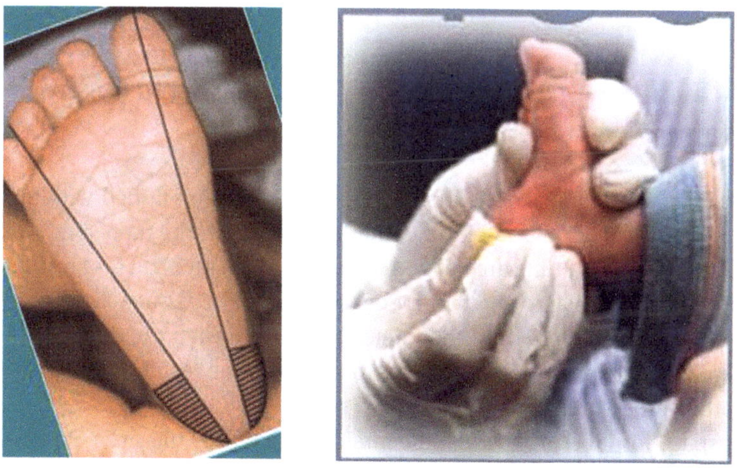

Fig. 2.2.3: Collection of capillary blood using heel lance.

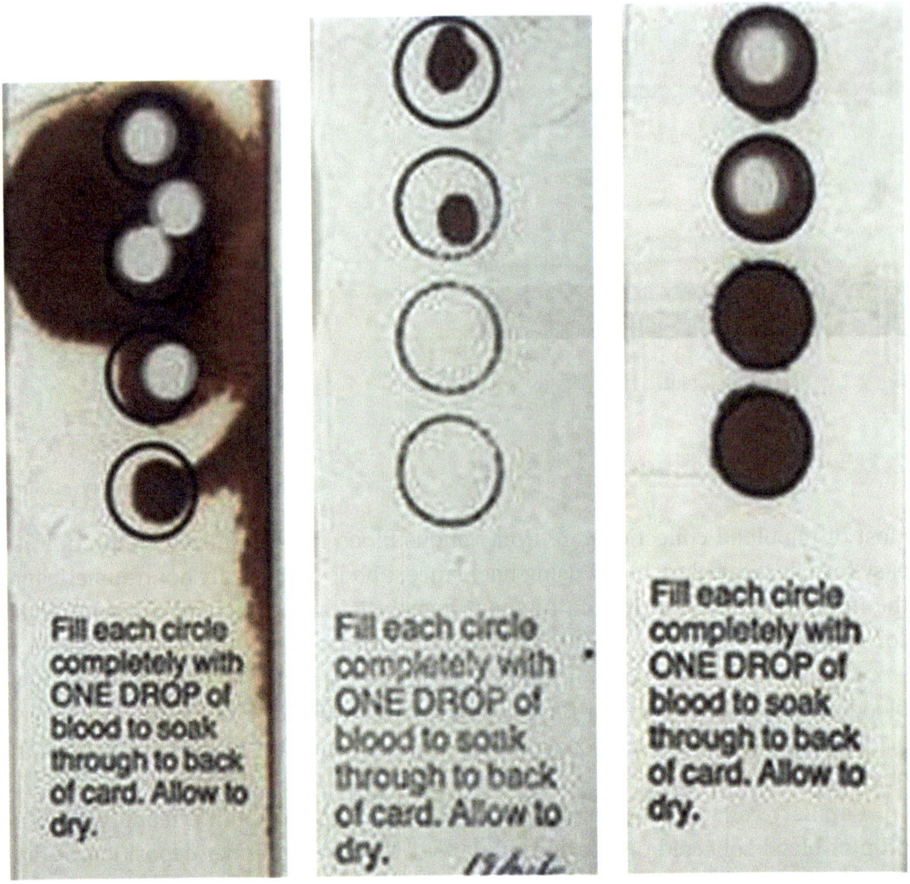

Fig. 2.2.4: Examples of poor blood spot samples received as compared to the appropriate spots.

2.2.5 Identity monitoring: identification of samples from newborn with information system

At birth, a newborn has no name; so, the first sample collected is identified with the mother's name. Some hours or days later, the father is involved in the name and the samples collected at the second time are labeled with the new name. The first results are identified with a different name and cannot be followed as necessary. The name written on the request form differs from the name reported through the Laboratory Informatics System (LIS). This could lead to errors in following the results or misidentification if a rigorous procedure is not instituted.

The reduced space for labeling the blood containers increases the risks of misidentification, even more where plasma or serum has to be separated and aliquoted. Therefore, analyzers with a direct sampling from different primary pediatric devices requiring only a small sample, dead space should be preferred. A direct sampling leads to reduced time consumption and to eliminate errors attributable to manual sample transfer and sample identification.

The geometric configuration of tubes for use with automatic analyzers and robotics has to be investigated before acquisition of systems and devices. There is a risk of misidentification by inversion where the support tube is not integrated to the device as shown Fig. 2.2.5.

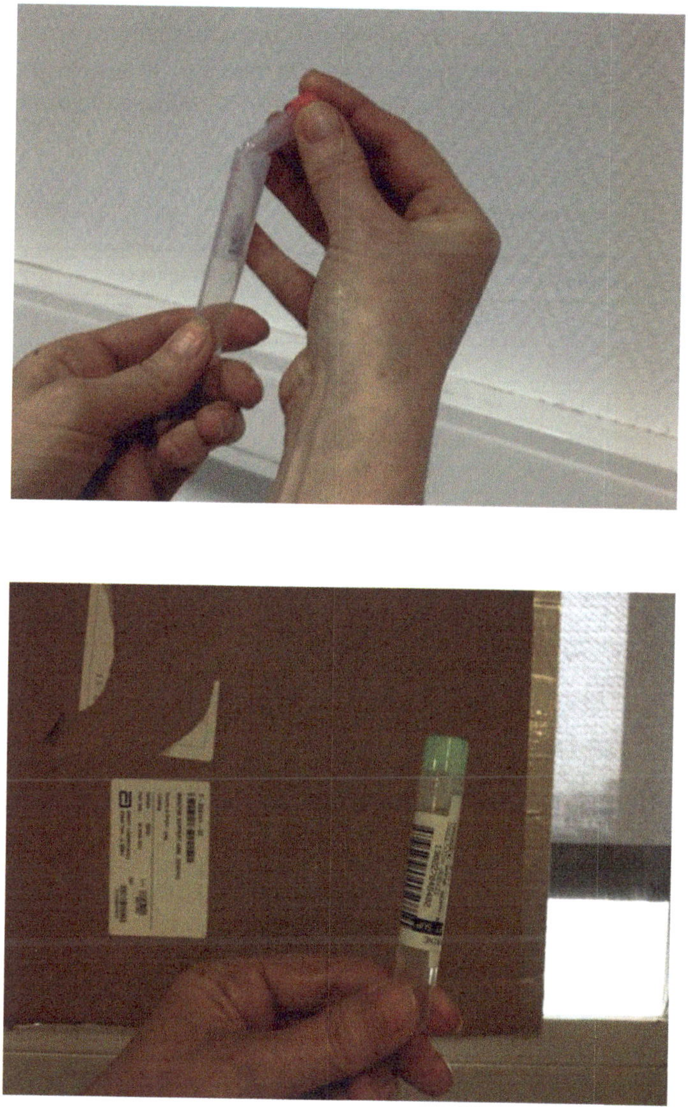

Fig. 2.2.5: Risk of misidentification by inversion: the support tube is not integrated to the device.

2.2.6 Order of filling the tubes

There is a risk of contamination from the anticoagulant of the first tube to the second tube.

The appropriate order of filling the tubes is to be observed regardless of the mode of collection: capillary, venous or catheter. The risk of contamination is less with evacuated tubes than with the tubes that are open. Contamination can also occur through contact from the micro drops of the caps with a compress or from contact with interior of the tube with the syringe. Due to reduced volume, the specimen is much more affected with contamination by EDTA, when the appropriate order for sampling is not fulfilled, leading to false hyperkalimia and hypocalcemia. In this case, the results are erroneous so that re-collection is necessary leading to a delay in arriving at medical decisions (Table 2.2.2).

Table 2.2.2: Consequences of the order of improper filling of the tubes.

Order of collection and/or filling of the tubes	Additives	Consequences
1. Tube without additives	No	
2. Hemostasis	Citrate liquid	Volume to be controlled
3. Biochemistry: electrolytes	Dry Li heparin	Heparin imported Prothrombin time N Activated clotting time ↓
4. Blood cell count	EDTA tri K liquid	EDTA complex ↓ Ca^{2+}, Mg^{2+}, iron imported K =>↑ K

2.2.7 Specimen rejection policy

The following criteria will be used for rejection of newborns and pediatric samples:

- Unlabeled, mislabeled or incompletely labeled specimens will be discarded by the laboratory and new collection will be requested.
- Discrepancy between the patient identification on the test requisition (request) and on the specimen label will cause specimen to be discarded by the laboratory and a new collection will be requested.
- Inappropriate container and additives
- Insufficient volume
- Sample(s) deemed unacceptable due to the requirements of test methodology; i. e. hemolysis, lipemia, in adequate samples, wrong tube or wrong specimen type
- Suspected specimen contamination due to discrepant result
- Preanalytic handling and transport requirements not met

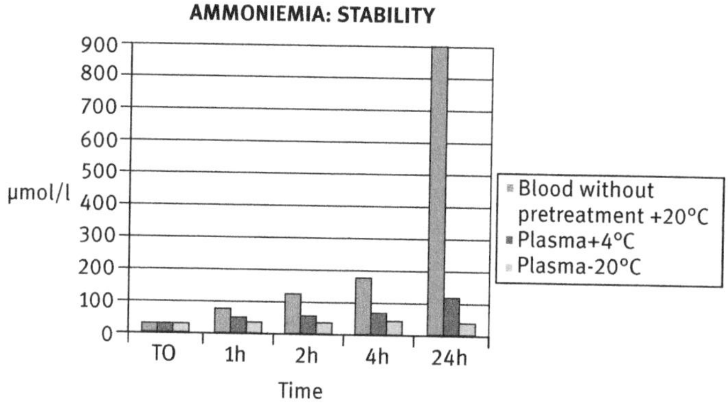

Fig. 2.2.6: Study of stability of ammonia in blood samples according to transport conditions.

2.2.8 Impact of preanalytical conditions on analytical process

- Appropriate method and analyzers are necessary for small volumes (reduced dead space). Usually, for check up, a volume less than 100 µL plasma or serum is required for infants.
- The method used has to deal with the interference of bilrubin which is very frequently increased
- Hemolytic samples are frequent: e. g. method for bilirubin assay
- A great deal of the requested test (usually in hospital practices, 50 % of the request) requires results with a short turnaround time (<60 min).

2.2.9 Impact of preanalytical conditions on postexamination process

The pediatric status impacts on the postanalytical process because the results must be interpreted by comparison with reference ranges according to age and sex. Providing age-specific reference ranges is necessary for an appropriate interpretation of the results by the pediatricians (e. g. creatinine, see Table 2.2.3).

Information about the time of sampling according to the fasting state for example is usually useful for an appropriate interpretation of the results (Fig. 2.2.7 for ketone bodies measurement in plasma)

Communication between laboratory personnel and pediatricians is required [11, 12].

2.2.10 Conclusion

To manage and improve the quality of the preexamination phase is essential for newborns and pediatric patients' care. The tests for pediatric patients necessitate appropriate devices, methods and analyzers.

Table 2.2.3: Plasma creatinine according to children's age

Newborn <4 months	30–90 µmol/L
Child <1 year	20–50 µmol/L
Chid <12 years	30–70 µmol/L
Adolescent	40–85 µmol/L
Adult male	65–120 µmol/L
Adult female	50–100 µmol/L

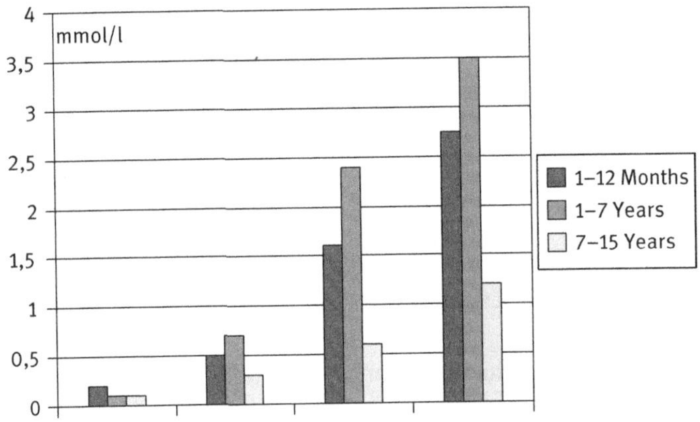

Fig. 2.2.7: Plasma ketone bodies according to fasting time.

A close cooperation among pediatricians, manufacturers and medical laboratory responsible is necessary.

References

[1] ISO EN Standard 15 189: Medical laboratories – particular requirements for quality and competence. GENEVA: ISO. 2012.

[2] Lal MK, McClelland J, Phillips J, Taub NA, Beattie RM. Comparison of EMLA cream versus placebo in children receiving distraction therapy for venipuncture. Acta Pædiatr 2001; 90:154–9.

[3] Strehle E-M. Making the invisible visible: near-infrared spectroscopy and phlebotomy in children. Telemedicine and e-Health 2010; 16: October.

[4] Kim MJ, Park JM, Rhee N, Je SM, Hong SH, Lee YM et al. Efficacy of VeinViewer in pediatric peripheral intravenous access: a randomized controlled trial. Eur J Pediatr 2012; 171:1121–25.

[5] Strehle E-M. Novel vein viewing system assists with venipuncture in children. J Pediatric Intensive Care 2012; 1:1–2.

[6] Baxter AM, Cohen LL, Lawson ML, McElvery H, von Baeyer CL. An integration of vibration and cold relieves venipuncture pain in a pediatric emergency department. Pediat Emerg Care 2011; 27:1151–56.

[7] Lima-Oliveira G, Lippi G, Salvagno GL, Montagnana M, Picheth G, Guidi GC. Quality impact on diagnostic blood specimen collection using a new device to relieve venipuncture pain. Indian J Clin Biochem 2013; 28:235–41.

[8] Lima-Oliveira G, Lippi G, Salvagno GL, Campelo MDR, Adala Tajra KS, dos Santos Gomes F, et al. A new device to relieve venipuncture pain can affect hematology test results. Blood Transfus 2014; 12 Suppl1:6–10.

[9] Shah VS, Ohlsson A. Venepuncture versus heel lance for blood sampling in term neonates (Review). The Cochrane Library 2011; 10:1–26.

[10] Howie SRC. Blood sample volumes in child health research: review of safe limits. WHO 2011; 89:46–53.

[11] Shaw JLV, Marvasti TB, Colantonio D, Adeli K. Pediatric reference intervals: challenges and recent initiatives. Crit Rev Clin Lab Sci 2013; 50:37–50.

[12] Couderc R, Vassault A. Pediatric clinical chemistry: why is it different? Clin Biochem 2014; 47:747–8.

Walter G. Guder, Michael P. Cornes

2.3 Venous Blood Sampling (Phlebotomy)

2.3.1 Materials used for venous blood sampling

Before venous blood sampling (phlebotomy) is performed, all materials required have to be collected and confirmed to be in date including the medical order requesting laboratory tests. The medical request should have the patient's name, address, date of birth and a unique identification number as well as the required tests as a minimum. They may be put together on a container suitable for transport to the patient's room or phlebotomy location.

The decision regarding size and type of tube is to be decided according to the recommendations of the working group on preanalytical quality (1 and Table 2.3.1), national [2, 3] international [4] or hospital specific guidelines in agreement with the recommendations of the respective organization like the European Working Group on preanalytical phase [5]. The present recommendations were largely taken from a compendium on preanalytics published in German [6], WHO and CLSI guidelines [7].

After all materials are prepared the person performing phlebotomy goes to the patient. He/she identifies the patient using positive identification wherever possible (i. e. ask them to state name, etc. and not to confirm what you state). The number of patient identifiers should be defined in local guidelines but should be at least 2. Patients should also have a patient ID bracelet. If the patient is unconscious, a nurse, relative or friend should identify the patient and their name recorded. If the person is not identical with the one named in the prepared samples, the request has to be restarted and the identity of the patient as well as the request confirmed by comparison.

Table 2.3.1: Materials needed for phlebotomy

Single use disposable latex free tourniquet
Prepackaged gauze pads
Antiseptic spray or liquid (70% isopropyl alcohol)
Needles and holders if needed
Tubes with patient's labels needed stored upright and dry
(heparin plasma tube,serum tube, EDTA tube, citrate tube, specialized tubes [like CTAD-tube] bandage and/or plaster)
Safety container for disposal of needles and other materials
Syringes and injection material, if needed
The tubes can be labeled with the patient's identification labels before or after sampling but must be labeled in the presence of the patient

2.3.2 How to perform phlebotomy

After the request has been assured the following order of actions has been recommended [7]:

- **Identify patient.**
- **Sanitize hands.**
- **Verify patient's diet restrictions and ensure they are suitable for the tests.**
- **Assemble supplies and put on gloves.**
- **Reassure patient.**
- **Position the patient in specific phlebotomy chairs with arms** or if lying the arm should be extended and fully supported.
- **Verify paperwork and tubes.**
- **Ensure the patient's hand is closed.**
- **Select vein site** taking note of any scarring, hematomas, intravenous therapy, mastectomy and any other contraindications.
- **Cleanse venipuncture site.**
  Cleanse by spraying appropiate antiseptic spray on the skin at and around the intended place of venipuncture. Prevent contamination of sample by allowing to dry or/and cleansing with a gauze by circular motions from center to periphery. Do not use alcohol-containing cleansers when ethanol is to be measured.
- **Apply tourniquet.**
  Tourniquet should be applied 7.5–10 cm above the site of puncture. Tourniquet pressure should be between the diastolic and systolic blood pressure to ensure further blood supply. Superficial veins become more prominent. Once blood flow commences, the tourniquet should be released.
- **Inspect needle and other equipment.**
- **Perform venipuncture by filling the tubes.**
  The needle, usually fixed to the holder or tube, is carefully inserted into the vein at a 30° angle, holding the whole needle unit between thumb and forefinger, using the thumb as a guide to the skin. Change hand position as soon as the needle is in the vein. In case vacuum tubes are used, put the first tube into the holder with your right hand thumb by keeping the holder tight by your left hand. When blood flow commences the tourniquet can be removed and the patient's hand relaxed.
  If tourniquet remains more than 2 min, the concentration of high molecular weight analytes increases (including cell counts). This is due to pressure-dependent changes of water and low molecular weight molecules into the extravascular space and the interstitium.
  If needed, more tubes may be filled by the same procedure following the order recommended (Table 2.3.2). This rule considers avoidance of contamination of tubes with possibly disturbing constituents of other tubes. If a citrate tube for

Table 2.3.2: Recommended order of draw of samples during phlebotomy [2, 3, 7]

1.	Blood culture tube
2.	Coagulation tube
3.	Serum tube with or without clot activators, with or without gel
4.	Heparin tubes with or without gel
5.	EDTA tubes
6.	Glycolytic inhibitor tubes
7.	Other tubes (like trace element tube)

measuring coagulation tests is the only tube, it is suggested to fill a "neutral tube" first to avoid contamination with tissue thromboplastin from the puncture site. **Tubes should be filled to recommended levels** to ensure accuracy of results.
- **Remove needle.** After the tubes are filled, the needle can be slowly removed and disposed of in sharps bin next to patient after activating safety procedures as per manufacturer's guidance.
- **Place gauze over puncture and ask patient to hold it in place with moderate pressure for at least 30 seconds.** The patient should not bend his/her arm.
- **Bandage patient's arm:** In patients with decreased coagulation, a plaster with gauze may close the vessel with slight pressure.
- **Ensure there is no excessive bleeding.**
- **Mix tubes by inversion as per manufacturer's guidance.**
- **Chill specimen (only for special tests).**
- **Time-stamp paperwork.**
  Time of sampling is important in interpreting results with circadian rhythm and in assays who are dependent on the timely action of therapy to know if sample was taken before or after drug dosage, massage or other treatment.
- **Send properly labeled tubes to proper laboratories.**

2.3.3 Influences and interferences with impact on laboratory results during phlebotomy

When using butterfly devices, minor changes have been observed [8]. These seem acceptable, if this version is described in the phlebotomy guideline [9].

The needle perforates the vein

If the needle perforates the vein on the opposite site of injection, it may bleed into the arm tissue. Remove the needle immediately and bandage the site with a pressure plaster to prevent further bleeding. For further sampling of blood or intravenous injections use the other arm only for the next 6 hours.

Sampling with catheter

When sampling with a venous catheter, the few milliliters of blood representing 1–2 volumes of the catheter should be discarded to avoid contamination with the anticoagulant used. A coagulation tube should be filled then before a heparin containing tube is collected.

References

[1] Guder WG, Fiedler W, daFonseca-Wollheim F, Schmitt Y, Töpfer G, Wisser H, Zawta B. German united Society of Clinical Chemistry and Labratory Medicine. Quality of Diagnostic Samples 4th ed 2015, Oxford BD Diagnostics.

[2] Nicolac N, Supak-Smolciç V, Simundiç A-M, Celap I. Croatian. Society of Medical Biochemistry and Laboratory Medicine: national recommendations for venous blood sampling. Biochem Med 213; 23:22–54.

[3] Gurr E, Arzideh F, Brandhorst G, Gröning A, Haeckel R, Hoff T et al. Musterstandardarbeitsanweisung Präanalytik. J Lab Med 2011; 35:55–60.

[4] World Health Organisation (WHO) Guidelines on drawing blood: best practices in phlebotomy. 2010; http://whqlibdoc.who.int/publications/2010/9789241599221_engl.pff

[5] Simundic A-M, Cornes M, Grankvist K, Lippi G, Nybo M. Standardization of collection requirements for fasting blood samples. For the working group on preanalytical phase (WG-PA) of the european federation of clinical chemistry and laboratory medicine (EFLM). Clin Chim Acta 2013, http://dx.doi.org/10.1016/j.cca.2013.11.008

[6] Guder WG, Hagemann P, Wisser H, Zawta B. Fokus Patientenprobe, Kompendium Präanalytik. CD 2007. Heidelberg, Basel, Schwechat BD (Becton Dickinson AG).

[7] Clinical Laboratory Standards Institute (CLSI); Procedures for the collection of diagnostic blood specimens by venipuncture. Approved standard. 6th edn, CLSI Wayne, PA 2007, Document H3–A6.

[8] Lippi G, Salvagno GL, Brocco G, Guidi GC. Preanalytical variablity in laboratory testing: influence of the blood drawing technique. Clin Chem Lab Med 2005; 43:319–25.

[9] Lippi G, Mattiuzi C, Guidi GC. Laboratry quality improvement by implementation of phlebotomy guidelines. MLO Med Lab Obs 2006; 38:6–7.

Walter G. Guder, Peter Hagemann

2.4 Arterial Sampling of Blood*

2.4.1 When is arterial sampling indicated

Whenever arterial sampling is indicated, particular care should be taken [1]. In general, arterial sampling is only needed when venous or capillary blood samples cannot provide the information required to support a medical decision in the treatment of a patient. For example, the respiratory component of acid–base homeostasis and blood gases cannot be derived from results obtained from venous or capillary blood samples.

2.4.2 Deciding on the puncture site

Decision about the puncture site includes several criteria:
- Accessibility of vessel
- Age of the patient
- Prevention of diagnostic and medical risks. For instance, prevention of edematous areas, preferential use of existing arterial bypasses like those existing in dialysis patients.

In adults, the femoral artery represents the most often used arterial vessel for diagnostic purposes, followed by the brachial artery and the radial artery.

In children, head arteries are used and, during the first two days after birth the umbilical artery can be used. Here capillary sampling is often preferable. In children below 4 years of age the femoral artery should be used to prevent infectious and bleeding risks.

Table 2.4.1: Materials needed for arterial puncture [2]*.

– Self filling tubes adequate for arterial sampling, containing buffered heparin and ion buffered solution [1]
– Needles above 25 gauge size in safety holders for risk free disposal
– Sterile gauze pads used for disinfection
– Suitable antiseptic solution (like 70 % alcohol)
– Gauze pads and bandage for closing the vessel after puncture
– Adequate material for patient identification on sample
– Local anesthetic solution
– A container keeping the sample at 1–4°C after sampling

* translated from Guder et al [2].

2.4.3 Performance of arterial puncture

Preparing arterial puncture: Prepare the self-filling tube with a volume not higher than needed. In case of a higher volume syringe reduce volume to the needed size. Localize arteries to be punctured and support stable position with two fingers. Hold arm whose brachial or radial artery is intended to be punctured on a stable support.

Arterial sampling: Locate the needle into the direction of blood stream, keeping the bevel of the needle upwards.

Puncture with an angle of about 45° to the skin surface. Femoral punctures should be performed vertically. A winged infusion set is helpful in observing the incoming pulsative blood flow.

When the needle is inside the artery the tube/syringe is filled quickly. Hold needle and blood container closely fixed to prevent contralateral puncture of artery and withdraw needle quickly, simultaneously pressuring a dry gauze sponge over the puncture site.

After the needle is withdrawn, keep pressure on the site of puncture for at least 5 min. To ensure stagnation of bleeding and prevent formation of hematoma, inspect site of puncture every minute after releasing pressure.

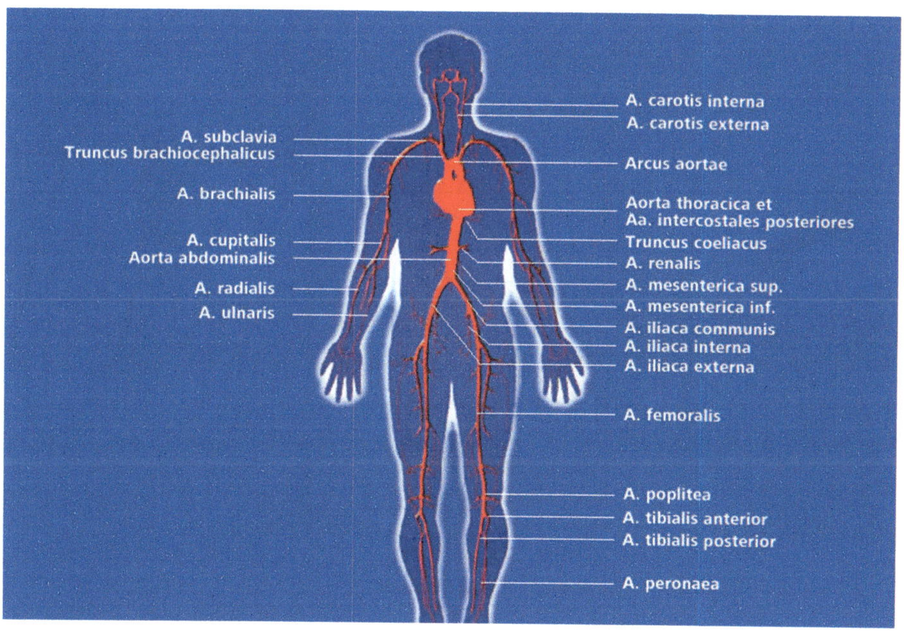

Fig. 2.4.1: Localization of arteria ulnaris and brachialis.

To remove air bubbles formed eventually, put syringe vertically on a gauze pad and eliminate bubbles through the needle before the tube is tightly closed and the pat for patient identification kept fixed on sample.

Sufficient mixing of sample with anticoagulants is to be ensured by rolling sample by vertical turning around in the hand surface. For samples intended for posting, the blood container is to be closed safely and storage and transport on ice cubes to be ensured.

The needle is to be disposed safely into an appropriate needle container.

2.4.4 Sampling from an arterial cannulation

Arterial indwelling catheters permit serial sampling of arterial blood without the need for repeated arterial punctures. Besides, arterial blood pressure and blood gases can be measured continuously. The puncture is usually done at the radial or ulnar artery, more seldom at the brachial artery. Central pulmonary arteries and veins are cathetered for controlling central blood circulation.

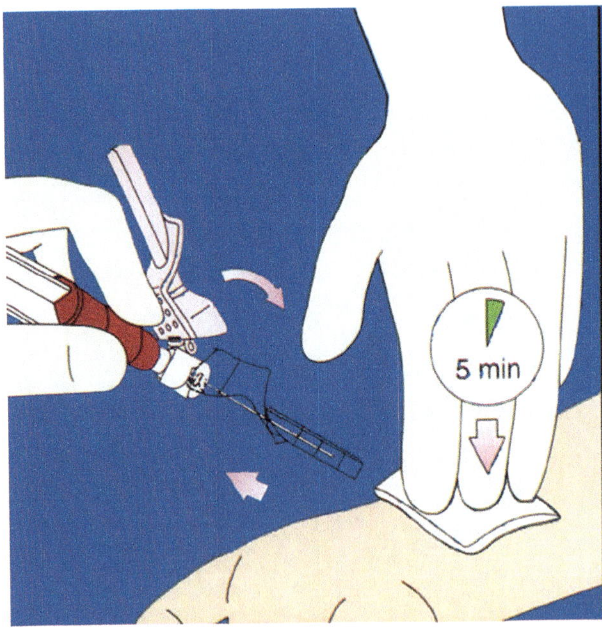

Fig. 2.4.2.1: Removal of sample from and pressure on artery.

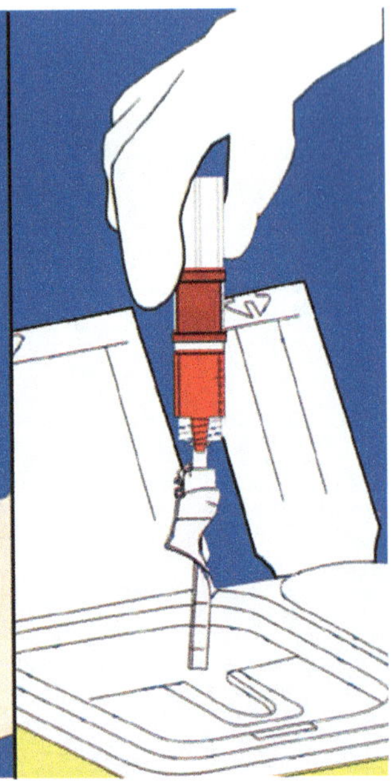

Fig. 2.4.2.2: Arterial sample handling and safety container for needles used for blood sampling.

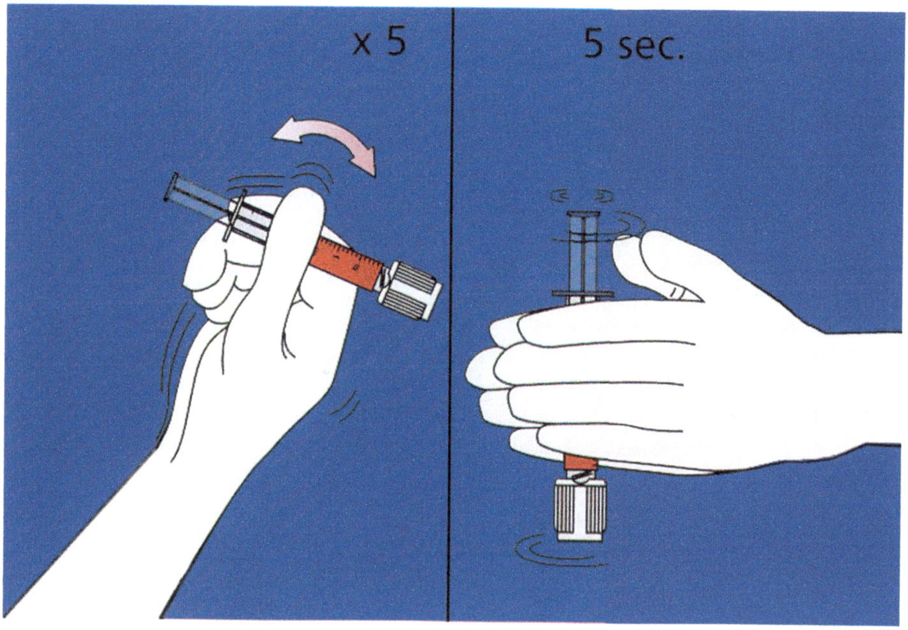

Fig. 2.4.2.3: Mixing of sample.

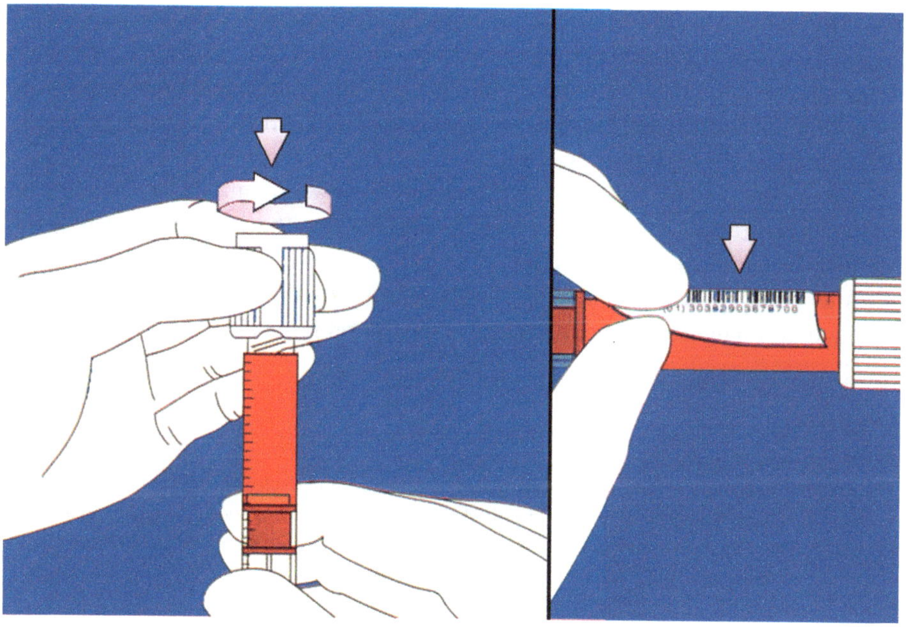

Fig. 2.4.2.4: Removing of air bubbles and position of identity marker.

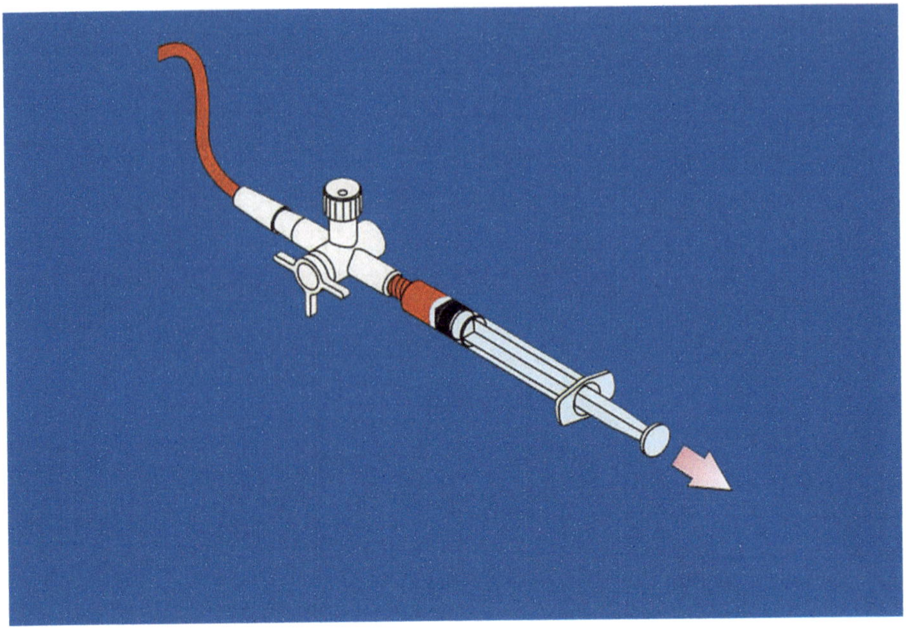

Fig. 2.4.3: Arterial catheter with three way cannula.

Before samples are collected, the catheter is to be cleaned using the isotonic salt solution. This is usually done by aspiration of the six-fold volume of the catheter with a syringe that is opened by a three-way cannula. This sample is discarded. Only then the diagnostic sample is collected by filling a self filling tube or syringe containing the appropriate anticoagulants and buffered ions. This syringe or vessel is optimally positioned in a 45° angle to the catheter.

When the sample container is filled the three-way cannula is closed. The sample tube is also closed with an appropriate stopper. The catheter as well as the cannula is cleaned by isotonic saline.

After air bubbles are eliminated (see above) the closed sample container is transported to the analytical site usually close to the sampling site.

References

[1] Clinical Laboratory Standards Institute (CLSI). Procedures and devices for the collection of arterial blood specimen. Approved standard 4th ed., CLSI Wayne PA: 2004, document H11–A4.

[2] Guder WG, Hagemann P, Wisser H, Zawta B. Fokus Patientenprobe. Kompendium Präanalytik CD 2007. Heidelberg, Basel, Schwechat BD (Becton-Dickinson AG).

Michael P. Cornes, Walter G. Guder

2.5 Capillary Sampling of Blood

2.5.1 Indication of capillary blood sampling

Capillary blood sampling is always indicated if small amounts of blood are needed. Likewise, capillary sampling is a good alternative when there is reduced accessibility to arterial blood. This is the case in situations where venous blood is not appropriate to obtain the required analytical information. For these reasons capillary blood is the major type of sample in pediatric medicine. In adults, the determination of acid–base status and blood gases, lactate and glucose is most commonly done from capillary samples. Following the introduction of analyzers that allow the simultaneous determination of other analytes at point of care, capillary samples are used increasingly for analysis in clinical chemistry, immunochemistry and serology.

2.5.2 Sites of sampling capillary blood

In selecting the site of sampling of capillary blood one has to consider age and gender-specific, anatomic and patient-specific psychological aspects. In newborns and neonates arterial blood can be obtained from the medial and plantar surface of the heel [1, 2]. Figure 2.5.1 shows the optimal area on the heel. Here bone cannot be injured by puncture if puncture does not exceed 2 mm depth. Children below 6 months of age

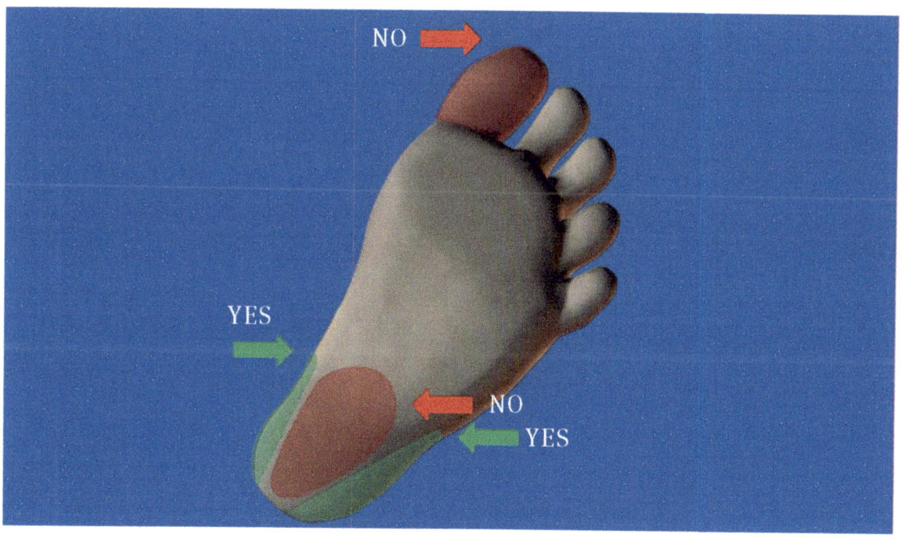

Fig. 2.5.1: Puncture sites to collect capillary blood in newborns and babies [1].

Table 2.5.1: Conditions influencing heel-prick location (adapted from WHO guideline [3].

Condition	Heel prick	Finger prick
Age	Birth to 6 months	> 6 months
Weight	3–10 kg	>10 kg
Placement of lancet	Medial or lateral plantar surface	On the side of the ball of the finger perpendicular to the lines of the fingerprint
Recommended finger	N/A	Second and third fingers

can also be punctured on the point of the big toe. The fingertips should not be punctured at this age.

In children >6 months and having sufficient blood flow and in adults the palmar surface of the distal phalanx of fingers should be used. The middle fingers are the preferred sites of puncture (see Table 2.5.1).

Sites of capillary blood collection in adults

For the analysis of capillary blood glucose, earlobes as well as finger tips can be used, if performed by well-trained personnel and not prevented by earrings or other ornaments. This is also indicated when fingers cannot be used for other reasons like reduced peripheral blood flow. Figure 2.5.2 shows which finger tips are most appropriate.

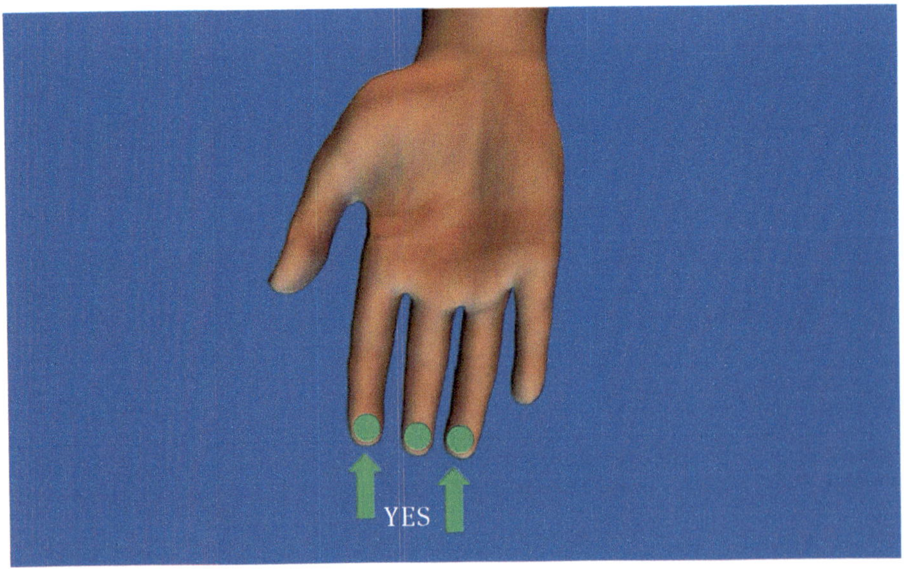

Fig. 2.5.2: Capillary blood sampling [1].

2.5.3 Materials needed for capillary blood sampling

Table 2.5.2 shows the list of materials that must be prepared before capillary blood sampling is performed [2].

2.5.4 Performance of capillary blood sampling

Preparation: After the patient has been appropriately identified as detailed in Chapter 2.3, select the puncture site and warm up skin by using a towel warmed up in hot water, infrared lamp or by microwave. This can increase blood flow up to seven-fold. Using warm hands or shawls carries the risk of contaminating sample or causing hemolysis. Keep puncture site in a clean safe position.
Sampling: After the site of puncture has been cleaned with disinfectant, the site is dried with sterile gauze. Using the lancet right angular to the skin inserting it with a short, precise, continuous prick. Semi automated lancettes can be inserted by pressing on the button. The first drop of blood that flows after puncture should be discarded by wiping with gauze, since this may be contaminated with tissue fluid. By application of gentle pressure but without "milking" or massaging the area around the puncture site allow free-flowing drops of capillary blood into the properly labeled microtube device. Samples are collected in the reverse order to venous blood collection to avoid platelet clumping, i.e. haematology followed by clinical chemistry then blood bank samples.

Table 2.5.2: Materials needed for capillary blood sampling.

- *Safety Lancets* are to be selected according to the site punctured and the amount of blood needed. The intended depth of puncture is to be decided. For newborns 1.4 and children 1.4–1.6 mm penetration depth lancets are to be choosen, whereas adults can be punctured with 1.8–2.5 mm lancets. Semiautomatic lancets facilitate this decision
- *Disinfectant*: 70% isopropanol in water
- *Capillary tubes*: depending on the intended investigations, serum tubes, buffered heparinated as well as citrated and EDTA-tubes are available
- *Iron mixing bar* ("flea") and magnet to mix blood in capillaries together with closures to allow air-free sampling for blood gas analysis
- *Filter-paper* or similar material used for newborn screening
- Container and carrier for capillary blood samples
- *Gauze* with desinfectant
- Bandage or plaster
- *Warm moist towel* and creme to warm up skin
- *Cooling bag*, when transport for blood gas analysis is intended
- *Gloves* and appropriate container
- *Identification labels* to identify each patient's sample
- *Safety container* to discard lancets and other material

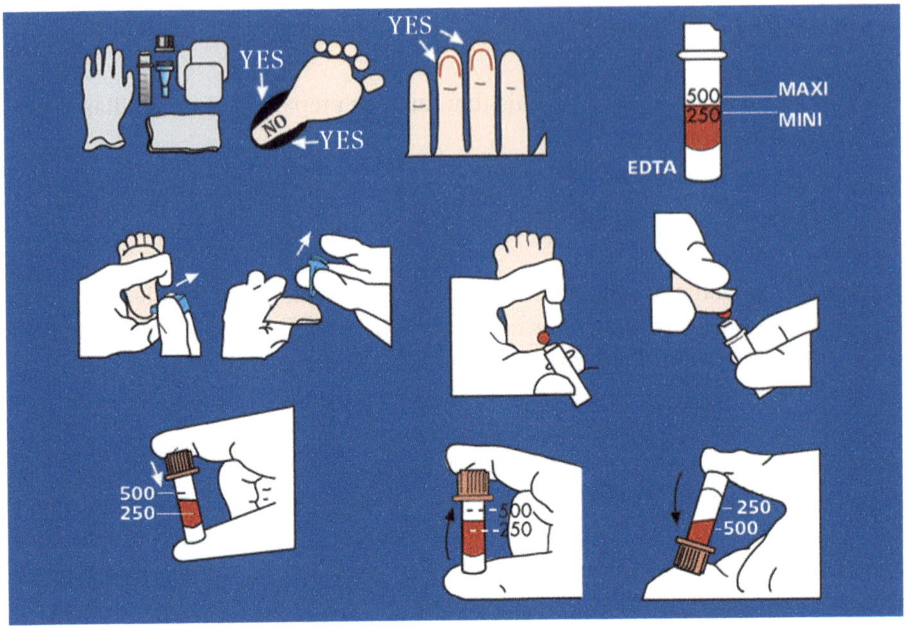

Fig. 2.5.3: Sampling capillary blood into appropriate containers [1].

After capillary tubes are appropriately filled, the tubes are closed and samples mixed by inverting them gently according to manufacturer recommendations. Samples should not be shaken.

If the puncture did not provide a sufficient volume of blood, a second puncture site should be selected to fill up the vessels. In case of blood gas determination, formation of air bubbles must be prevented at all stages of sampling, storage and transport. Air bubbles are to be eliminated by carefully inverting samples.

After confirmation of sample identity on the label, the safely closed samples may be transported to the point of analysis together with the intended request.

2.5.5 Typical complications during capillary sampling

To ensure that the capillary sample represents the arterio-venous mixture of the patient's circulation, any errors changing the results of intended examination should be avoided.

These include:

- Use of the first blood drop after puncture may contaminate the sample with tissue factors from the skin or tissue.
- Contamination with infusion solutions (like glucose). Do not puncture at a site where infusion liquids may have been in contact with the skin.

- Applying too high pressures to the side of puncture site leading to hemolysis and contamination with extravascular fluid.
- Too little amount of blood creates problems when mixed with a well defined amount of anticoagulant. This is particularly important in coagulation tests from capillary blood.

Moreover, the depth of penetration is to be observed in relation to the age of patients and site of puncture. For such reasons fingers should not be punctured in newborn children. After puncture, cleaning with gauze followed by air drying may be sufficient. Plasters and bandages may hurt the skin more than the puncture itself.

References

[1] Guder WG, Hagemann P, Wisser H, Zawta B. Fokus Patientenprobe, Kompendium Präanalytik. CD 2007. Heidelberg,Basel, Schwechat BD (Becton Dickinson AG).

[2] Clinical Laboratory Standards Institute (CLSI) Procedures and decises for the collection of diagnostic capillary blood specimens. Approved standard 5th edn , CLSI Wayne PA 2004,Document H4–A5.

[3] World Health Organisation (WHO). Guidelines on drawing blood: best practices in pflebotomy. Geneva:2010.

Walter G. Guder, Sheshadri Narayanan

2.6 Plasma or Serum? Which Anticoagulant to Use?

2.6.1 Introduction

Before anticoagulants were developed as additives in blood collection tubes, serum was the standard sample of blood utilized for analysis in clinical chemistry, immunology and serology.

Serum is obtained from whole blood by separating the extracellular fluid after the completion of coagulation process. Serum can therefore be regarded as an artifact [1]. By definition it is devoid of clotting factors but enriched with cellular components of platelets and metabolic products.

Plasma is the virtually cell-free supernatant following centrifugation of anticoagulated blood. This has been possible by collecting blood into an anticoagulant vessel. After this has become a standardized procedure, plasma can be used as standard sample in analytical procedures. The question, however, as to which anticoagulant should be used and when anticoagulated plasma is the preferable sample is the subject of this chapter.

2.6.2 Anticoagulants to be used

The first anticoagulant used for diagnostic purposes was oxalate, used either as sodium or lithium salt. After citrate and EDTA were introduced, this was only seldomly used, because it may lead to wrong results with haematological and hemostasiological examinations [2]. Table 2.6.1 gives a brief overview on the present state of use of anticoagulants. Currently, heparinated blood is the preferred source for plasma used in clinical chemistry and immunology, whereas citrate is routinely used in hemostaseology and EDTA in hematology. However, EDTA may be recommended also for sensitive analytes in plasma, because stability is superior in this medium. For detailed information on all presently needed analytes see table in Annex [3].

For some analytes there are diagnostically relevant differences between the results obtained from serum and those obtained from plasma. This may be due to several physiological and technical reasons as follows:

- The analyte may be consumed during clotting: fibrinogen, platelets, glucose.
- The analyte may be released from cells during clotting: potassium, lactate dehydrogenase, phosphate, ammonia, lactate (likewise formed from glucose during clotting), but also neurone specific enolase, dopamine and serotonin [4, 5].
- The anticoagulant may interfere with the assay or contaminate with its cations: Lithium (heparinate) with flame photometry, when calibrated with lithium.

Table 2.6.1: Anticoagulants and their application in diagnostic examination of blood.

Anticoagulant	Recommended cation	Usual final concentration	Application	Comment
Ethylenediamine-tetraacetate (EDTA)	Na_2 salt K_2 salt K_3 salt	4.1–6.8 mmol/L (1.2–2.0 g/L)	Blood cells, Plasma for sensitive analytes	K_2-EDTA recommended by ICSH, See Annex
Citrate	Na_3-citrate	10.9 mmol/L (3.2%) 12.9 mmol/L (3.8 %)	Coagulation assays	Buffered pH 5.5–5.6 recommended
Heparinates	Na, K, or Li- salts	12–30 kiU/L, 30–50 kiU/L	Plasma for Clinical Chemistry	
Hirudin	----	10 mg/L	“universal”	Not widely used
Oxalate				Almost obsolete

– Methodology used (monochromatic versus bichromatic measurement) [6] may cause interference. This includes interference of fibrinogen with some heterogeneous immunoassays.

Therefore before changing from serum to plasma in the routine laboratory, the different advantages and disadvantages have to be considered.

2.6.3 Advantages of using plasma

The following aspects support the preferential use of plasma versus serum in laboratory medicine:

Time saving: Plasma samples can be centrifuged directly after sample collection, unlike serum, in which coagulation is completed after 30 min,

Higher yield: 15 to 20 % more in volume of plasma than of serum can be isolated from the same volume of blood.

Prevention of coagulation-induced interferences: Coagulation in primary and secondary tubes that were already centrifuged, may block suction needles of the analysers when serum tubes are used. This is prevented by using anticoagulants.

Prevention of coagulation-induced influences: The coagulation process changes the concentrations of numerous constituents of the extracellular fluid beyond their maximum allowable limit (7, 8). The changes are induced by the following mechanisms:

a. Increase in the concentrations of platelet components in serum as compared to plasma (e. g. potassium, phosphate, magnesium, aspartate aminotransferase, lactate dehydrogenase, serotonin, neurone-specific enolase, zinc). Release of amide-NH_3 from fibrinogen induced by action of clotting factor XIII.

b. Decrease in the concentration of constituents in serum as a result of cellular metabolism and the coagulation process (glucose, total protein, platelets).
c. Activation of the cell lysis of erythrocytes and leukocytes in noncoagulated blood (cell-free hemoglobin, cytokines, receptors).

Lower risk of hemolysis and thrombocytolysis: In healthy individuals, free hemoglobin is about 10 times less concentrated in plasma than in serum. Likewise, platelets remain intact in vitro; there is hence no pseudohyperkalemia in plasma samples [8].

Certain constituents should only be measured in plasma (e. g. neuron-specific enolase, serotonin, ammonia) to obtain clinically relevant results.

2.6.4 Disadvantages of plasma over serum

The addition of anticoagulants may interfere with certain analytical methods or change the concentration of the constituents to be measured:

Contamination with cations: NH_4^+, Li^+, Na^+, K^+.

Disturbing platelets: Even if platelet potassium is not released into plasma, when heparinated blood is centrifuged, it was observed that platelets often remain above the cellular layer and even above the plasma separator and can be the cause of increased potassium concentrations if the lower part of the centrifuged plasma is used for the flame photometric potassium determination, or in detergent containing sampling needles with ion selective electrodes [9].

Assay interference caused by metals complexing with EDTA and citrate: Inhibition of alkaline phosphatase activity by zinc binding, inhibition of metallo-proteinases, inhibition of metal-dependent cell activation in function tests, binding of calcium (ionized) to heparin [10].

Interference by fibrinogen in heterogeneous immunoassays [11]. Thus lower TSH and T_3- values results were observed in some immunoassays, which were no longer visible in later test sets or in free T_3 assays [12].

Inhibition of metabolic or catalytic reactions by heparin: Taq polymerase, as needed in the polymerase chain reaction (PCR), is strongly inhibited by low doses of heparin [13, 14]. This can only be eliminated by application of heparinase or precipitation of mRNA with Lithium chloride 1.8 mmol/L final concentration [15].

Changes in the distribution of ions between the intra-cellular and extra-cellular space: Cl^-, NH_4^+ ions change their extracellular concentrations when EDTA or citrate is used as anticoagulant [1].

Serum electrophoresis results are disturbed by fibrinogen: An extra peak appears in protein electrophoresis, when plasma is used instead of serum. This additional peak between β and γ-phase can lead to misinterpretation as M-gradient and can be prevented only after pretreatment with a fibrinogen precipitating agent activating coagulation in plasma.

2.6.5 Recommendations

The table of the Annex indicates sample materials that are recommended for a specific test. The table also contains information on the utility of other sample materials as long as the results measured by that method do not exceed the maximum allowable deviation of measurement [16] as defined by the biological variation [17]. A maximum deviation of 10 % is assumed as being acceptable for a constituent if deviation of measurement is not defined [18].

References

[1] Guder WG, Narayanan S, Wisser H, Zawta B. Diagnostic Samples: From the Patient to the Laboratory. 4th updated ed. Weinheim: Whiley-Blackwell 2009.

[2] Hallmann L. Klinische Chemie und Mikroskopie. 1st ed. Stuttgart: Thieme 1939.

[3] Guder WG, Fiedler GM, da Fonseca Wollheim F, Schmitt Y, Töpfer G, Wisser H, Zawta B. Quality of Diagnostic Samples, 4st completely revised ed. Oxford: BD-Europe 2015.

[4] Gross J, Ungethüm U, Moller R, Priem F, Heldt J, Ziebig R et al. Preanalytical factors influencing the measurement of NSE levels in blood. J Lab Med 1995; 18:286–9.

[5] Guder WG. Thrombocytes as interfering factors in clinical chemistry (editorial comment). J Clin Chem Clin Biochem 1990; 28:445.

[6] Hallbach J, Hoffmann GE, Guder WG. Overestimation of albumin in heparinized plasma. Clin Chem1991; 37:566–8.

[7] Guder WG, Ehret W, da Fonseca Wollheim F, Heil W, Müller-Plathe O, Töpfer G, et al. Serum, plasma or whole blood? Which anticoagulant to use? J Lab Med 1998; 22:297–312.

[8] Lutomski DM, Bower RH. The effect of thrombocytosis on serum potassium and phosphorus concentration. Am J Med Sci 1994; 307:255–8.

[9] Guder WG, Hoffmann GE. Analytische und medizinische Aspekte der flammenphotometrischen Elektrolytbestimmung – Ergebnisse einer Kurzevaluierung des Elektrolytgerätes EFIX 5055. Lab Med 1992; 15:14–18.

[10] Boink ABTJ, Buckley BM, Christiansen TF, Covington AK, Maas AHJ, Müller-Plathe O, et al. IFCC recommendations on sampling, transport and storage for the determination of concentration of ionized calcium in whole blood, plasma and serum. Eur J Clin Chem Clin Biochem 1991; 29:767–72.

[11] Voit R. Plasma-Serum-Unterschiede und Lagerungsstabilität klinisch chemischer Messgrößen bei Verwendung von Plasmatrennröhrchen. Dissertation München, Ludwig-Maximilian-Universität 1993.
[12] Narayanan S. Analytical and specimen considerations including specimen stability for selected analytes. Conference report "Präanalytik" Plymouth: Becton Dickinson 1983, pp. 23–25.
[13] Holodny M, Kim S, Katzenstein D, Konrad M, Groves E, Merigan TC. Inhibition of human immunodeficiency virus gen amplification by heparin. Clin Microbiol 1991; 29:676–9.
[14] Neumaier M, Braun A, Wagener C. Fundamentals of quality assessment of molecular amplification methods in clinical diagnosics. Clin Chem 1998; 44:12–26.
[15] Jung R, Soondrum K, Wimmer M, Neumaier M. Rolle der Präanalytik bei molekularbiologischen Untersuchungen von humanen Zellen. J Lab Med 1999; 23:275–9.
[16] Bundesärztekammer. Richtlinie der Bundesärztekammer zur Qualitätssicherung laboratoriumsmedizinischer Untersuchungen. Dtsch Ärztebl 2008; 105:A341–55; and 2011; 108:43–6.
[17] Ricos C, Alvarez V, Cava F, Garcia-Lario JV, Hernandez A, Jiminez CV, et al. Current databases on biological variation: pros, cons and progress. Scand J Clin Lab Invest 1999; 59:491–500.
[18] Engstadt CS, Guttenberg TJ, Osterut B. Modulation of blood cell activation by four commonly used anticoagulants. Thromb Hemost 1997; 77:690–6.

Walter G. Guder, Joris Delanghe

2.7 Spot or Timed Urine – Preanalytical Aspects of Urinalysis

2.7.1 Historical introduction

Urine has been used as a diagnostic material since the commencement of medicine in human history. 5000 years ago in the Egyptian civilization, specialists knew that urine of pregnant women could be used to diagnose the state of pregnancy. Likewise, physicians in Ancient India possessed the knowledge about the sweet taste of urine in patients with diabetes mellitus; the ancient Hippocratic medicine was the basis of the Middle Age uroscopy, which used urine turbidity, color, smell and taste for diagnostic and prognostic conclusions (Fig. 1.1.1). There was a special container for carrying urine sample called matula, covered by canned or web basket to protect urine from sunlight [1].

The scientific background for the use of urine as material for medical examination was created by the modern knowledge about the function of the kidneys and the application of chemical and microscopic procedures in the 19th century [2]. From this the "Urinalysis" developed, consisting of chemical tests (today in the form of test strips) and microscopic examination of the urinary sediment [3, 4].

Despite its existence over many centuries, the sampling and transport of urine for diagnostic procedures is still the most vulnerable part of the total diagnostic process as a source of errors in results. Therefore preanalytical processes became the center of international recommendations [5, 6, 7].

2.7.2 When is urine of diagnostic value?

Urine samples are easy to obtain. Compared to blood sampling, no bodily puncture is needed. Therefore urine is always to be preferred, when the diagnostic evidence of investigation is of similar value compared to that of a blood sample.

Urine is always to be the preferred sample, when the analyte to be measured sufficiently represents the status in the patient's body. Urine is often used for screening purposes as basic diagnostic performance. Thus one teststrip of sufficient sensitivity can exclude kidney diseases, diabetes mellitus and bleeding processes in the urinary tract at the same time.

In special diagnostic conditions, on the other hand, urine samples are often equally or even more useful compared to blood samples. This is true, if pregnancy is to be ascertained or drugs to be detected, which are excreted into urine. Many inborn metabolic dysfunctions can be detected if a pathological metabolite is excreted into urine.

2.7.3 Urine collection

2.7.3.1 Patient preparation

The laboratory is responsible for the correct information regarding optimal patient preparation and best sampling procedure [8]. Informing the patient goes far beyond only explaining the practical aspects of urine sampling. The effects of possible biological factors such as dietary intake, diuresis, exercise and other influences, should be emphasized. If necessary, illustrated instructions for sampling may be provided [7]. This may include informations about the first morning urine, washing of the outer genitals with water and time of collection, if timed urine is to be collected.

2.7.3.2 Types of urine sample

Interpretation of test results is only possible when kind of sample as well as time of collection and other patient conditions are known. The type of urine sample depends largely on the indicated examination. Table 2.7.1 summarizes the definitions and main areas of application of different types of urine samples. Samples are differentiated by the kind of sampling, time of collection and sampling technique [7, 9].

The choice of urine type should be documented. Likewise, the time of collection and its volume also should be documented. The techniques of urine sampling are to follow regulations described in the quality journal [10].

Table 2.7.1: Definition and range of application of different urine samples [9].

Type and definition of urine samples	Time of collection	Area of application
Midstream urine: Urine after the first urine portion	First morning urine	Microbiological and morphologic examinations, teststrip
	Second morning urine (spot urine in the morning)	Test strip using correction field for density (conductivity) of urine, urine protein differentiation
	Spot urine	Same as second morning urine
Timed urine: Urine collected over defined times	24 hour urine overnight urine	Quantitative examination results related to collection time
Catheter urine: Bladder urine after insertion of sterile catheter through the urethra	At any time	Microbiological examinations
Suprapubic aspiration urine: Bladder urine after suprapubic puncture	At any time	Microbiological examinations

2.7.3.3 Materials for urine collection

Before urine is collected for diagnostic purposes, all needed materials and the accompanying request form should be collected. The decision about the type and size of container depends on the medical question and urine type. The following materials may be needed.

Midstream urine:
Urin collection container with cover
Urine tube
Holder with plastic funnel (sampling unit)
Timed urine:
Container designed for collecting 24 h urine

Once all the materials are prepared, the person responsible for sampling the urine visits the patient. If this is not the same person who prepared the sampling materials, he/she has to confirm the correctness of materials in relation to the requested investigation.

When asking which materials should be preferred, each has its own problems. For example, plastic containers can bind proteins or charged molecules [11], vacuum aspiration can cause decreases in cell and casts [12, 13].

2.7.3.4 Performance of urine collection

Preparation of patient

According to medical requests, time and type of urine collection is to be defined and the patient informed verbally or, in specific cases, in written form. During this process, the patient is to be informed about the reason for urine collection and the collection technique. Written information is of special help when timed urine is to be collected. Materials are to be shown and their use explained. The following texts describe the details for different types of urine sample.

Midstream urine

> **Women:**
> Wash your hands with water and soap and dry them thoroughly.
> Hold the clean urine container ready without contacting the inner surface with your fingers.
> When sitting in the toilet, spread your labia and clean them with running warm water. Dry-wipe genital surface with a paper towel without using disinfectants. Dry with toilet paper.
> When urine flows, let the first pass into the toilet (or respective vessel) and fill at least half of the urine collection container with the next urine portion. Allow excess urine to pass again into the toilet or urine container.

When urine collection is finished, part of urine may be transfered from the collection container into the urine vessel. Ensure that the vessel is marked with the patient's name and date and time of urine collection is documented.

Men:
Wash your hands with water and soap and dry them thoroughly.
Hold the clean urine container ready without contacting the inner surface with your fingers.
Clean your penis by withdrawing the foreskin. Clean end of your penis with towel dipped in warm water. Do not use any disinfectant. Dry with a paper towel.
When urinating let the first pass into the toilet or respective vessel. The middle portion is to be urinated into the urine collection container to fill at least half of the vessel. Allow excess urine to pass again into the toilet or urine container.
When urine collection is finished, part of urine may be transfered from the collection container into the urine vessel. Ensure that the vessel is marked with the patient's name and date and time of urine collection is documented.

If you have any problems please contact the person in service to help you.

Children, who can urinate themselves:
After appropriate information about the meaning of a midstream urine, children can collect urine sample themselves using potty chair. This can be made easier if the collection vessel is put into the potty chair. Elder children can be treated like adults.

24 hour timed urine

Before collecting 24 h timed urine, patient should be informed verbally and/or by written procedures and presentation of the collection vessels and if needed handling of stabilizers. After the agreement, a written description may be handed to them, containing the following text.

In the frame of your medical examination you are asked to collect a 24 hour urine. This was ordered because investigations of urine allows conclusions regarding your medical treatment. If you are not admitted in the hospital, please select a day, when you can visit the same over 24 hours and leave the collection vessel at this place to ensure a continuous collection of your urine.
PLEASE READ THE FOLLOWING INFORMATION CAREFULLY; BEFORE YOU START THE COLLECTION TIME.
Collection of your urine probably needs the use of stabilizers to keep the concentration of the measured substances during collection and transport stable. The person taking care of you will inform you about the use of stabilizers. These are given into the sample container before or after the first sample of urine is collected.
Empty your bladder at a certain time selected by you and write down the time. This will be the start time of your collection. From now on, each urine you pass is let into the collection vessel. Empty your bladder exactly 24 h after beginning urine collection into the container. Then close the vessel and make sure that your name and time of collection is clearly documented on the container. Deliver your collection container at the place indicated.
Our staff will help you in case you have questions or problems.

If only a small portion of the timed urine is needed, ensure complete mixing of the whole collection container before separating part into a urine vessel with a sampling unit. Otherwise urine constituents may be unequally distributed in the container. Therefore the quantitative detection of constituents may not be accurate. The urine vessel sent to the laboratory has to be labeled with the patient's identification data and the volume and time of collection (24 h). Only then the right calculation of excretion rate from the measured concentration is possible. When these procedures are completely done, the rest of timed urine can be eliminated.

Transport and storage of urine samples

Time between the sampling and performance of the examination procedure is a critical reason for the relatively unreliability of urine results. This is because changes in concentration of urine constituents can appear, making the measured result useless for medical use. Therefore a standardized organization of transport and storage time is needed as well as documentation of storage temperature. Stabilization and adequate timing of transport are of special importance.

Nearly all the tests performed in spot and/or midstream urine do not need any stabilizing agent, if examination is performed in 24 h time after voiding and samples stored at 4–8° C. Table 2.7.2 illustarates all the stability data for most common urine examinations. These data were collected over many years by the working group of the preanalytical phase of the German United Society for Clinical Chemistry and Laboratory Medicine (DGKL) [6].

Table 2.7.2: Recommended sample stability, stabilizers and storage conditions regarding urine analytes.

Analyte	Stability in urine at −20°C	4–8°C	20– 25°C	Stabilizer	Recommended Sample, Comments	References
Albumin	6 m	1 m	7 d			19, 20, 21, 22, 23
Aluminium	1 y	7 d	3 d			24
5(δ)-Aminolevulinic acid	1 m	4 d	1 d	pH 6–7, stabilized with 0.3 % $NaHCO_3$	Drugs ↗ Light ↘	25, 26
Amphetamine	1 y					27
Amylase	>3 w	>10 d	2 d		Saliva contaminates ↗↗	28
Bence Jones protein (immunoglobulin light chains k, λ)	6 m	1 m	7 d			22, 23

Table 2.7.2: (continued)

Analyte	Stability in urine at -20°C	4–8°C	20– 25°C	Stabilizer	Recommended Sample, Comments	References
Calcium	>3 w	4 d	2 d	Acidify, pH <2	Crystallization at cool temperature	29
Catecholamines Norepinephrine Epinephrine Dopamine	Unstabilized 20 d Stabilized 1 y	 4 d 1 y	 4 d 3 w	Acidify, pH <2.5–5 (9 mL 20% HCl in 24 h urine) or EDTA (250 mg/L) and sodium metabisulfite (250 mg/L)		30, 31, 32
Citrate	4 w*		1 d*	*pH <1,7	Unstable in native urine	33
Cocaine metabolite Benzoylecgonine	4 m	3 w		pH 5, ascorbic acid		27, 34, 35
Codeine	1 y					27
Copper	1 y	7 d	3 d			24
Cortisol, free	1 w	1 w	2 d	10 g/L boric acid		29, 36, 37
C-peptide	2m	6 d	19 h			38, 39
Creatinine	6 m	6 d	2 d			25, 29
C-terminal telopeptide (ß-crosslabs®)	1 y	5 d	1 d		UV light ↘	40
Cystine (Cysteine)	> 1 y*	3 m*	7 d*	*Stabilized in HCl		33
Ethanol		30d				41, 42
Glucose	2 d	2 h↘	2 h ↘	10 mmol/L azide	Bacteria decrease stability	25, 29
5-Hydroxyindoleacetic acid	2 d	2 d	2 h	Acidify		25, 26
Hydroxyproline	5 d	5 d	5 d			25
Immunoglobulin G (IgG)	Unstable	1 m	7 d			19, 20, 22, 23

Table 2.7.2: (continued)

Analyte	Stability in urine at −20°C	4–8°C	20–25°C	Stabilizer	Recommended Sample, Comments	References
Iron	>1 y	7 d	3 d			29
Lysergic acid diethylamide (LSD)	2 m	1 m	1 m			27, 42
α_2-Macroglobulin		7 d	7 d			43
Magnesium	1 y	3 d	3 d	Acidify, pH < 2		29, 33
Methanephrines			8 d			32
α_1-Microglobulin	6 m	1 m	7 d			19, 20, 22, 23
Morphine	1 y					27, 35, 44
Myoglobin	>12d*	12d*	12d*	*pH >8.0	Unstable at acid pH	45
N-Acetyl-β, D-glucosaminidase (β-NAG)	1 m	7 d	1 d			46
Neutrophil gelatinase associated lipocalin (NGAL)		7 d	1 d			
N-telopeptides (NT_x)	4 w	5 d				40
Osmolality	>3 m	7 d	3 h			29
Oxalate	>4 m (at pH 1.5)	unstable↘	<1 h	pH <2, HCl 1 vol %, thymol 5 mL/L	Vitamin C ↗	33
pH		unstable↗			Increase by NH_4 formation	29
Phosphate, inorganic		6 m (pH <5)	2 d at pH <5.0	1 vol % thymol, 5 mL/L, pH <5	Precipitates at alkaline pH	29, 33
Porphobilinogen	1 m*	7 d*	4 d*	*pH 6–7 by $NaHCO_3$	Acid pH↘ Light↘	25, 26

Table 2.7.2: (continued)

Analyte	Stability in urine at −20°C	4–8°C	20– 25°C	Stabilizer	Recommended Sample, Comments	References
Porphyrines	1 m*	7 d*	4 d*	*0.3% $NaHCO_3$, pH 6–7 Use amber colored containers	Light ↘	25, 26, 47
Total porphyrine Uroporphyrine Heptacarboxyporphyrine Hexacarboxyporphyrine Pentacarboxyporphyrine Coproporphyrine Tricarboxyporphyrine Dicarboxyporphyrine						
Potassium	1 y	2 m	45 d			29
Protein	1 m	7 d	1 d			29
Pyridinolines	>1 y	1 w	3 d		UV light ↘↘	48, 49, 50
Sediment of urine		1–8 h	1–2h	20 g/L polyethylene glycol and ethanol (“Saccomanno’s fixative”)	Do not freeze	7, 29, 51, 52.
Acanthocytes		2 d	1 d*		*>300 mosmol/kg	
Bacteria		24 h↗	1–2 h↗***		**pH <6.5 ***pH >7.5	
Casts (hyaline and others)			2 d	Osmolality >300 mosmol/kg		
Epithelial cells			3 h			
Erythrocytes		1–4 h	1 h, 24 h*			
Leukocytes		1–4 h	24 h** <1 h ↘***			

Table 2.7.2: (continued)

Analyte	Stability in urine at −20°C	4–8°C	20– 25°C	Stabilizer	Recommended Sample, Comments	References
Sodium	1 y	45 d	45 d			29
Test-strip fields						7, 29, 51
Erythrocytes		1–3 h	4–8 h		* >300 mosmol/kg	
Leukocytes		1 d*	1 d ↗		** Unstable at pH >7.5	
Nitrite		8 h	4 h			
Protein			2 h**			
Transferrin	4 w	1 w	7 d			20
Urea	4 w	7 d	2 d	pH <7		29
Uric acid	Unstable		4 d	pH >8	Precipitation at pH <7	29, 33
Vanillyl mandelic acid (VMA)	>1 y	>7d	7d at pH 3–5	pH <5		25, 26, 29

Use of preservatives

Alkaline pH, low relative density and low osmolality can induce a rapid lysis of some urine particles after collection [3]. Addition of stabilizers usually prevents metabolic changes of urine analytes and overgrowth of bacteria. Recently, the value of preservatives for maintenance of semiquantitative and qualitative assessment of urine cultures was again demonstrated, especially when the sample transport times exceeded 2 h [14]. However, preservatives may affect some chemical properties and alter the appearance of particles. An appropriate label carrying a hazard symbol should give information dealing with any preservative [3, 5, 7].

Recommended stabilizers are given in Table 2.7.2 [6]. The correct preservative to specimen ratio should be determined when samples are preserved for transport and analysis [15]. The minimal urine volume needed in order to obtain correct results has been determined for two different preservative containing systems [16]. Lyophilized formulations should be chosen among the conservative preservatives as there is no risk of sample dilution or spillage. Also, containers supplemented with boric acid alone or in combination with formic acid or other stabilizing media are used [7, 17]. As shown recently [8], white cells, casts, epithelial cells and bacteria are well preserved, whereas red cells tend to shrink and are less stable. After no preservative seems ideal for all tests required from one sample, the fact that 24 h urine is only seldomly needed, helps to solve this problem. Thus spot urine in the morning is of equal value when proteinuria is differentiated. This is possible by relating concentration to creatinine [7]. The albumin/creatinine ratio, as ratios of tubular and postrenal protein markers were shown to be of equal value [17, 18] without proteinase inhibitor.

References

[1] Guder WG, Kutter D, Bonini P. From uroscopy to molecular analysis – improving diagnostic information from urine analysis. Editorial. Clin Chim Acta 2000; 297:1–3.
[2] Büttner J. Urina ut signum: Zur historischen Entwicklung der Urinuntersuchung. In Guder WG, Lang H (Hrsg) Pathobiochemie und Funktionsdiagnostik der Niere. Merck-Symposium 1989, Heidelber, Springer-Verlag, 1991.
[3] Fogazzi GB, Ponticelli C, Ritz E. The urinary sediment. An Integrated View. New York: Oxford University Press 1999.
[4] Voswinkel P. From uroscopy to urinalysis. Clin Chim Acta 2000; 297:5–16.
[5] Clinical and Laboratory Standards Institute (CLSI). Urinalysis and Collection, Transportation and Preservation of Urine Specimens. Approved guideline 3rd ed. CLSI-Document GP16–A3, Available at :http://www.clsi.org/.assessed 2014.
[6] Guder WG, Fiedler M, da Fonseca Wollheim F, Schmitt Y, Töpfer G, Wisser H, Zawta B. The Quality of Diagnostic Samples 4th ed 2014 BD Diagnostics Oxford UK.
[7] Kouri T, Fogazzi G, Gant V, Hallander H, Hofmann W, Guder WG. European Urinalysis Guidelines Scand J Clin Lab Invest 2000; 60:suppl 231.
[8] Delanghe J, Speeckaert M. Preanalytical requirements of urinalysis. Biochemia Medica (Zagreb) 2014; 24:89–104.
[9] Guder WG, Hagemann P, Wisser H, Zawta B. Fokus Patientenprobe; Kompendium Präanalytik. CD BDHeidelberg 2007.
[10] ISO/EN/DIN 15189 Medical laboratories – particular requirements for quality and competence Geneva, Bruxelles, Berlin 2007.
[11] Hara F, Shiba K. Nonspecific binding of urinary albumin on preservation tube. Jpn J Clin Chem 2003; 32:28–9.
[12] Delanghe J, Langlois M. Pre-analytical effects of vacuum aspiration on urinalysis. Proceedings of the 5th symposium : Preanalytical Phase in Patient Care and Hospital Management. Pratolino 1999.
[13] Langlois M, Delange JR, Steyaert R S, Everaert KC, DeBuyzere ML. Automated flow cytometry compared with an automated dipstick reader for urinalysis. Clin Chem 1999; 45:118–22.
[14] Eisinger SW, Schwartz M, Dam L, Riedel S. Evaluation of the BD Vacutainer Plus Urine C&S Preservative tubes compared with nonpreservative urine samples stored at 4° C and room temperature. Am J Clin Pathol 2013; 140:306–13.
[15] Stankovic AK and DiLauri E. Quality improvement in the preanalytical phase: focus on urine specimen workflow. Clin Lab Med 2008; 28:339–50.
[16] Nickander KK, Shanholtzer CJ, Peterson LR. Urine culture transport tubes: effect of sample volume on bacterial toxicity of the preservative. J Clin Microbiol 1982; 15:593–5.
[17] Gillespie T, Fewster J, Masterton RG. The effect of specimen processing delay on borate urine preservation. J Clin Pathol 1999; 52:95–8.
[18] Hofmann W, Rossmueller B, Guder WG, Edel HA. A new strategy for characterizing proteinuria and hematuria from single pattern of defined proteins in urine. Europ J Clin Chem Clin Biochem 1992; 30:707–12.
[19] Hofmann W, Guder WG. Präanalytische und analytische Faktoren bei der Bestimmung von IgG, Albumin, a1-Mikroglobulin und Retinol-bindendem Protein im Urin mit dem Behring Nephelometer System (BNS). J Lab Med 1989; 13:470–8.
[20] Ledue TB, Collins MF, Craig WY, Ritchie RF. Effect of storage time, temperature and preservative on IgG, transferrin, albumin and alpha-1-microglobulin levels in urine. Clin Chem 2000; 46Suppl.: A 48 (abstract).
[21] Guy M, Borzomato JK, Newall RG, Kalra PA, Price CP. Protein and albumin-to-creatinine ratios in random urines accurately predict 24 h protein and albumin loss in patients with kidney diseases. Ann Clin Biochem 2009; 46:468–76.

[22] Tencer J, Thysell H, Andersson K, Grubb A. Stability of albumin, protein HC, immunoglobulin G, kk-, ll-chain immunoreactivity, orosomucoid and a1-antitrypsin in urine stored at various conditions. Scand J Clin Lab Invest 1994; 54:199–206.

[23] Tencer J, Thysell H, Andersson K, Grubb a. Long-term stability of albumin, protein HC, immunoglobulin G, kappa- and lambda-chain immunoreactivity, orosomucoid and alpha-1-antitrypsin in urine stored at −20° C. Scand J Urol Nephrol 1997; 31:67–72.

[24] Schmitt Y. Influence of preanalytical factors on the atomic absorption spectrometry determination of trace elements in biological samples. J Trace Elem Electrolytes Health Dis 1987; 1:107–14.

[25] Thomas L (ed.). Clinical Laboratory Diagnosis. Frankfurt: TH-Books, 1998; Labor und Diagnose, 7th ed. Frankfurt TH-Books, 2008.

[26] Young DS. Effects of preanalytical variables on clinical laboratory tests, 2nd ed. Washington: AACC Press, 1997.

[27] Dugan S, Bogema S, Schwartz RW, Lappas NT. Stability of drugs of abuse in urine samples stored at −20 °C. J Anal Toxicol 1994; 18:391–6.

[28] Lorentz K. Approved Recommendation on IFCC methods for the measurements of catalytic concentration of enzymes. Part 9. IFCC method for α-amylase (1,4-α-D-glucan 4-glucanohydrolase, EC 3.2.1.1). Clin Chem Lab Med 1998; 36:185–203.

[29] Hagemann P, Scholer A. Aktuelle Urindiagnostik, Rotkreuz: Labolife 2011.

[30] Boomsma F, Alberts G, van Eijk L, Man in't Feld, Schalekamp MADH. Optimal collection and storage conditions for catecholamine measurement in human plasma and urine. Clin Chem 1993; 39:2503–8.RS.

[31] Morello R, Boos KS, Seidel D. Standardized preanalytics of urinary marker molecules and on line SPE-LC-ECD analysis of catecholamines in urine and plasma. 2nd Pharmaceutical Sciences World Congress (PSWC) 2004, Kyoto, Japan.

[32] Willemsen JJ, Ross HA, Lenders JWM, Sweep FCG. Stability of urinary fractionated metanephrines and catecholamines during collection, shipment and storage of samples. Clin Chem 2007; 53:268–72.

[33] Hesse A, Claßen A, Röhle G. Labordiagnostik bei Urolithiasis. Stuttgart: Wissenschaftliche Verlagsgesellschaft, 1989.

[34] Hippenstiehl MJ, Gerson B. Optimization of storage conditions for cocaine and benzoylecgonine in urine: a review. J Anal Toxicol 1994; 18:104–9.

[35] Levine B, Smith ML. Stability of drugs of abuse in biological specimens. Forensic Sci Rev 1990; 2:148–56.

[36] Keiichi M, Suday A, Kawasaki J. Effect of urine pH, storage time and temperature on stability of catecholamines, cortisol, and creatinine. Clin Chem 1998; 44:1759–62.

[37] Wilding P, Zilva JA, Wilde CE. Transport of specimens for clinical chemistry analysis. Ann Clin Biochem 1977; 14:301–6.

[38] Evans MJ, Livesey JH, Ellis MJ, Yandle TG. Effect of anticoagulants and storage temperatures on stability of plasma and serum hormones. Clin Biochem 2000; 34:107–12.

[39] McDonald TJ, Knight BA, Shields BM, Bowman P, Salzmann MB, Hattersley AT. Stability and reproducibility of single sample urinary C-peptide/creatinine ratio and its conversion with 24 h urinary C-peptide. Clin Chem 2009; 55:2035–39.

[40] Lippi G, Brocco G, Salvagno GL, Montagnana M, Guidi GC, Schmidt Gayk H. Influence of sample matrix on the stability of ß-CTX at room temperature for 24 and 48 hours. Clin Lab 2007; 53:455–9.

[41] Gibitz HJ, Schütz H. Bestimmung von Ethanol im Blut. Mitt. XX der Senatskommission für klinisch toxikologische Analytik. Weinheim VCH 1993.

[42] Levine B, Smith ML. Stability of drugs of abuse in biological specimens. Forensic Sci Rev 1990; 2:148–56.

[43] Guder WG, Boesken WH. Hämaturie und α2-Macroglobulin, in Guder WG Nolte J Das Laborbuch, 2. Aufl. München: Elsevier Urban und Fischer 2009. pp 789–91.
[44] Evans WE, Oellerich M, Holt DW. Drug monitoring. Leitfaden für die klinische Praxis. 2nd ed Wiesbaden: Abbot Diagnostics Division 1994.
[45] Wu AHB, Laios I, Green S, Gornet TG, Wong SS, Parmley L, et al. Immunoassays for serum and urine myoglobin: myoglobin clearance assessed as a risk factor for acute renal failure. Clin Chem 1994; 40:796–802.
[46] Matteucci E, Giampetro O. To store urinary enzymes: How and how long? Kidney Int 1994; Suppl 47:58–9.
[47] Guder WG, Lammert F. Porphyrine, ihre Vorläufer und Derivate. In Guder WG, Nolte J. Das Laborbuch.2nd ed. München: Urban und Fischer, Elsevier 2009; pp 973–6.
[48] Schmidt-Gayk H. Marker der Knochenresorption. In Guder WG, Nolte J. Das Laborbuch.2nd ed. München: Urban und Fischer, Elsevier 2009; pp 909–15.
[49] Walne AJ, James IT, Perret D. The stabilityof pyridinium crosslinks in urine and serum. Clin Chim Acta 1995; 240:95–7.
[50] Withold W. Monitoring of bone turnover. Biological, preanalytical and technical criteria in the assessment biochemical markers. Eur J Clin Chem Clin Biochem 1996; 34:785–99.
[51] Koivula T, Groenroos P, Gävert J, Icen A, Irjala K, Penttillä I et al. Basic urinalysis and urine culture: Finnish recommendations from the working group on clean midstream specimens. Scand J Clin Lab Invest 1990; 50 (Suppl 200):26–33.
[52] Kouri T, Vuotari L, Pohjavaara S, Laippala P. Preservation of urine for flow cytometric and visual microscopy testing. Clin Chem 2002; 48:900–5.

Walter G. Guder

2.8 When are other Body Fluids to be Analyzed?

Urine was the first body fluid to be used for the visual, and later chemical analysis, of health status in human history. Centuries later, blood (plasma or serum) became the preferred body fluid to be analyzed. In special cases, however, other body fluids are more suitable to answer diagnostic questions. These will be briefly discussed in this chapter.

2.8.1 Cerebrospinal fluid (CSF)

When diagnosing diseases of the central nervous system, examination of cerebrospinal fluid (CSF) is critical in providing the required information [1].

Important conclusions about the site and mechanisms of inflammatory and degenerative diseases of the central nervous system can be drawn by comparing CSF/serum (plasma) ratios [2]. Indication of CSF investigation should be done by an expert physician in neurology and/or psychiatry. Several preanalytical questions arise concerning CSF investigation, and these are discussed below.

2.8.1.1 Which puncture site?

The puncture site is generally between the 3rd and 4th lumbar vertebrae. In some cases, CSF is obtained from the suboccipital area (e. g. in newborns), or directly from the ventricular area (e. g. when a stent is inserted to affect release from high intracerebral pressure). Suboccipital puncture is no longer recommended in adults because of the increased risk involved. Since the composition of CSF changes along its flow from the cerebral to the lumbar region, the puncture site is of great importance for the interpretation of results and should therefore be documented.

Time of puncture is of additional importance while planning a CSF sampling [3]. The time of antibiotic treatment is to be considered. If the success of this therapy is to be evaluated, CSF obtained 2–3 days after finishing the antibiotic treatment is meaningful. In addition, other medical actions can be of impact. Thus injecting contrast media to document cerebral vessels can interfere with some analytes. Therefore not only the day, but the hour of puncturing CSF should also be documented.

2.8.1.2 Materials needed to obtain cerebrospinal fluid [4]

To prevent postpuncture headache caused by the lumbar puncture, special needles have been developed. These “atraumatic needles”, as designed by Sprotte and

Whitacre are characterized by an outer diameter of 0.7 mm and exhibit a pencil-shaped tip and a special site of opening. It is closed by an inserted mandrin.

In addition, several sterile colorless sample tubes are to be kept ready to collect CSF portionwise. These tubes should allow a clear view and sterile closure. They should withstand freezing and permit thawing. They should neither contain preservatives nor anticoagulants.

Further, a disinfectant and a local anesthetic agent together with a syringe, cotton dip and plaster to attend the puncture site should be available.

2.8.1.3 CSF sampling procedure

In preparing the puncture, the fasting patient should be seated or asked to lie on his/her side on a flat support. The puncture site should be marked and the area disinfected.

Application of a local anesthetic is desirable for the patient's comfort, but is not always needed for the experienced puncturer. Puncture after local anesthetic is effective, if puncture is saggital and sloping upwards with 20° angle.

After the needle has been positioned, cerebrospinal fluid flows spontaneously after the mandrin has been recommended. The first drops should be discarded (about 0.5 mL) to prevent contamination with skin components and capillaries punctured. Then CSF is collected in several portions into sterile vessels, which are opened shortly before being filled separately for microbiological, cytological and clinical chemical investigations. Investigations for clinical chemistry and cytology tests are possible from the same sample. By the use of small needles (22-gauge) a slow flow of CSF is secured, thus avoiding headache caused by rapid decrease of pressure in CSF area. When visible color changes or turbidities are observed, the follow up of samples is to be numbered to allow diagnostic conclusions about the source and mechanisms of local (like bleedings) and cerebral causes.

To prevent errors by contamination, the following sequence is recommended, when three tubes are obtained:

1. Tube 1: Clinical chemical and immunological investigations
2. Tube 2: Microbiological investigations including cultures
3. Tube 3: Cytological investigations including differentiation of tumor cells and other pathological cells

If contamination of the first tube is suspected, the second may be used for microbiological and the third for clinical chemical and cytological investigations, using the sediment for microscopic and the supernatant for chemical and immunological tests.

After the successful sampling of CSF, the canula is removed and the puncture site treated by pressing a plaster on it. The patient should keep resting in lying position for at least 30 min to prevent further CSF effluents.

2.8.1.4 What amount of CSF is required?

Despite the fact that less sample is needed for routine analysis, the amount required depends on the clinical question. Thus, the detection of pathological cells and bacteria increases with the volumes investigated. This is to be considered against the fact that increasing volumes taken increase the risk of side effects like postpuncture syndrome (headache). This is to be expected in children where less CSF is available and even the recommended three tubes would cause severe headaches.

For microbiological investigations 2 mL are recommended, 1 mL in young children and babies. In contrast, cytological investigations need 5–10 mL as optimal volume in adults. Total volume should not exceed 20 mL in adults and 2 mL in young children. These volumes can be replaced in a few hours by normal new CSF formation.

2.8.1.5 What measures of transport and storage of the CSF samples?

CSF samples obtained should be transported to the laboratory as rapidly as possible.

Maximal allowable transport times depend on the intended investigation. Stability is not only different for different analytes, but likewise different in normal, compared to pathological cases. Therefore, transport times depend on the analytes to be investigated in CSF. The following general recommendations have been published, which may help to ascertain the right treatment of these precious samples independent of the intended investigation (Table 2.8.1).

In addition, the following information may be useful:

When bacterial meningitis is suspected, meningococci in CSF are stable at room temperature for 3 days, whereas other disease-causing germs decrease considerably after 6 hours' storage at room temperature. Implanting culture bottles can prevent sensitive germs from dying.

Table 2.8.1: Recommendations regarding transport and storage times of cerebrospinal fluid samples.

Expected transport times	transport- and storage conditions
< 1 h	Closed at room temperature
Up to 3 hours	Closed and cooled to 4–8 °C without additives
Up to 3 hours for microbiological investigations	Closed at room temperature with parallel culture bottle of whole blood.
>3 hours	Centrifuge cells, implant culture bottle if needed, transfer supernatant into freezable polypropylene tubes
Long term storage	Freeze supernatant after centrifugation at −70 °C, cellular dry smear stable.

Granulocytes are the most sensitive cells in CSF. Therefore, a dry smear should be prepared in 3 hours' time.

Of the diagnostically relevant analytes, lactate and glucose change most rapidly. Depending on the number of leucocytes present in CSF, stability is limited to 30 min–2 h. Cooling to freezer temperature can only double the time of stability.

Table 2.8.2 gives the known times of stability of diagnostically relevant analytes in CSF.

2.8.2 Puncture fluids (ascites, pleural fluid, amniotic fluid, synovial fluid) [6–12]

Ascites, pleural and pericardial fluids and synovial fluid are investigated for the following medical reasons:

- Identification and differentiation of body fluids, for instance, is the fluid from the nose nasal secretion or CSF, is the liquid amniotic fluid or urine?
- Diagnostic examination, if the liquid sample is a transsudate or exsudate, benign or is it a malignant effusion or chylus or so called pseudochylus?

The following questions may be answered before sampling is performed:

Native Sample: What analyte should be measured? Clinical chemical analytes including tumor markers and gram staining needed, bacterial culture intended? The following samples are recommended:

EDTA-Sample:	Cell count, cell smear
Heparinized sample:	pH-measure, tumor cells
NaF-stabilized sample:	Lactate

Table 2.8.2: Stability of CSF components at different temperatures (5).

Analyte	at −20 °C	at 4–8 °C	at 20–25 °C
Albumin	>1 year	2 months	1 day
Glucose	>1 month	3 days (sterile)	5 hours
Immunoglobulines A,G, M	unstable, since tendency to precipitate	7 days	1 day
Lactate	Several months	1 hour	30 min
Granulocytes	Not recommended	3–5 hours	1–2 hours
Total protein	>1 year	6 days	1 day
Oligoclonal bands	unstable after 3 months	7 days	1 day
Tumor cells	Not recommended	1–12 hours	3–5 hours
Bacterial culture	Not recommended	depending on germ	1–24 hours

Citrated sample:	Fibronectine
Is venous blood needed for comparison?	When synovial fluid is used for enzyme and substrate examinations. Fluid/serum/plasma ratio possible
Stabilizer?	Add 25 mg hyluronidase/mL synovial fluid and incubate at 37 °C over 5 min.

2.8.2.1 Special sampling techniques

- When oto- or laryngorhoe is suspected, collect **secreted liquids** in tubes with a coined bottom if aspiration is needed. When only low amounts are expected, liquids may be absorbed on to small sterile sponges.
- **Amniotic fluid** aspiration: reject the first 1–2 ml because of possible contaminations with maternal tissue or cells.
- Sampling of **tears** with glass capillaries from the lower "tear lake" by a little swamp, which is inserted into the lower conjunctiva space.
- **Duodenal juice** for the detection of lambliae are to be microscopically examined instantly.
- **Bronchoalveolar lavage:** Use siliconized glass or polyethylene tubes to prevent absorption of macrophages and other diagnostically important constituents. Lavage liquid may be filtered through sterile gauze before being analyzed. When posted, airdried cytocentrifuged sample preparations may be advisable.

2.8.2.2 Sample preparation

Cellular diagnostic examinations should be performed instantly in up to 2–5 h time, since cells, crystals, glucose, lactate and viscosity are unstable. If this is not possible, cells may be fixed in the native liquid by adding 50 % ethanol in a 1:1 ratio. At low cell numbers, a cytocentrifuged sample should be made for storage. At higher cell numbers, a manual smear may be produced or a diluted cytocentrifuge preparation done for transport and storage.

Joint effusion or punctates: Whenever instant examination is not possible, joint fluid samples should and can be stored at –20 °C, when enzyme- and other metabolites are to be measured later.

Veneous blood for comparison with punctured samples should be obtained within approximately 0.5 h.

2.8.3 Saliva

Oral fluid (OF) as a sample matrix offers significant advantages: collection can be performed at almost any location, is noninvasive and it can be collected under direct observation, thus reducing the risk of adulteration and substitution.

Oral liquid is composed of fluid from one single (gland-specific sputum) or all salivary glands (mixed sputum), oral mucosa and gingiva. For routine use, mixed sputum in oral fluid is used. Different techniques to collect sputum from OF are described [14].

Saliva can be used to analyze different analytes like steroids and drugs. According to Gorodischer [15] 85 % of all parents and 50 % of all children prefer collection of oral fluid compared to venous blood sampling. Compared to blood samples, this material offers different advantages. The simplicity of sampling makes sputum the ideal sample for selftesting at home or onsite testing, or in cases sampling of blood is complicated (like in newborns). Limiting factors of oral fluid sampling are the viscosity and sometimes the sufficient amount of sample.

Because of these advantages, saliva is also used for controlling drugs of abuse [16] by the police as well as for checking doping control in sports [17]. Here the medical needs to analyse saliva samples as well as the technical performance may be of relevance [4, 18, 19]. Data on sputum/blood concentration ratios for different drugs are also available [20].

Sample authenticity can be proven via the determination of biomarkers. Depending on concentration and the cut-off defined, it is possible to detect drugs in OF for the same detection time window as in urine [14, 20, 21]. OF is acidic compared to serum. In most cases, drugs of abuse possess an alkaline pKa. If the pH of the collection device is acidic, the drugs still will be enriched during the collection process. The un-ionized drugs will be "trapped" and stabilized for further analysis. In OF, the nonprotein bound parent drugs are mostly analyzed. One exception is cannabis. The analysis of the component tetrahydrocannabinol carbonic acid (THC) in OF is not only for medical reasons, but also to control oral contamination of this administered drug. Analysis of drugs transferred from blood as well as of drugs adhering to the buccal mucosa is possible with a liquid based saliva collection by rinsing the whole oral cavity. For a definite result, in particular for forensic applications, a correct definition of the collected amount of oral fluid within the sample is of some significance. The most precise method is a photometrical measurement which enables the quantification of saliva and a recalculation of contained drugs of abuse to 1 mL of collected sample [16, 22].

This makes it possible to monitor therapies and abstinence programs in OF quite accurately. Modern chromatography enables detection of parallel consumption of different drugs of abuse within minutes. Additionally, trends in the abuse of new drugs can be detected more easily and rapidly.

Beside hormone and drug testing, the analysis of DNA in OF will gain more acceptance and is under investigation. Bacterial DNA tests will provide an overview on the current status of oral microflora. It might be possible to recognize periodontitis and possible risks arising from rejection of implants earlier [23]. In OF, human genomic DNA can be collected quite easily for use in the analysis of single nucleotide polymorphisms in future [24].

2.8.3.1. Methods to collect oral liquid (saliva) samples

Before sputum is to be collected, patients should abstain from eating and alcoholic drinks for 12 hours. Smoking should be avoided for 30–60 min [25, 26].

The following basic procedures were recommended:
- **Collecting free floating sputum**
  (Proband in sitting position allows the oral fluid collected in the mouth to float over the lower lips into the vessel.
- **Spitting saliva into a collecting vessel**
- **Aspirating saliva:** aspirate oral fluid from the inner mouth ground by a platic syringe.
- **Collect with cotton rolls:** put four cotton rolls connected by a thread into the mouth above the **tongue into the mouth.** To collect from the lower part put three cotton rolls connected by a thread below the tongue. On the left and right side metalline strips are shown, which are wrapped in polyethylene foil with a rectangular opening in the area of the orificium of the ductus parotidis. Leave all parts in position for about 9 minutes. Saliva secretion may be stimulated by slight chewing for up to 60 sec. In addition, saliva secretion can be stimulated by using citric acid impregnated into the cotton rolls. After sputum is absorbed in the cotton, these are taken into a Salivette^{R} container and centrifuged to obtain liquid saliva.

Collecting sputum from individual glands
- Insert a Lashley-case into Stenon's or Warthon's duct and fix it under light pressure. Guide sputum by a low size tube into a suitable vessel.
- Alternatively, a silicon tube may be inserted and the sputum aspirated carefully.

Different sampling systems are provided to collect sputum. None fulfills the ideal conditions for all analytes to be measured. The different collection devices, containing absorbing material either contain substances interfering with the analytical procedures or give incomplete recoveries for some analytes. The recoveries described are between 59–107 % [20]. Table 2.8.3 gives an overview about the present saliva-collection devices available.

Table 2.8.3: Commercially available oral fluid/saliva collection devices based on the adsorption principle (modified from [4]).

Product name	Manufacturer	Adsorptive material	Sampling procedure
Salivette®	Sarstedt, Germany	Cotton rolls	Saliva container useful for storage and centrifugation
Salivette®	Sarstedt, Germany	Cotton rolls containing citric acid	Saliva container useful for storage and centrifugation
Salivette®	Sarstedt, Germany	Polyester rolls	Saliva container useful for storage and centrifugation
Saliva Sampler™	Stat sure Diagnostic Systems USA	Cushion at handle with volume indicator	Pressing saliva through a separate filter into a container with buffer
Orapette®	Trinitiy Biotech	Silc cotton rolls	Pressing by turning the piston
Ora Sure® =	Epitope USA	Cushion at the "lolli-pop"-handle	Centrifugation tube with anti-microbial buffer
Accu Sorb ™	Avitar Technologies	Polymer foamy plastic – finger fixed at the cap of container	Pressing with the fingers ("milking") into plastic vial
Oral Screen™	Avitar Technologies	Polymer foamy square	Pressing by pressure against a surface
Q. E. D.® Saliva Alcohol Test	STC Technologies	Cotton dipper with stick	Pressing by pressure against a surface
Clin Rep®	Recipe	Cotton wool rolls with citric acid preparation (1–3 %) with wooden stick	Centrifugation container with filter insert (0.45 μm pore size)
Saliva Sampler™	Stat Sure Diagnostic Systems USA		
SCS saliva collection system	Greiner Bio One, Austria		
UltraSal-2™, VersiSal®, Oragene™	Oasis Diagnostics USA		

Saliva samples can be stored up to 12 h at room temperature. When rapid analysis of samples is not possible, they should be stored at –20 °C. During freezing, mucins are expected to precipitate; they can be separated after redissolving by centrifugation [26].

References

[1] Henry JB. Clinical Diagnosis and Management by Laboratory Methods. 19th ed Philadelphia:-Saunders, 1996.

[2] Felgenhauer K, Beuche W. Labordiagnostik neurologischer Erkrankungen. Liquordiagnostik und –zytologie. Diagnose- und Prozessmarker. Stuttgart: Thieme 1999.

[3] Guder WG, Fiedler M, daFonseca-Wollheim F, Schmitt Y, Töpfer G, Wisser H, Zawta B. Quality of Diagnostic Samples. 4th ed. Oxford: BD Diagnostics 2015.

[4] Guder WG, Narayanan S, Wisser H, Zawta B. Diagnostic Samples: From the Patient to the Laboratory. Weinheim: Wiley –VCH-Verlag 2009, pp. 32–33.
[5] Guder WG, Hagemann P, Wisser H, Zawta B. Fokus Patientenprobe, Kompendium Präanalytik. CD Heidelberg: BD (Becton-Dickinson) 2007.
[6] Satz N. Laborchemische Untersuchungen im Aszites. Schweiz Med Wschr 1991; 121:536–47.
[7] Jüngst D, Gerbes AL, Martin R, Paumgartner G. Value of ascitic lipids in the differentiation between cirrhotic and malignant ascites. Hepatology 1986; 6:239–43.
[8] Schölmerich J, Volk BA, Köttgen E, Hasler C, Wilms H, Billmannn P, Gerok W. Aszites. Neue Aspekte zur Diagnostik und Therapie. Dtsch med Wschr 1985; 100:512–8.
[9] Gerbes AZ, Paumgartner P. Diagnostik des Aszites. Dtsch med Wschr 1994; 119:1507–11.
[10] Loddenkemper H. Diagnostik der Pleuraergüsse. Dtsch med Wschr 1992; 117:1487–91.
[11] Hamm H, Fabel H. Chylotorax und Pseudochylotorax. Dtsch med Wschr 1989; 114:2017–22.
[12] Kleesiek K. Gelenkerkrankungen. Med Welt 1980; 31:1609–17.
[13] Berkman N, Kramer MR. Diagnostic test in pleural effusion. Postgrad Med J 1993; 69:12–8.
[14] Heltsley R, DePriest A, Black DL, Crouch DJ, Robert T, Marshall L et al. Oral fluid drug testing of chronic pain patients. II Comparison of paired oral fluid and urine specimens. J Anal Toxicol 2012; 36:75–80.
[15] Gorodischer R, Burtin P, H wang P, Lewine M, Koren G. Saliva versus blood sampling for the therapeutic drug monitoring in children; patient and parental preferences and economic analysis. Ther Drug Monit 1994; 16:437–43.
[16] Wendy M. Bosker AB, Huestis MA.: Oral fluid testing for drugs of abuse; Clin Chem 2009; 55:1910–31.
[17] Lippi G, Banfi G, Botrè F, de la Torre X, De Vita F, Gomez-Cabrera MC, Maffulli N, Marchioro L, et al. Laboratory medicine and sports:between Scylla and Carybdis. Clin Chem Lab Med 2012; 50:1309–16.
[18] Haeckel R, Walker RF, Colic C. Reference ranges for mixed saliva collected from the literature. J Clin Chem Clin Biochem 1989; 27:249–52.
[19] Haeckel R, Hänecke P. Application of saliva for drug monitoring. An in vivo model for transmembrane transport. Europ J Clin Chem Clin Biochem 1996; 34:171 – 91B.
[20] Verstraete AG. Detection times of drugs of abuse in blood, urine and oral fluid. Ther Drug Monit 2004; 26:200–5.
[21] Vindenes V, Ytredal B, Oiestad EL, Waal H, Bernhard JP, Moerland JG, Christophersen AS. Oral fluid is an alternative for monitoring drug abuse: detection of drugs in oral fluid byliquid chromatography-tandem mass spectrometry and comparison of results from urine samples from patients treated with methadone or buprenorphine. J Anal Toxicol 2011; 35:32–9.
[22] Anizan S, Huestis A. The potential role of oral fluid in antidoping testing: Clin Chem 2014; 60:307–22.
[23] Haririan H, Andrukhov O, Bertl K, Lettner S, Kierstein S, Moritz A, Rausch Fan X. Microbial analysis of subgingival plaque samples compared to that of whole saliva in patients with periodontitis. J Periodontol 2014; 85:819–28.
[24] Paar C, Euko D, Zahel B, Mayr R, Berg J. Reliable analysis of single nucleotide polymorphism of lactate persistence LPH (-13910)C/T from saliva derived DNA: validation of a standardized saliva collection system. Clin Lab 2014; 60:1977–82.
[25] Choo RE, Huestis MA. Oral fluid as a diagnostic tool. Clin Chem Lab Med 2004; 42:1773–87.
[26] Samyn N, Laloup M, De Boeck G. Bioanalytical procedures for determination of drugs of abuse in oral fluid. Anal Bioanal Chem 2007; 388:1437–53.

Ana-Maria Simundic

2.9 Who is Doing Phlebotomy in Europe?

2.9.1 Introduction

Phlebotomy is the most common invasive procedure in health care, available worldwide. It is also the most common source of preanalytical errors. Errors in phlebotomy may compromise sample quality, affect laboratory test results and cause unnecessary delays. Moreover, errors in phlebotomy can cause patient discomfort and harm and jeopardize patient and personnel safety, by putting them at risk of different injuries to a various extent and blood-borne infections. Controlling and minimizing all potential sources of error is a demanding and challenging task [1].

- Major risks for healthcare workers related to phlebotomy are [2]: Sharps injuries which can cause blood borne infections in health-care workers.
- Disease transmission through blood-borne and airborne infection, mainly associated with noncompliance with the basic principles of infection control (vaccination of healthcare professionals, sanitizing and washing hands, patient isolation, etc.).

Major risks for patient injuries related to phlebotomy are [2]:

- Fainting which may (if patient is falling from a venipuncture chair) potentially result in bone fractures and even some serious head injuries. It is the most common adverse event related to phlebotomy.
- Hematoma due to the subcutaneous accumulation of the blood in the proximity of the venipuncture site.
- Nerve damage (punctured or nicked nerve) which may lead to permanent disability.
- Arterial puncture or laceration which leads to bleeding into the venipuncture area with subsequent severe pain and possibly even nerve injury due to the tissue compression.

Phlebotomy therefore requires solid theoretical knowledge and adequate practical skills. Knowledge and skills are acquired through education and training. By education, personnel involved in phlebotomy is learning about good laboratory practice, potential risks associated with any violation of the recommended procedure and related consequences and effects to the patient and health-care worker safety. Unfortunately education and training of phlebotomy personnel are lacking both worldwide and across Europe.

2.9.2 Existing international recommendations and guidelines for phlebotomy

The document published by the Clinical Laboratory Standards Institute (CLSI) H3–A6 entitled 'Procedures for collection of diagnostic blood specimens by venipuncture' has a very detailed description of the recommended phlebotomy procedure and all related steps [3]. This document is published in 2007 and is currently under revision. Seventh edition is expected to be published soon. Another important document is the guidelines issued by World Health Organization (WHO) in 2010 [4].

Those international documents are mostly evidence based (where evidence is available) and quite comprehensive. Compliance to these guidelines ensures standardization of phlebotomy procedures and thus minimizes error risk, both for the patient as well as the health care personnel.

However, English language in Europe is spoken as a native language only in the UK and Ireland. Hence, these international guidelines are not well suited to be directly applied to the local practice in the rest of the European countries where English is not a native language.

As recognized by the European survey on national guidelines, education and training for phlebotomy performed by the European Federation of Clinical Chemistry and Laboratory Medicine (EFLM) working group for the preanalytical phase (WG-PA), it is of national as well as local interest to adopt an international phlebotomy guideline and adapt (modify) it for application in the specific cultural and organizational environment. This modification should respect many national and/or local issues such as language, culture, education and training of personnel, legislation, healthcare setting, economy, standard of care and many others. Once adopted and adapted, local guidelines should serve as the basis for nationwide education and standardization of phlebotomy procedures [5]. Implementation of the guidelines are challenging and there are many reasons for the low level of compliance such as the lack of knowledge and theory behind the recommended steps as well as the lack of support from the hospital and lab management.

2.9.3 National guidelines for phlebotomy in European countries

Presently, only a minority of European countries (i. e. one quarter) have produced their national guidelines for phlebotomy. Those countries are Croatia, Ireland, Italy, Slovenia, Spain, Sweden, The Netherlands and the UK. Germany has published a standard operation procedure preexamination also covering phlebotomy [6]. In most of those countries guidelines were issued by national societies in laboratory medicine. Production and publication of guidelines are driven by the respective governmental bodies in a much lesser extent, as was the case, for example in Spain and Sweden.

Obviously, there is a room for improvement in that area and effort should be made to ensure that patients are receiving the same level of quality and standard of care across Europe. National societies should either alone or in collaboration with national health authorities take the leading role in guideline production and implementation.

2.9.4 Education of phlebotomists

There is a large heterogeneity in terms of the type of the personnel involved in phlebotomy across Europe. Moreover, there is a remarkable difference in their level of background training and education needed to become qualified for phlebotomy and opportunities for life-long learning.

In the majority of European countries phlebotomy is performed almost equally often by nurses and laboratory staff. Only some countries offer education as phlebotomist. In addition physicians are responsible for any kind of blood sampling in some European regions (Germany, Austria). Laboratory staff involved in phlebotomy is of different level of background education, from laboratory technicians to even specialists in laboratory medicine.

Nurses perform phlebotomy less often in outpatient than in inpatient settings, whereas laboratory personnel are involved in phlebotomy more often in outpatient settings. In the majority of countries in Europe nurses are almost exclusively responsible for performing phlebotomy for hospital inpatients.

However, there are also some specific and quite unique situations. For example, in Austria and Germany phlebotomy for hospital inpatients is very often performed by medical doctors, whereas in the rest of Europe medical doctors are rarely involved in phlebotomy.

In Denmark, phlebotomy is performed by laboratory technicians both for inpatients and outpatients in the majority of the health-care facilities.

Specialized phlebotomists are recognized as professionals and are involved only in phlebotomy in several countries in Europe (Belgium, Ireland, The Netherlands and the UK), whereas this profile does not exist in the remaining part of Europe.

And last, but not the least, in some countries even nonmedical staff, i. e. administrative staff is sometimes performing phlebotomy, mostly for outpatients.

2.9.5 Specific training for phlebotomy

The level of education of personnel involved in phlebotomy varies substantially throughout Europe. In most of the countries across Europe, nurses and laboratory technicians have 4–5 years of high school education, followed by 2–5 years of college or university education. In some countries, nurses and laboratory technicians become qualified for phlebotomy right after high school.

Whether or not phlebotomy is included in the training varies substantially depending on the type of profession. While in some European countries specific training for phlebotomy is not a part of the education required to become qualified, in majority of the European countries specific training in phlebotomy (in the duration of at least 5 h or more) is provided and is mandatory for nurses and laboratory technicians.

In many countries, phlebotomy training for medical doctors and even specialists in laboratory medicine is limited to less than 5 hours. Unfortunately, administrative staff receives most often only up to one hour of training, which is especially problematic, since they are nonmedical personnel with no knowledge and understanding of the phlebotomy procedure and associated risks.

Strikingly enough, a specific training for phlebotomy is not provided in some European countries, neither as a part of the education required to become qualified as a nurse or laboratory technician, nor as a separate training or educational resource.

Specific training for phlebotomy as a continuous educational resource is available only in a minority of European countries. In those countries the training periods are longer (up to one week) and mostly mandatory for nurses, laboratory technicians and specialized phlebotomists, whereas they tend to be much shorter (up to 1 day) and are usually optional for all other professionals.

Such courses are almost always provided by healthcare institutions or governmental bodies regardless of the profession. Although not so often in the past, training courses are recently becoming much more often provided by blood collection system suppliers.

2.9.6 The way forward

Patients should be receiving the same level of care across Europe. This is not debatable. Unfortunately, currently this is not the case. Phlebotomy procedures are not uniformly implemented, personnel involved in phlebotomy is of different level of education, has different level of knowledge, skills and expertise. In order to improve the quality of phlebotomy in Europe, it is necessary to ensure that international phlebotomy guidelines are adopted and successfully implemented in all European countries which do not have their own guidelines. Professional societies in laboratory medicine should take the leading role and be at the forefront in this process by becoming engaged in the development of basic training programs and continuous education, with subsequent assessment of competence and certification of healthcare and phlebotomy staff.

In addition, professional associations should be raising awareness for the need to improve the quality of phlebotomy and continuously encouraging standardization of phlebotomy practices across Europe.

References

[1] Simundic AM, Lippi G. Preanalytical phase – a continuous challenge for laboratory professionals. Biochem Med 2012; 22:145–9.

[2] Johnson L. Phlebotomy - It's A Risky Business. National Center for Competency Testing. USA, 2013.

[3] Clinical and Laboratory Standards Institute (CLSI). Procedures for collection of diagnostic blood specimens by venipuncture; approved guideline – 6th ed. CLSI document H3-A6. Wayne, PA: C L S I; 2007.

[4] World Health Organization. WHO guidelines on drawing blood: best practices in phlebotomy. Available at: http://whqlibdoc.who.int/publications/2010/9789241599221_eng.pdf Accessed on 10 August, 2014.

[5] Simundic AM, Cornes M, Grankvist K, Lippi G, Nybo M, Kovalevskaya S, Sprongl L, Sumarac Z, Church S. Survey of national guidelines, education and training on phlebotomy in 28 European countries: an original report by the European Federation of Clinical Chemistry and Laboratory Medicine (EFLM) working group for the preanalytical phase (WG-PA). Clin Chem Lab Med 2013; 51:1585–93.

[6] Gurr E, Arzideh F, Brandhorst G, Gröning A, Haeckel R, Hoff T, Roggenbuck L, Schumann G, Wolters B, Wosniok W. Musterstandardarbeitsanweisung Präanalytik (Exemplary standard operation procedure pre-examination). J Lab Med 2011; 35:55–60.

3. Biological Variables Influencing Laboratory Results

Walter G. Guder and Sheshadri Narayanan

3.1 Age and Gender Differences – Unavoidable Influences on Laboratory Results

Endogenous factors like **age, gender, race** and other **genetic** influences belong to intrinsic influences which will impact on the concentration of several analytes.

3.1.1 Age

Several analytes exhibit age-dependent concentrations in body fluids which have to be taken into consideration while comparing data of the same patient retrospectively. These changes are found more often and in a greater number of analytes in childhood than in adult stage [1].

Figure 3.1.1 provides examples for the age dependence of concentrations of some analytes in the newborn period, during childhood and at adult age [2, 3]. Thus, compared to adults, hemoglobin concentration and erythrocyte count are significantly higher in newborns when compared to adults. Physiologically the increased pO_2 a few days after birth leads to increased erythrocyte degradation. The increased hemoglobin degradation leads to increased bilirubin formation. The newborn's liver lacks the enzymes involved in the glucuronidation of bilirubin and as such unconjugated bilirubin in plasma is increased. Uric acid concentration, in contrast, is comparable to that in adults after birth. However, whithin a few days after birth, a significant decrease in uric acid concentration is observed.

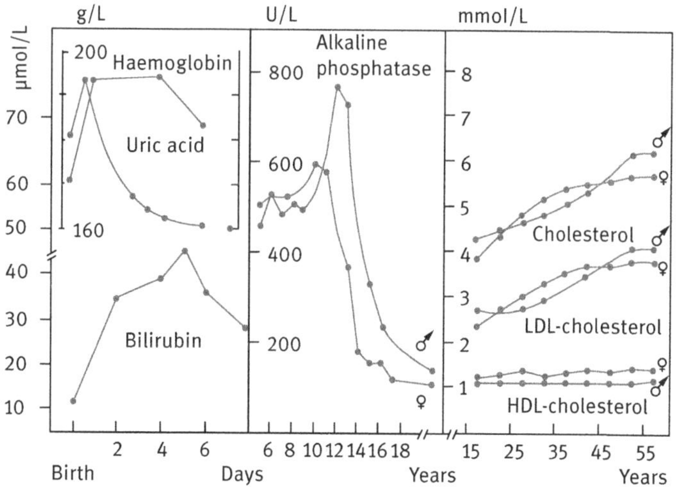

Fig. 3.1.1: Age dependence of various substrates and enzyme activity. Alkaline phosphatase was measured at 30 °C (86 °F) [2].

Other examples of age-dependent concentrations include alkaline phosphatase in serum which peaks during the growth phase and then declines to the adult status, mirroring osteoblast activity. On the other hand, total and LDL-cholesterol increase during adult phase. In addition to the age dependence, a small gender difference of these analytes may also be observed. These gender differences in turn change as a function of age.

3.1.2 Gender

In analogy to the outer phenotypic signs and the gender specific hormone patterns, several clinical, chemical and hematology markers exhibit gender-specific reference ranges (Fig. 3.1.2). Examples of gender differences are the serum concentrations of creatine kinase and creatinine. Both markers depend on muscle mass which in general is more pronounced in men. Since these differences disappear after intense training, these gender-specific differences may not be relevant in old patients with low muscle mass.

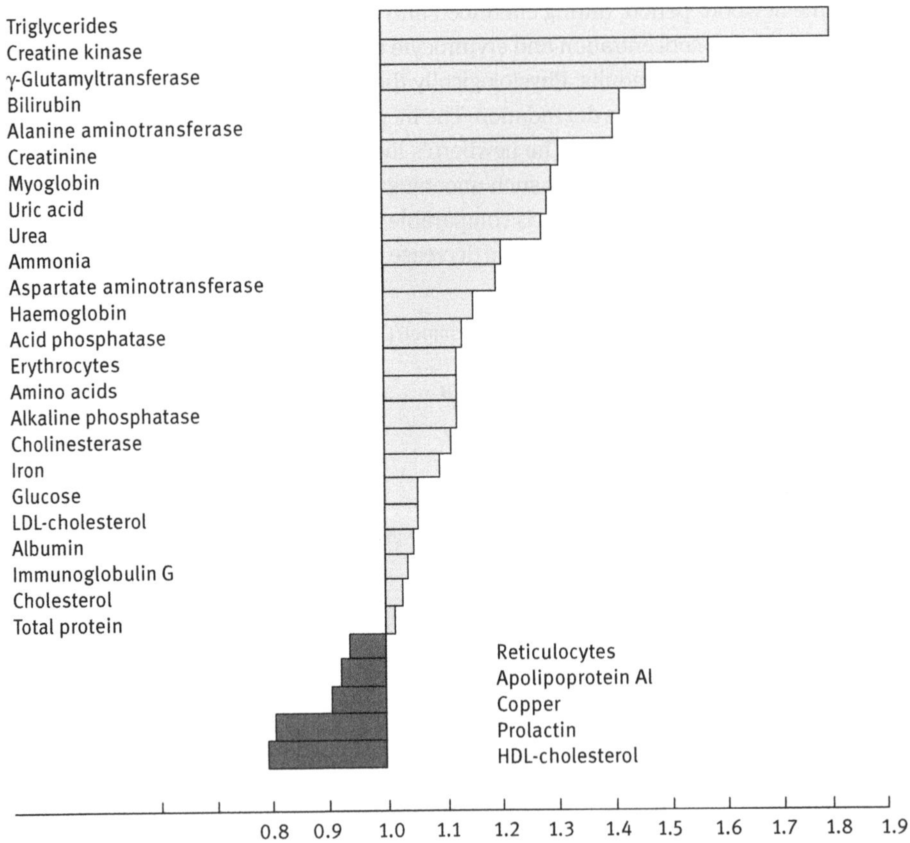

Fig. 3.1.2: Male-female differences related to the mean values of reference ranges in females [2].

Gender-specific differences likewise exist in urine excretion rates [4, 5].

3.1.3 Race

Figure 3.1.3 illustrates examples of analytes, which seem to demonstrate race dependence. A significant difference in creatine kinase activity has been described in black and white people. This difference was not evident between Hispanic, Asian and European whites. The difference was not due to differences in age, height or body weight [6]. Interestingly, the difference of creatinine is similar to that of creatine kinase. Since both analytes are dependent on muscle mass, aquired differences may account for this apparant race difference. The difference in creatinine has led to different formulas for the calculation of glomerular filtration rate in blacks as well as white persons [7]. On the other hand, black Americans of both genders have lower leucocyte numbers compared to white Americans. This difference is due to a higher granulocyte number in white people. In contrast, hematokrit, hemoglobin and lymphocyte counts are similar in both ethnic groups [8]. Monocyte counts in whites exceeded that in black people [9].

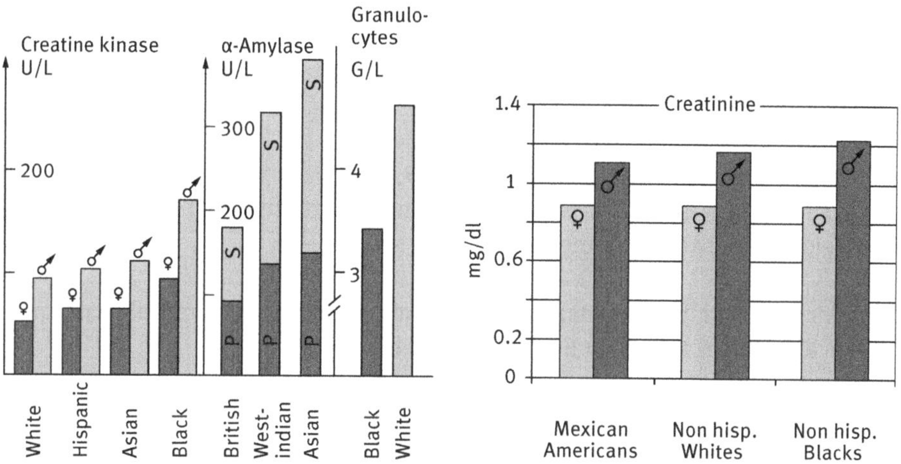

Fig 3.1.3: Influence of race/ethnicity on creatine kinase, creatinine, amylase (P = pankreatic-, S = salivary gland – isoenzyme) and granulocyte counts in blood (modified from [10]).

Another example is the striking difference of amylase activities in blood between British and West Indian as well as Asian originating patients, living in Britain [11]. The difference was more significant with the salivary isoenzyme. If interpreted with the British reference ranges of this method, 50 % of probands from West Indians would be declared as having elevated amylase activity. Significant racial differences have also been reported for serum concentration of vitamin B_{12} (1.35 times higher in black compared to white people) [12] and of Lp(a) (two times higher in black compared to white people) without any difference in arteriosclerosis risks nor in mortality [13, 14].

3.1.4. Other genetic variables

Here genetic variables are considered, which have no medical impact by not having any injurious effect. As an example familiar analbuminemia may serve as example. The missing albumin in plasma has no clinical consequences. Persons having this defect get neither edemas nor other clinical symptoms, but may disturb diagnostic interpretation whenever albumin is either the diagnostic analyte for risk (as plasma albumin in intensive care or albuminuria in checking for diabetic nephropathy). Another example is the inherited low choline esterase activity in pseudocholine esterase variants, which have no impact in narcoses using succinyl-bis-choline.

3.1.5 Conclusions and recommendations

The abovementioned examples clearly illustrate that only academic knowledge on the effect of diseases is not adequate to interpret laboratory findings. It is recommended to have all of the following data available when a laboratory result is to be assessed and interpreted:

- Age and sex of patient providing the sample
- Genetic origin of patient (country, ancestors)
- Reference ranges available should be controlled regarding the patient's origin to decide their usability for the person under study.

Note: Also correct laboratory data can lead to false conclusions, if unchangeable preanalytical variables are not sufficiently considered.

References

[1] Soldin SJ, Wong EC, Brugnara C, Soldin OP. Pediatric Reference Intervals, 7th ed. Washington DC: AACC Press 2013.

[2] Guder WG, Narayanan S, Wisser H, Zawta B. Diagnostic Samples: From the Patient to the Laboratory. 4th updated ed. Weinheim: Wiley-Blackwell, 2009.

[3] Heil W, Ehrhardt V. Reference Ranges for Adults and Children. Preanaytical Considerations. Mannheim: Roche Diagnostics, 9th. Ed 2008.

[4] Lamb Ej, Nonan KA, Burrin JM. Urine-free cortisol excretion: evidence of sex-dependence. Ann Clin Biochem 1994:31: 455–8.

[5] Hagemann P, Scholer A. Aktuelle Urindiagnostik. 1911 Rotkreuz: Labolife

[6] Harris, EK, Wong ET, Shaw ST jr. Statistical criteria for separate reference intervals: race and gender groups in creatine kinase. Clin Chem 1991; 37:1580–2.

[7] Stevens LA, Coresh J, Schmid CH, Feldmann HI, Froissart M, Kusek J et al. Estimating GFR using serum cystatin alone and in combination with serumn creatinine: a pooled analysis of 3.418 individuals with CKD. Am J Kidney Dis 2008; 51:395–406.
[8] Karayalcin G, Rosner F, Sawitsky A. Pseudoneutropenia in american negroes. Lancet 1972; 1:387.
[9] Bain B, Seed M, Godsland I. Normal values of peripheral blood white cell counts in women of four different ethnic groups. J Clin Pathol 1984; 37:188–93.
[10] Guder WG, Hagemann P, Wisser H, Zawta B. Fokus Patientenprobe, Kompendium Präanalytik. CD-Rom; Heidelberg:BD,2007.
[11] Tsianos EB, Jalali MT, Gowenlock AH, Braganza JM. Ethnic "hyperamylasaemia":clarification by isoamylase analysis. Clin Chim Acta 1982; 124:13–21.
[12] Saxena S, Carmel R. Racial difference in vitamin B_{12} levels in the United States. Am J Clin Pathol 1987; 88:95–7.
[13] Guyton JR, Dahlen GH, Patsch W, Kautz JA, Gotto AM jr. Relationship of plasma lipoprotein Lp(a) levels to race and to apolipoprotein B. Arteriosclerosis 1985; 5:265–72.
[14] Heyden S, von Eckardstein A, Schulte H, Schneider K, Assmann G. Raised lipoprotein (a) in hypercholesterinaemic black students compared to age matched whites in North and South Dakota. Int J Epidemiol 1994; 23:301–6.

Walter G. Guder, Sheshadri Narayanan*

3.2 Variables during Sampling

When organizing the preanalytical phase, certain influences during the preparation and performance of sampling are of special relevance. These can be standardized and thereby reduce their influence. Thus information of patients before samples are collected and education of persons involved in sampling procedures can reduce or even prevent negative influences on and misintepretaion of laboratory results. The following influences may serve as examples:

- Influence of exercise
- Influence of time during the day and during biological cycles
- Influence of diagnostic and therapeutic procedures
- Influence of body posture
- Influence of extended tourniquet application

3.2.1 Exercise

While interpreting laboratory findings, it is sometimes important to know if the subject had exercised prior to sampling, since this can have major influence on the concentration of some of the analytes. Two types of exercise have to be distinguished. First, static or isometric exercise of brief duration and high intensity which utilizes the energy (ATP and creatine phosphate) already stored in muscles (e. g. bodybuilding, household tasks) and second, dynamic or isotonic exercise of lower intensity and longer duration (e. g. running, swimming, cycling [to the physician's office!]) which utilizes ATP produced by aerobic or anaerobic pathways.

These changes depend on the state of training. Acute changes of analytes during exercise are due to volume shifts between the intravasal and interstitial compartments, volume loss by sweating (e. g. electrolytes, proteins, blood cell counts) and changes in hormone concentrations (e. g. increase in the concentrations of epinephrine, norepinephrine, glucagon, somatotropin, cortisol, ACTH and decreased concentrations of insulin) [2, 3]. These changes in hormone levels may in turn alter the leukocyte count to more than 25 G/L (>25 x 10^9/L) as well as increase glucose concentrations. Figure 3.2.1 shows changes in analyte concentrations induced by participating in a marathon run [4, 5]. The extent of change depends on a variety of individual and/or environmental factors (e. g. training status, air temperature and intake of electrolyte and carbohydrate-containing liquids during the actual run). In nontrained persons heavy exercise can cause muscle tissue degradation with increases in creatine kinase, creatinine and creatine [6] (Fig. 3.2.1).

* Partially published with Hermann Wisser and Berndt Zawta [1]

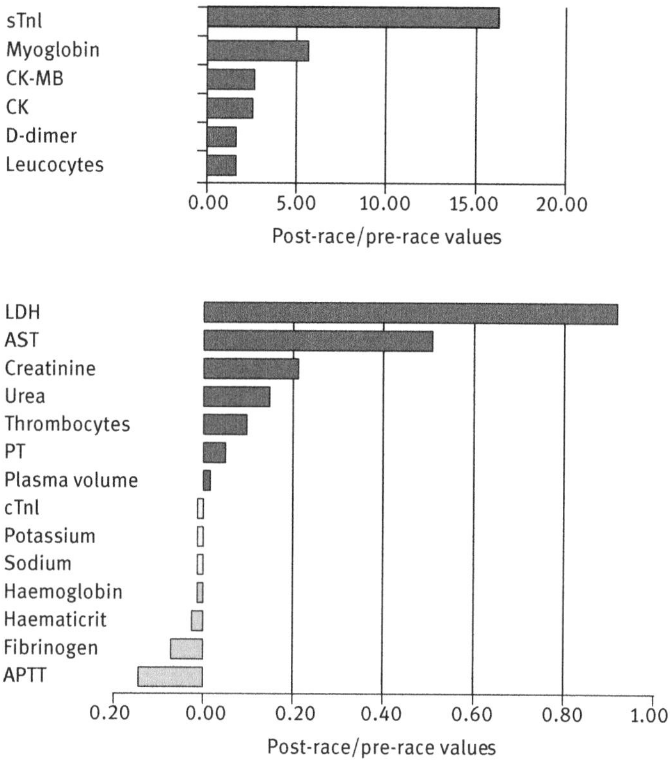

Fig. 3.2.1: Effects of a marathon run on biochemical and hematological parameters. Blood was drawn 1–3 days before the race and within one hour of completion [4, 5, 6].

The changes observed (e. g. increased albumin) can in part be attributed to the above-mentioned volume shift from intravasal to the interstitium or to loss of volume by sweating, but the small increase in plasma volume indicates only a minor hemodilution in most runners [4]. The increased uric acid concentration in serum is a consequence of reduced urinary excretion due to increased lactate concentration. Hypoxia-mediated creatine kinase (CK) increase depends on the training status and hence shows a high degree of individual variability. The less physically fit an individual is the more pronounced the increase in CK. Training increases both the number and the size of mitochondria which is associated with increased capacity of the oxidative enzyme system. This effect in turn increases the capacity of the muscle to metabolize glucose, fatty acids and ketone bodies in aerobic pathways. As a consequence, mitochondrial CK-MB increases to more than 8 % of the total CK activity without evidence of altered myocardial function. Well-trained individuals have a higher percentage of total activity in terms of the CK-MB of skeletal muscle compared to untrained persons. In the study of Smith et al. [4] with amateur runners the percentage CK-MB of the total CK post exercise never exceeded 5 %. Several other analyte concentrations likewise depend on muscle mass and training status. Thus, plasma creatinine, myoglobin, LDH and sceletal troponin increased, but cardiac troponin (cTnI) was never elevated. Urinary creatinine and creatine excretion increased. Lactate formation after exercise decreases in trained compared to untrained athletes. Hematological parameters increase following exercise too. The increase in leukocytes is caused mainly by granulocytes. Coagulation is influenced by activation of clotting (decrease of prothrombin time (PT) and activated partial thromboplastin time (aPTT), fibrinolysis, increase of D-dimer) and increase of platelets. Vigorous exercise may cause erythrocytes or other blood cells to be excreted in urine. These exercise-induced changes, however, usually disappear within a few days.

3.2.2 Time of sampling

Changes occurring in specimens due to the time factor should be taken into account in the preanalytical phase. Three questions are essential in this context:
- When should a sample be taken?
  - Time of day
  - Time after last sample
  - Time after last meal
  - Time after administering a drug, etc.
- When do I require the result of the specimen collected?
- Can results be compared with the results obtained at a different time in daily, monthly and yearly rhythms, either from the same patient or from a reference population?

For the sake of clarity, we can differentiate between linear time, going from the past to the future, and cyclic time; both of these can influence the results of laboratory tests (Fig. 3.2.2).

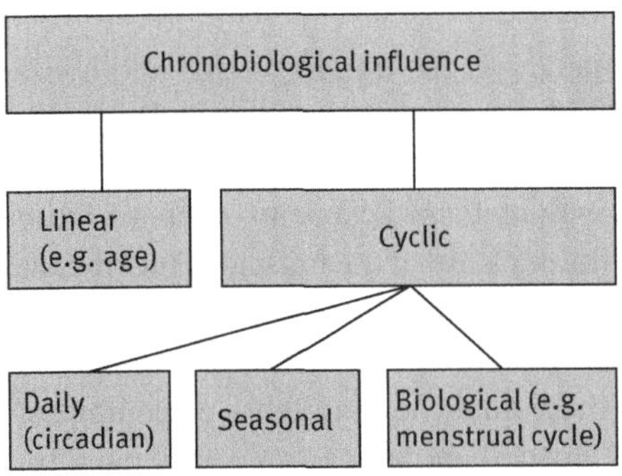

Fig. 3.2.2: Linear and cyclic chronobiological influence.

Influence of circadian rhythm [7]

Several analytes tend to fluctuate in terms of their plasma concentration over the course of a day (Table 3.2.1). Thus, the concentration of potassium is lower in the afternoon than in the morning, whereas that of cortisol decreases during the day and increases at night (Fig. 3.2.3). The cortisol rhythm may well be responsible for the poor results obtained from oral glucose tolerance testing in the afternoon. For this reason, reference intervals are actually obtained between 7 and 9 a.m.

The circadian rhythm can also be influenced by individual rhythms concerning meals, exercise and sleep. These influences should not be confused with real circadian changes. In some cases, seasonal influences also have to be considered. Thus, total triiodothyronine (T_3) is 20 % lower in summer than in winter [8] whereas 25 OH-cholecalciferol exhibits higher serum concentrations in summer [9].

Analytes may change during the menstrual cycle [7]

Analytes can also exhibit statistically significant changes due to the biological changes that occur in the hormone pattern during menstruation. Thus, the aldosterone concentration in plasma is twice as high before ovulation than in the follicular phase. Likewise, renin can show a pre-ovulatory increase. Even cholesterol exhibits a significant decrease during ovulation. In contrast, phosphate and iron decrease during menstruation.

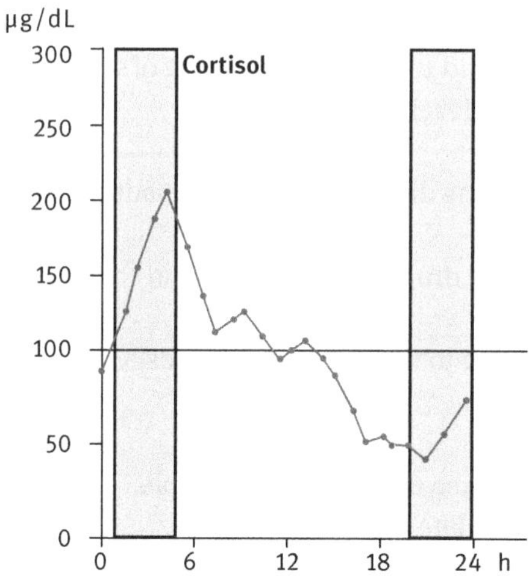

Fig. 3.2.3: Daily variation of plasma concentrations of cortisol (shaded area = sleep period).

Table 3.2.1: Diurnal variation of selected analytes (P = plasma; U = urine) [10].

Analytes	Maximum (time of day)	Minimum (time of day)	Amplitude (percentage of daily mean)	Analytes	Maximum (time of day)	Minimum (time of day)	Amplitude (percentage of daily mean)
Corticotropine (ACTH)	6–10	0–4	150–200	Norepinephrine (P, U)	9–12	2–5	50–120
Cortisol (P,U)	5–8	21–3	180–200	Hemoglobin	6–18	22–24	8–15
Testosterone	2–4	20–24	30–50	Eosinophils	4–6	18–20	30–40
TSH	20–2	7–13	5–15	Iron (P)	14–18	2–4	50–70
Thyroxine (T_4)	8–12	23–3	10–20	Potassium (P)	14–16	23–1	5–10
Somatotropin	21–23*	1–21	300–400	Phosphate (P)	2–4	8–12	30–40
Prolactin	5–7	10–12	80–100	Sodium (U)	4–6	12–16	60–80
Aldosterone	2–4	12–14	60–80	Phosphate (U)	18–24	4–8	60–80
Renin	0–6	10–12	120–140	Volume (U)	2–6	12–16	60–80
Epinephrine (P)	9–12	2–5	30–50	Body temp.	18–20	5–7	0.8–1.0°C

* Start of sleeping phase

3.2.3 Influence of diagnostic and therapeutic procedures

Timing with regard to diagnostic and therapeutic processes

In order to prevent disturbing preanalytical influences, the timing of sampling has to be organized to take place before interfering diagnostic procedures. Likewise, interfering drugs should be administered after collecting a blood sample. On the other hand, in drug monitoring (see Chapters 3.5, 4.5 and 10.8) the exact timing of sampling is essential for correct interpretation of the drug level.

Important rules for the timing of sampling:

- Samples should be collected before interfering diagnostic and therapeutic procedures are performed.
- In drug monitoring, consider the peak after drug administration and the steady state phase before the next dose.
- Always document the exact time of sampling in the charts and requests.

Warning:

- A sample taken at the wrong time can be worse than taking no sample.
- A sample whose analytical results arrive too late is a wasted sample.

The following diagnostic and therapeutic measures can result in both in-vivo (frequent) and in-vitro (less common) effects on laboratory tests [11, 12]:

- Operations
- Infusions and transfusions

- Punctures, injections, biopsies, palpations, whole-body massage
- Endoscopy
- Dialysis
- Physical stress (e. g. ergometry, exercise, ECG)
- Function tests (e. g. oral glucose tolerance test)
- Immunoscintigraphy
- Contrast media, drugs
- Mental stress
- Ionizing radiation

Operations

Changes in serum enzyme activities are frequently so pronounced that specific targeting of an organ is no longer possible. The elevation in acute phase proteins (e. g. C-reactive protein (CRP), fibrinogen) at the beginning of the postoperative phase is accompanied by a decrease in albumin; this cannot be explained alone by hemodilution.

Transient elevations in urea concentration in serum/plasma (up to 60 mg/dL or 10 mmol/L) as well as a decrease in cholesterol are very frequent in the first postoperative days whilst the creatinine concentration remains normal. This may be due to protein breakdown subsequent to gastro-intestinal tract surgery as well as to bleeding in the lumen of the bowel, e. g. in the case of a stress ulcer.

Therapeutic drugs

Nearly unavoidable are influences and interferences on diagnostic analyte concentrations due to therapeutic treatment with drugs. The influences of drug therapy are of different nature, as described in Chapter 3.5. Besides the intended therapeutic drug effect (like decreasing blood glucose concentration after an oral antidiabetic drug) concentration of other analytes can be altered by physiological side effects of the drug (like an increase in plasma lactate) without one being aware of this cause. Besides these side actions, drugs can cause interferences in vitro (see Chapter 4.4). The complex method- and concentration dependence of these drug interferences has led to the development of data banks in internet and special books providing information on drug interactions with analyte's different measurement procedures.

Infusions, transfusions

Hemolysis and hence the concentrations of free hemoglobin and potassium, as well as the activity of lactate dehydrogenase in plasma obtained from conserved blood, increase with the age of the transfused conserved material.

Table 3.2.2: Infusions/transfusions as interfering factors and/or contaminants of laboratory diagnostic tests.

Infusion/Transfusion	Analyte affected	Trend	Comments, Mechanism
Dextran	Thrombin time, reptilase time	↓	5–10 sec slower
	von Willebrand factor	↓	
	Total protein in serum, plasma	↑	Biuret, method-dependent (turbidity, flocculation, greenish coloration)
	Urea, serum	↓	
	Blood grouping serology		Pseudoagglutination
γ-Globulin	Serological determinations during virus-mediated and bacterial infections		False positive
Electrolytes	Potassium, sodium, magnesium	↑	Contamination
Glucose	Glucose	↑	Contamination
Glucose	Inorg. phosphate, potassium,	↓	Insulin
	Bilirubin, amylase	↓	Up to 15 %, particularly in neonates
Fructose	Uric acid	↑	Metabolic effect
Citrate	pH value in blood	↓	
(blood transfusion!)	Coagulation tests	↓↑	Inhibition

Contamination of laboratory samples by infusion solutions is the commonest and often the most relevant form of preanalytical interference in the hospital [13, 14] (Table 3.2.2).

Blood should never be collected proximal to the infusion site.

Specimens should be collected from the opposite arm. A certain period of time should be allowed to elapse following infusion therapy (Table 3.2.3).

Special disturbing possibilities exist when blood is transfused. Here results from blood samples can be artificially altered even after the equilibration phase has elapsed. Thus depending on the age of blood that is transfused, increasing amounts of extracellular fluid with higher potassium, free hemoglobin and lactate dehydrogenase activity can be observed. On the other hand, plasma constituents which were pathologically high could be diluted by infusion of normal blood (like tumor markers, pathological blood cells and poisons). Likewise coagulation tests can give different results after transfusion of anticoagulated blood.

It is recommended that the laboratory be informed of when and what type of infusions was administered and when blood samples were collected.

Table 3.2.3: Recommendations for scheduling infusions and blood sampling.

Infusion	Earliest time of blood sampling in hours after cessation of infusion
Fat emulsion	8
Carbohydrate-rich solutions	1
Amino acids and protein hydrolysates	1
Electrolytes	1

Sampling from catheters

Three techniques of blood sampling are described in the literature: the widely used discard method, the reinfusion method and the push-pull or "mixing" method [15]. The infusion should be terminated 1–2 minutes before sampling and the sample should never be collected proximal to the infusion site. If not possible, sampling from the opposite arm is recommended.

If samples are to be taken from intravenous and intraarterial infusion catheters, the cannula should be rinsed with isotonic saline commensurate with the volume of the catheter. The first 5 mL of blood should be discarded before a blood sample is taken.

Sampling for coagulation tests from heparin-contaminated catheters is particularly critical. For heparin-dependent methods (thrombin time, aPTT), it is recommended that an amount of blood equivalent to twice the volume of the catheter be discarded; the blood first taken after this should be used for nonhemostaseological investigations and the subsequently obtained citrated blood only used for determining heparin-insensitive analytes: Prothrombin time, reptilase time, fibrinogen according to Clauss, AT III, fibrin monomers. It is important that before transferring blood to the sampling vessel containing sodium citrate solution there is no lengthy pause during which the blood in the catheter is allowed to "stand".

Mental stress

The importance of mental stress on laboratory results is frequently underestimated (anxiety prior to blood sampling, preoperative stress, etc.). Increased secretion of hormones (aldosterone, angiotensin, catecholamines, cortisol, prolactin, renin, somatotropin, TSH, vasopressin) and increased concentrations of albumin, fibrinogen, glucose, insulin, lactate and cholesterol have been observed.

3.2.4 Sampling in the supine or upright position?

It is a well-known fact that **body posture** influences blood constituent concentrations. This is caused by different mechanisms. First, the effective filtration pressure

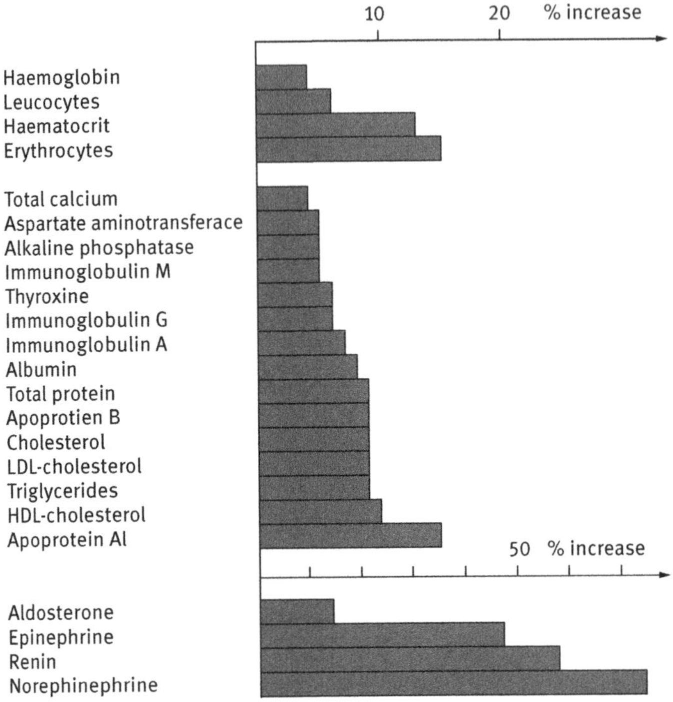

Fig. 3.2.4: Increase (%) of plasma concentration of various analytes when changing from supine to an upright position [19, 20].

(e. g. the difference between capillary pressure and colloidal osmotic pressure in plasma) increases in the lower extremities when changing from the supine to the upright position. As a consequence, water is moved from the intravasal compartment to the interstitium; thus reducing the plasma volume by about 12 % in normal individuals. Blood particles with a diameter of more than 4 nm are restrained by membranes and cannot follow this volume shift. A change from the upright to the supine position leads to a decrease in the effective filtration pressure and hence to a volume shift in the reverse direction [16]. A change in plasma volume leads to an apparent concentration change in cells, macromolecules and protein-bound small molecules. Most low molecular weight compounds show no change in their apparent concentrations when changing from the upright to the supine position. As osmolality is mainly mediated by such compounds, they are only modestly affected by changes in plasma volume (1–2 %). Because of partial protein binding, the concentrations of free and bound calcium are affected in a different manner. Whilst the concentration of free calcium is independent of posture, total calcium increases by 5–10 % when changing from the supine to the upright position [17]. Other changes are due to altered blood pressure which in turn causes secretion of vasoactive compounds. In addition, the metabolic consequences of regulatory changes due to postural changes may alter body fluid composition. The effects of posture on analytes in venous anticubital blood are shown in Fig. 3.2.4. As expected from the described mechanism, most cellular and macromolecular analytes decrease between 5 and 15 % compared to the supine position. These effects can be more pronounced in patients with a tendency to edema (cardiovascular insufficiency, liver cirrhosis). Reduction in plasma volume induces a decrease in blood pressure which in turn leads to increased secretion of renin, aldosterone, norepinephrine and epinephrine.

An example of the metabolic changes brought about due to posture is the urinary excretion of calcium which increases during long-term bed rest (see Fig. 13–2 in Guder et al 2009 [1, 18].

Recommendation: Blood sampling should be performed at least 10 min after a preceding phase of rest in recumbant position. In comparing laboratory results, sample is to be obtained under identical conditions regarding body position.

3.2.5 Extended Tourniquet application

What happens when a tourniquet is kept on during sampling? A tourniquet is usually applied to facilitate finding the appropriate vein for venipuncture (see Chapter 2.3). Using a pressure below the systolic pressure maintains the effective filtration pressure inside the capillaries. As a consequence, fluid and low molecular compounds are moved from the intravasal space to the interstitium. Macromolecules, compounds bound to protein and blood cells, do not penetrate the capillary wall so that their concentration apparently increases while the concentration of low molecular substances remains unchanged. Figure 3.2.5 shows the changes of different analyte concentrations [21]. The alterations of low molecular weight analytes observed following 6 min of constriction are in the range of ±3 % and therefore within the range of analytical imprecision. However, it has been shown that constriction of the forearm muscles causes an increase in serum potassium concentration. Therefore, during venipuncture for potassium determination, repeatedly clenching and unclenching a fist should be avoided and a superficial vein selected [22, 23]. The extent of changes in high molecular weight analytes depends on the duration of constriction. Figure 3.2.6 demonstrates changes of blood cell concentration due to sample drawing procedure. Hemoglobin, hematocrit (data not shown) and platelets show an increase, leucocytes, granulocytes and lymphocytes decrease. The latter effect is caused by venostasis and hypertension induced release of several mediators promoting leucocytes migration and penetration of the vascular wall [23]. Constriction times of one minute with subsequent release of the tourniquet have no consequences on plasma serum analyte concentrations and coagulation factors.

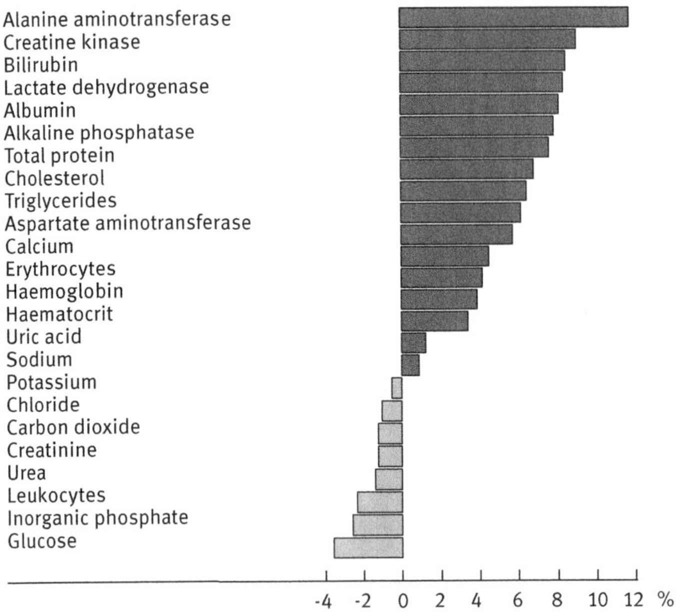

Fig. 3.2.5: Change (%) in serum concentration of various analytes after a tourniquet application time of 6 min [21].

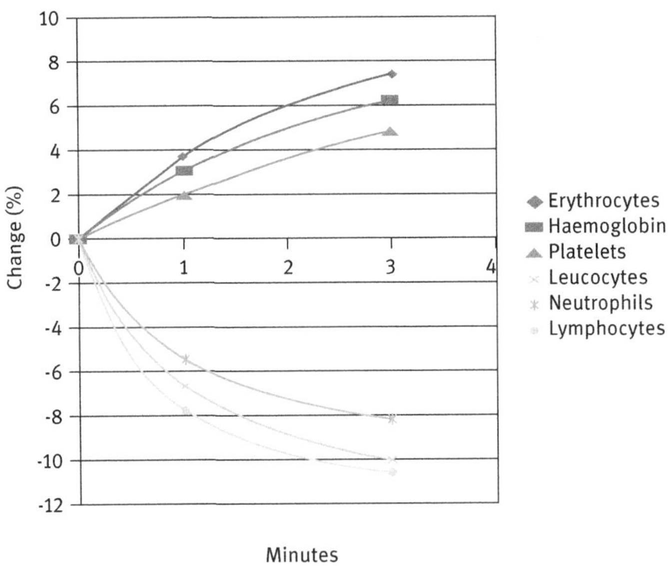

Fig. 3.2.6: Changes in hematological parameters after 60 mm Hg external pressure for 1 and 3 minutes [24].

To reduce the intra- and interindividual variance of laboratory results a standardized sampling procedure is an important prerequisite. The tourniquet application time should not be longer than one minute.

References

[1] Guder WG, Narayanan S, Wisser H, Zawta B. Diagnostic Samples: From the Patient to the Laboratory. 4th updated edn. Weinheim: Wiley-Blackwell 2009.
[2] Alexander S. Physiologic and biochemical effects of exercise. Clin Biochem 1984; 17:126–31.
[3] Röcker L, Schmidt HM, Matz W. Der Einfluss körperlicher Leistungen auf Laboratoriumsbefunde im Blut. Ärtzl. Lab. 1977; 23:351–7.
[4] Smith JE, Garbutt G, Lopes P, Tunstall Pedoe D. Effects of prolonged strenuous exercise (marathon running) on biochemical and haematological markers in the investigation of patients in the emergency department. Br J Sports Med 2004; 38:292–4.
[5] Stansbie D, Begley JP. Biochemical consequence of exercise. IFCC J 1991; 3:87–91.
[6] Sorichter S, Mair J, Koller A, Calzolari C, Huonker M, Pan B, et al. Release of muscle proteins after downhill running in male and female subjects. Scand J Med Sci Sports 2001; 11:28–32.
[7] Young DS. Effects of Preanalytical Variables on Clinical Laboratory Tests. 3rd edn. Washington DC: AACC Press 2007.
[8] Harrop IS, Ashwell K, Hapton MR. Circannual and within individual variation of thyroid function tests in normal subjects. Ann Clin Biochem 1985; 22:371–5.
[9] Nordin BEC, Wilkinson R, Marshall DH, Galagher JC, Williams A, Peacock M. Calcium absorption in the elderly. Calcif Tiss Res 1976; 21:442–51.
[10] Wisser H, Knoll E. Tageszeitliche Änderungen klinisch-chemischer Messgrößen. Ärtzl. Lab. 1982; 28:99–108.
[11] Hagemann P. Qualität im Arztlabor, Optimierung der Präanalytik, Berlin-Heidelberg: Springer, 1994.
[12] Keller H. Klinisch chemishe Labordiagnostik für die Praxis. 2nd ed. Stuttgart: Thieme 1991.
[13] Watson KR, O´Kell T, Joyce JT. Data regarding blood drawing sites in patients receiving intraveneous fluids. Am J Clin Pathol 1983; 79:119–21.
[14] Zawta B. Infusionstherapie und Labordiagnostik. GIT LaborMedizin 1994; 17:171–3.
[15] Frey AM. Drawing blood samples from vascular access devices. J Inf Nurs 2003; 26:285–93.
[16] Röcker L, Schmidt HM, Junge B, Hoffmeister H. Orthostasebedingte Fehler bei Laboratoriumsbefunden. Med Lab 1975; 28:267–75.
[17] Renoe BW, MacDonalds JM, Ladenson JH. Influence of posture on free calcium and related compounds. Clin Chem 1979; 25:1766–9.
[18] Deitrick JE, Whedon GD, Shorr E. Effect of immobilization upon various metabolic and physiological functions in men. Am J Med 1948; 4:3–36.
[19] Felding P, Tryding N, Hyltoft-Petersen P, Hørder M. Effects of posture on concentrations of blood constituents in healthy adults: practical application of blood specimen collection procedure recommended by the Scandinavian Commiteee of reference values. Scand J Clin Lab Invest 1980; 40:615–21.
[20] Miller M, Bacharik PS, Cloey TA. Normal variation of plasma lipoproteins: postural effects on plasma concentrations of lipids, lipopropteins and apolipoproteins. Clin Chem 1992; 38:569–74.
[21] Junge B, Hofmeister H, Feddersen HM, Röcker L. Standardisierung der Blutentnahme. Dtsch Med Wschr 1978; 103:260–5.

[22] Don BR, Sebastian A, Cheitlin M, Christiansen M, Schambelan M. Pseudohyperkalemia caused by fist clenching during phlebotomy. N Engl J Med 1990; 322:1290–2.
[23] Skinner SL, Adelaide MB. A cause of erroneous potassium levels. Lancet 1961; i:478–80.
[24] Lippi G, Salvagno GL, Montagnana M, Franchini M, Guidi GC. Venous stasis and routine haematologic testing. Clin Lab Haem 2006; 28:

Sheshadri Narayanan, Walter G. Guder*

3.3 Food, Drinks and Smoking

3.3.1 Influences of diet

Before sampling blood, the disturbing influences of food and drinks should be excluded. Diet and drinking are major factors influencing a number of analytes.

From a practical point of view, one should distinguish acute effects from those observed over a longer period. A critical question in daily routine is whether a standard meal affects target analytes. Figure 3.3.1 shows the percentage change in different analyte concentrations as a function of food intake [1, 2]. Effects of 5 % or less may be neglected (below <1.005 in Fig. 3.3.1), since they are clinically irrelevant. Therefore, samples for these analytes do not require strict food deprivation. Not only do triglycerides and glucose increase during the reabsorptive phase, the turbidity of the plasma/serum sample following reabsorption of lipids and present as chylomicrons can in addition interfere with other measuring procedures and cause apparent changes in analyte concentrations. The effect of food is therefore dependent on the composition of diet and the interference or influence dependent on the elapsed time between food intake and sampling.

The serum concentration of cholesterol and triglycerides are influenced by various factors as food composition [4], physical activity, smoking, consumption of alcohol and coffee [5]. Elevated levels of ammonia, urea and uric acid are observed during a high protein and nucleotide diet. The changes occurring after a standard carbohydrate meal (75 g glucose) are diagnostically helpful in testing glucose tolerance [6].

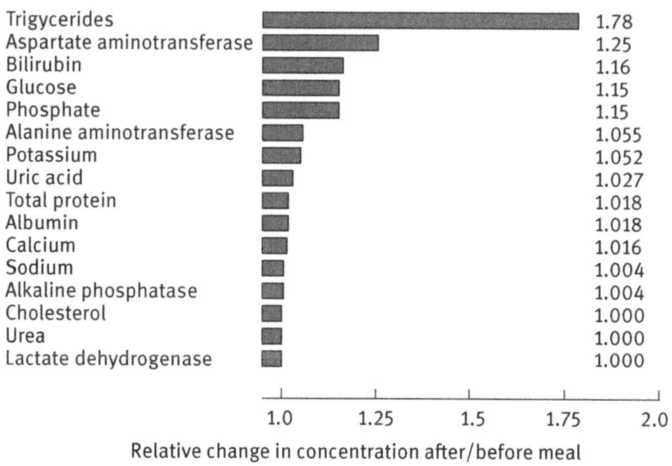

Fig. 3.3.1: Change in serum concentration of different analytes 2 h after a standard meal (from [3]).

* Partially published earlier with Hermann Wisser and Bernd Zawta [3].

Changes may be due either to an increase in reabsorption of the measured analyte (triglycerides, glucose) intestinal or liver metabolism of reabsorbed metabolites (VLDL), urea, ammonia) or regulatory changes due to food intake (uric acid, retinol binding protein, ketone bodies).

Recommendation: In order to avoid misinterpreation of laboratory results, sampling after 8 h better 12 h fasting and reduced activity (bedrest) is recommended as a standard procedure.

3.3.2 Starvation

On the other hand, malnutrition and starvation may alter analyte concentrations in a clinically relevant fashion. Early indicators of low-protein diet are reduced serum concentrations of transthyretin (prealbumin) and retinol-binding protein. Some alterations in clinical chemical analytes induced by starvation over 48 h are summarized in Fig. 3.3.2. Metabolic acidosis with a decrease of both pH and bicarbonate results from an increase in organic acids, mainly the ketone bodies (acetoacetic acid, 3-hydroxybutyric acid) [7].

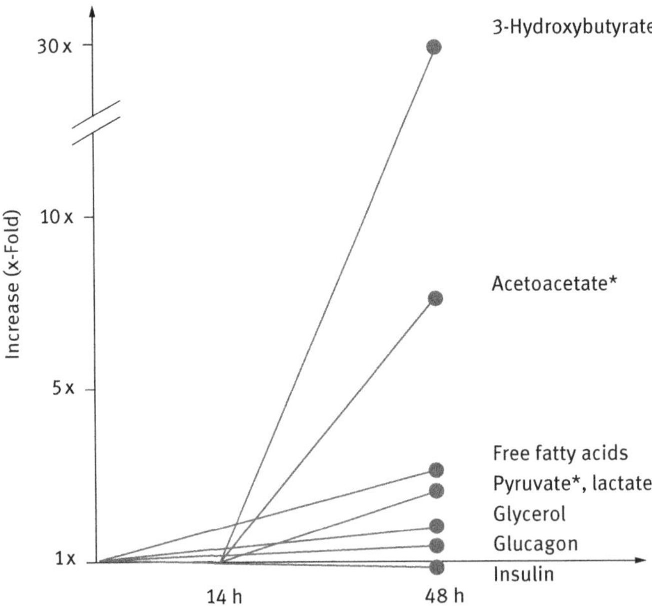

Fig. 3.3.2: Variation of several analytes in plasma/serum after 40–48 h of starvation (from [3]).

*Starting point after 14 h of starvation

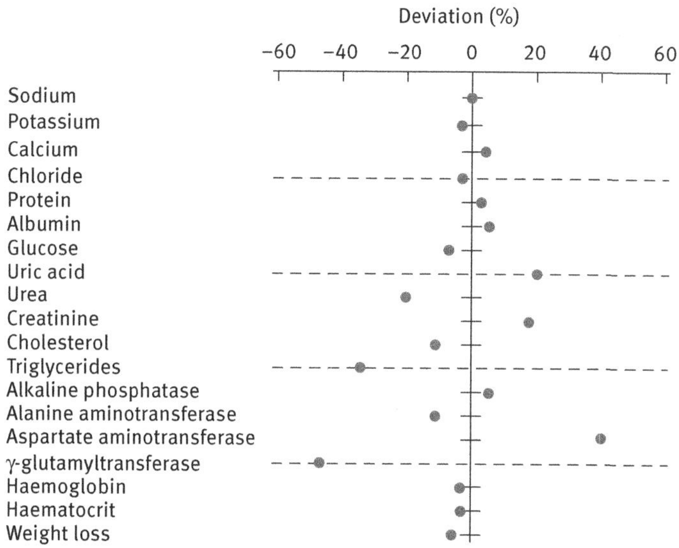

Fig. 3.3.3: Change (%) of clinical chemical analytes after 4 weeks starvation and a daily supply of 33 g protein, vitamines and electrolytes (from Ditschuneit et al [8], Wisser 1995 [9] in Guder et al 2009 [3]).

Changes in analyte concentrations induced by long-term starvation (4 weeks) are shown in Fig. 3.3.3 at the end of the starvation period in comparison to the initial values. The concentrations of blood cholesterol, triglycerides and urea are reduced. In contrast, creatinine and uric acid concentrations are elevated. The increase in uric acid concentration during starvation periods even requires treatment. The latter is due to reduced clearance of uric acid as a result of ketonemia [8]. It is readily apparent that long-term starvation is closely associated with reduced energy expenditure.

During long-term starvation, metabolism changes to endogenously stored metabolites. This explains the increase in fatty acids, ketones, lactate and glycerol, characterized by a decrease in insulin with increasing glucagon levels causing an increase in lipolytic activity in adipose tissue, followed by increased fatty acid degradation to form acetoacetate and 3-hydroxybutyrate. Degradation of muscle mass leads to increased creatinine and uric acid formation, whereas protein degradation increases urea excretion in urine. Only after extended starvation plasma proteins start to decrease. Urea formation is replaced by increasing excretion of ammonia, thus sparing nitrogen [10]. In addition, plasma calcium and phosphate decrease. Similar observations are made during catabolic states of various diseases (postoperative and polytraumatic conditions, malignant diseases in progredient stages).

Besides plasma alterations, urinary excretion of several compounds is likewise affected by long-term starvation. Urinary excretion of ammonia and creatinine is increased whereas that of urea, calcium and phosphate is reduced [11]. Changes in analyte concentrations brought about by long-term starvation are similar to those observed following surgical procedures or in patients with a catabolic status. Not only long-term, but short-term food restriction also induces significant changes of serum concentration of several biochemical and endocrine quantities. After an average weight loss of 1.8 kg during one week following changes were registered: triglycerides −25 %, free fatty acids +124 %, glycerol +74 % by increased lipolysis, ammonia −17 %, urea +11 %, uric acid +10 % by activated protein metabolism and insulin −42 %, corticotropine (ACTH) +41 %, cortisol −24 % and testosterone −34 % [12].

Recommendations: In measuring quantitative urinary excretion rates, excreted amounts per day are preferable to those per liter in order to eliminate variations in drinking habits and water excretion.

3.3.3 Effects of drinks before sampling

Whereas drinking of coffee or low amounts of alcohol is largely seen as part of normal life and therefore not worth reporting to the physician, are influences of these "luxuries of life" on various analyte concentrations very important to consider in laboratory results interpretations. For example, in muscle activity and diet effect acute and chronic effects are to be differentiated.

Caffeine

Caffeine is found in many constituents of food ingested daily. Despite its wide- spread use, the influence of caffeine on various analytes in clinical chemistry has not been investigated in detail. Thus coffee by caffeine action not only increases blood pressure, but likewise the hormones adrenaline and noradrenaline. This in turn increases lipolysis and renin activity in plasma.

Caffeine causes an increase of the glomerular filtration rate (GFR) and a decrease of the tubular reabsorbtion of electrolytes followed by inhibition of the renal adenosine receptors and inhibits phosphodiesterase and hence cyclic AMP degradation. Cyclic AMP in turn promotes glycogenolysis, thereby increasing blood glucose concentrations. In addition, the glucose concentration increases due to gluconeogenesis via epinephrine. Activation of triglyceride lipase leads to a threefold increase of non-esterified fatty acids [13]. Quantification of hormones and drugs bound to albumin is hampered by the fatty acid-induced displacement effect. Three hours after the intake of 250 mg of caffeine, plasma renin activity and catecholamine concentrations have been found to be elevated [14].

Studies intended to investigate these analytes should take caffeine consumption into account.

Alcohol

Alcohol consumption, depending on its duration and extent, may affect a number of analytes. These alterations are used in part for diagnosis and therapeutic monitoring. Among alcohol-related changes, acute and chronic effects should be considered separately. The acute effects (within 2–4 h) of ethanol consumption are decreased plasma glucose and increased lactate due to the inhibition of hepatic gluconeogenesis. Ethanol is metabolized to acetaldehyde and then to acetate. This increases hepatic formation of uric acid [15] and inhibits renal urea excretion thus causing an increase of uric acid in plasma [16]. Together with lactate, acetate decreases plasma bicarbonate, resulting in metabolic acidosis.

After one to several days of alcohol ingestion, γGT activities are induced, included in detoxification of acetaldehyde in liver and the biliary excretion of gluthathion, thus leading to an increase in plasma enzyme activity. Likewise the direct effect of ethanol degradtion to acetaldehyde in liver inhibits sialinisation of proteins formed in liver cells, which becomes measurable in the so called carbohydrate deficient transferrin (CDT). The long-term effects of ethanol ingestion include an increase in the serum activity of liver enzymes. The increase of γ-glutamyltransferase activity is caused by enzyme induction. Glutamate dehydrogenase as well as aminotransferases (AST, ALT) activities increase due to a direct liver toxic effect [20]. The increase in desialylated forms of proteins in blood (i. e. carbohydrate deficient transferrins) is due to an inhibition of enzymatic glycosylation during posttranslational processing of these proteins in the liver. In chronic alcoholism, serum triglycerides increase due to decreased plasma triglyceride breakdown. The increased MCV may be related to a direct toxic effect on the erythropoetic cells or a deficiency of folate [21]. The data in Fig. 3.3.4 do not take into account either the dose or the time dependency, which underly both the acute and the chronic effects. Increased urine ethanol excretion leads to a decrease formation of vasopressin with increasing diuresis. Enhanced diuresis is followed by increased secretion of renin and aldosterone [22].

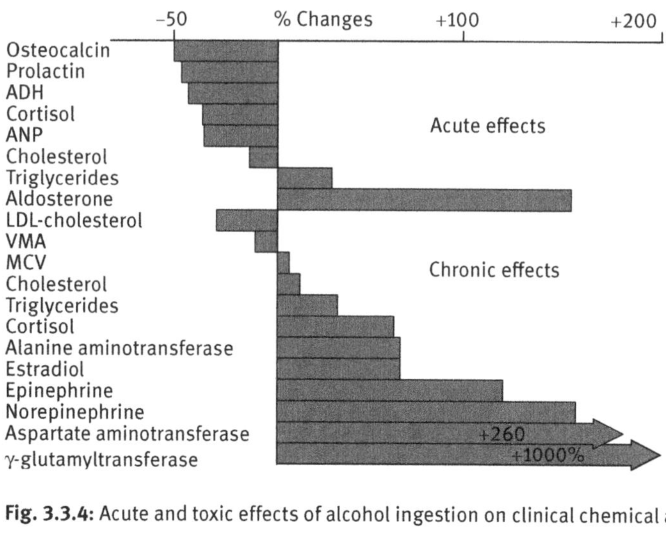

Fig. 3.3.4: Acute and toxic effects of alcohol ingestion on clinical chemical analytes [17, 18, 19].

Recommendations: The behavior of patients with alcoholic drinks should be documented in the clinical charts reagarding amount and times of ingestion to avoid misinterpretations of laboratory results. On the other hand, can a typical laboratory result constellation (increased γGT, MCV and CDT) not been used as proof of chronic consumption.

3.3.4 Smoking

Smoking leads to a number of acute and chronic changes in analyte concentrations, the chronic changes being rather modest. Smoking increases the plasma/serum concentrations of fatty acids, epinephrine, free glycerol, aldosterone and cortisol [19]. These changes occur within one hour of smoking 1–5 cigarettes. Alterations in analytes induced by chronic smoking include blood leukocyte count, lipoproteins, the activities of some enzymes, hormones, vitamins, tumor markers and heavy metals (Fig. 3.3.5) [19].

Four of the changes presented are of special clinical importance: the increase of the white cell count of neutrophils, lymphocytes and monocytes, and the concentration of carcinoembryonic antigen (CEA), a tumor marker. The letter marker needs to use different reference ranges for smokers and non smokers. The higher concentration found in smokers is caused by an increased synthesis and secretion of CEA in colon.

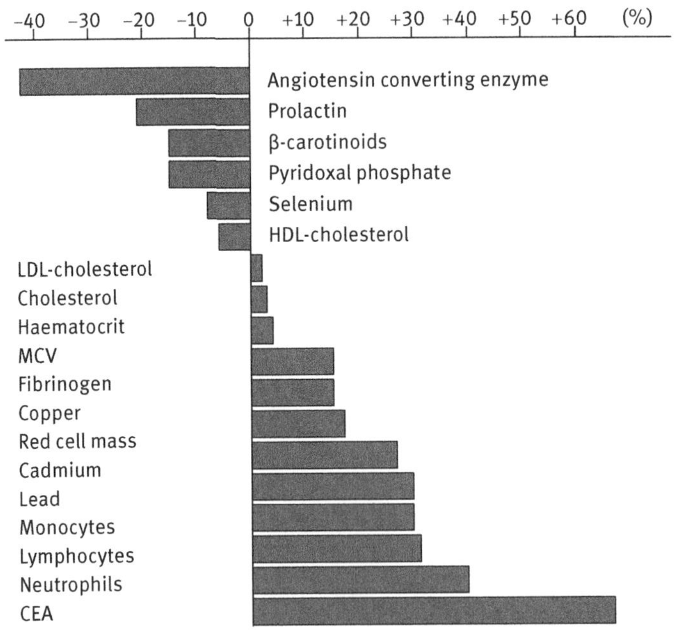

Fig. 3.3.5: Deviation (%) of blood analyte concentrations between current smokers and nonsmokers, chronic effects [19].

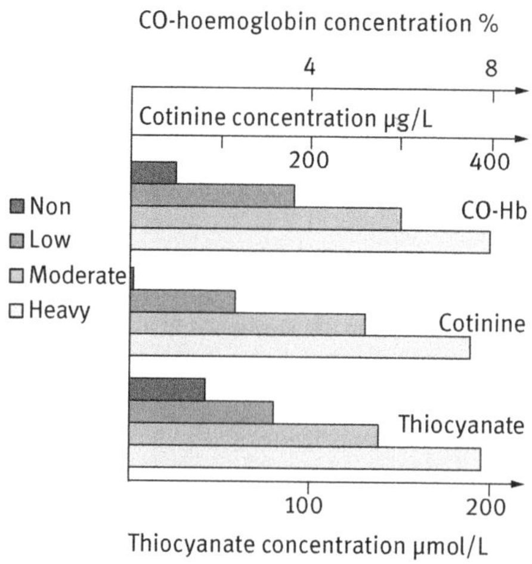

Fig. 3.3.6: Effect of smoking on different blood analytes caused by smoke constituents [13, 19].

The mechanism underlying these changes has not been fully elucidated. A large number of pyridine compounds, hydrogen cyanide and thiocyanate are found in tobacco smoke. They can account for concentration changes by direct or indirect effects. Decreased angiotensin converting enzyme activity (ACE) in smokers is believed to result from the destruction of lung endothelial cells with a subsequent reduction in the release of ACE into the pulmonary circulation and/or enzyme inhibition [23]. The extent of changes also depends on the amount, kind (cigarettes, cigars, pipes) and technique of smoking (with or without inhalation).

Moreover, smoking-induced changes are influenced by age and gender [16]. Specific for the uptake of tobacco smoke and nicotine are the metabolites of nicotine, cotinine and thiocyanate, which mirror the number of cigarettes smoked in blood. Better known is the marker CO-hemoglobin in smokers Fig. 3.3.6. shows the concentrations of cotinine, thiocyanate and carboxyhemoglobin, used as markers for the qualitative and quantitative assessment of smoking habits. Cotinine has the advantage of having a longer half-life (20–28 h) than nicotine, the parent compound (12–15 min) [24].

Recommendations: Smoking habits should be documented in the clinical charts. Smoking needs to be considered when interpreting result of laboratory tests sensitive to smoking habits.

References

[1] Cohn JS, McNamara JR, Cohn SD, Ordovas JM, Schaefer EJ. Postprandial plasma lipoprotein change in human subjects of different ages. J Lip Res 1088; 29:368–9.

[2] Steinmetz J, Panek E, Souriau F, Siest G. Influence of food intake on biological parameters. In: Siest (ed.), Reference Values in Human Chemistry. Basel: Karger 1973, pp. 195–200.

[3] Guder WG, Narayanan S, Wisser H, Zawta B. Diagnostic Samples: From the Patient to the Laboratory. 4th updated ed. Weinheim: Wiley-Blackwell 2009.
[4] Rifai N, Merill JR, Holly RG. Postprandial effect of a high fat meal on plasma lipid, lipoprotein cholesterol and apolipoprotein meaurements. Ann Clin Biochem 1990; 27:489–93.
[5] Evans K, Laker MF Intraindividual factors affecting lipid, lipoprotein and apolipoprotein measurement: a review. Ann Clin Biochem 1995; 32:261–80.
[6] Kerner W, Brückel J. Definition, Klassifikation und Diagnostik des Diabetes mellitus. Diabetologie 2013; 8:104–7.
[7] Lamers KJB, Doesburg WH, Gabreels FJM, Lemens WAJG, Romson AC, Wevers RA et al. The concentration of blood components related to fuel metabolism during prolonged fasting in children. Clin Chim Acta 1985; 152:155–63.
[8] Ditschuneit H, Ditschuneit H, Wechsler J. Probleme der ambulanten Nulldiätbehandlung. Dtsch Aerztebl 1979; 76:871–80.
[9] Wisser H. Einflussgrößen und Störgrößen. In Greiling H, Gressner AM (eds), Lehrbuch der Klinischen Chemie und Pathobiochemie. 3rd edn. Stuttgart: Schattauer 1995, pp. 37–49.
[10] Guder WG, Häussinger D, Gerok W. Renal and hepatic nitrogen metabolism in systemic acid base regulation. J Clin Chem Clin Biochem 1987; 25:457–66.
[11] Wisser H. Paeanalytische Faktoren bei der Harnanalytik. In Guder WG, Lang H. (eds), Pathobiochemie und Funktionsdiagnostik der Niere. Berlin: Springer, 1989, pp. 171–82.
[12] Degoutte F, Jouanel P, Bègue RJ, Colombier M, Lac G, Pequinot JM, Filare E et al. Food restriction, performance, biochemical, psychological and endocrinological changes in judo athletes. Int J Sports Med 2006; 27:9–18.
[13] Narayanan S.Preanalytical variables in blood sampling. Klin Chem Mitt 1993; 24:130–4.
[14] Robertson D, Frölich JC, Carr RK, Watson JTh, Holliefield JW, Shand DG et al. Effects of caffeine on plasma renin activity, catecholamines and blood pressure. N Engl J Med 1978; 298:181–6.
[15] Grunst J, Dietze G, Wicklmayr M, Mehnert H. Hepp KD, Eisenburg J. Einfluss von Ethanol auf den Purinkatabolismus der menschlichen Leber. Verh. D Gesellsch Innere Med 1973; 79:914–7.
[16] Statland BE, Winkel P. Effects of preanalytical sorces of variation. In: Gräsbeck R, Alström T (eds), Reference values in Laboratory Medicine. Chichester: Wiley 1981, pp. 127–37.
[17] Leppaluoto J, Vuolteenaho O, Arjamaa O, Ruskoao H. Plasma immunoreactive atrial natriuretic peptide and vasopressin after ethanol intake in man. Acta Physiol Scand 1992; 144:121–7.
[18] Ratge D, Brugger G, Wehr M, Bode JCh, Wisser H. Catecholamines in plasma and urine of patients with alcoholic liver damage under resting and exercise conditions. J Clin Chem Clin Biochem 1985; 23:447–52.
[19] Young DS Effects of Preanalytical Variables on Clinical Laboratory Tests. 3rd ed. Washington DC: AACC Press 2007.
[20] Freer DE, Statland BE. Effects of ethanol (0, 75 g /kg body weigth) on the acivities of selected enzymes in sera of healthy adults: 2. Interindividual variations in response of γ-glutamytransferase to repeated ethanol challenges. Clin Chem 1977; 23:2099–102.
[21] Rico H. Alcohol and bone disease. Alcohol Alcoholism 1990; 25:345–52.
[22] Bode JC. Die internistischen Folgeerkrankungen des Alkoholismus. In Kisker KP, Lauter H, Meyer JE, Müller C, Strömgren E (eds) Psychiatrie der Gegenwart. Abhängigkeit und Sucht. 3rd edn. Berlin:Springer 1987, pp. 206–41.
[23] Haboubi NAA, Bignell AHC, Haboubi BY. Serum angiotensin converting enzyme activity in cigarette smokers. Clin Chim Acta 1986; 154:69–72.
[24] Sepkovic DW, Haley NJ, Wynder EL. Thyroid activity in cigarette smokers. Arch Intern Med 1984; 144:501–3.

Sheshadri Narayanan, Donald S. Young

3.4 Effect of Herbs

3.4.1 Introduction

Herbs can influence the results of clinical laboratory test results in a variety of ways. They may alter the therapeutic effect of a drug by affecting either its metabolism, or uptake and transport. Herbs may have a beneficial effect such as protecting the liver, improving glycemic control, lowering plasma cholesterol concentration, relieving stress and improving urinary flow in patients with an enlarged prostate to cite a few. Herbs can also have a toxic effect affecting liver and kidney function [1].

In this chapter we will attempt to describe the mechanisms of herb-drug interactions that have an effect on therapy and illustrate these effects with specific examples. We include examples of herbs that have a beneficial effect as well as those which are toxic.

Finally we'll also address the effects of contaminants that creep into the formulation of herbs thus reinforcing the importance of quality control of herbal preparations.

3.4.2 Mechanism of herb–drug interactions

Herbs may either induce or inhibit the cytochrome P450 (CYP) family of enzymes involved in the metabolism of drugs. A variety of CYP isoforms such as CYP3A4, CYP2C9, CYP1A2, CYP2C19, CYP2E1 and CYP2D6 are involved in the metabolism of drugs. One major isoform CYP3A4, found both in the liver and intestine, is involved in the metabolism of a majority of prescription drugs. Genetic variants of these isoforms also have an effect on drug metabolism. In addition to the CYP isoforms, mutations in genes of specific enzymes such as uridine diphosphoglucuronyltransferase 1A1 (UGT1A1) can affect the metabolism of chemotherapeutic drug Irinotecan. Likewise, mutations in the gene coding for the enzyme vitamin K epoxide reductase complex subunit 1 (VKORC1) can affect the metabolism of warfarin.

Herbs that affect the activity of the efflux transporter P-glycoprotein (an ATP-dependent transport protein) present in the kidney and the gut can influence the concentration of a drug especially if the drug is a substrate for the efflux transporter thus affecting the drug's absorption and clearance.

Herbs can also influence the concentration of a drug by affecting its uptake by specific transporters such as organic anion-transporting polypeptide A (OATAP1A2) [1].

3.4.3 Examples of specific Drug-herb interactions

The following are some specific examples of widely used herbs as well as beverages that affect the metabolism of drugs commonly used in therapy.

St. John's wort: This herb is widely used to treat depression. The major active component of St. John's wort is hyperforin which activates the human nuclear pregnane X receptor (hPXR). The latter, in turn activates the expression of cytochrome P-450-3A (CYP3A) gene in the liver and small intestine [2]. The induction of CYP3A4 isoform by St. John's wort affects the metabolism of many different drugs. In addition to inducing the CYP3A4 isoform, St. John's wort also induces the CYP isoforms CYP 2C9, CYP1A2 and CYP2C19. The herb is also responsible for the expression of P-glycoprotein (P-gp), an efflux transporter present in the kidney and the gut.

St. John's wort, by inducing the CYP isoforms (CYP3A4, CYP1A2 and CYP2C9) involved in the metabolism of both of the enantiomeric forms of the oral anticoagulant warfarin (R and S), decreases their concentration in blood. The decrease in R-warfain is due to the induction of CYP3A4 and CYP1A2 isoforms by St. John's wort. S-warfarin which is two to five times more potent than the R-enantiomer is metabolized by the CYP2C9 isoform which is induced by St. John's wort. The decrease in R-and S-warfarin was reflected in the decrease of the INR (International Normalized Ratio) which is used to monitor warfarin therapy by the measurement of prothrombin time (PT). The mean half-life ($t_{1/2}$) of both R-and S-warfarin is also reduced by St. John's wort [3]. This effect of St. John's wort necessitates the adjustment of warfarin dosage upward in patients who were also ingesting this herb.

In addition to its effect on warfarin, St. John's wort affects the metabolism of a variety of drugs, some of which are listed in Table 3.4.1.

Ginseng: This herb is used as a restorative tonic and to increase resistance to stress. In Chinese medicine ginseng is used to treat gastric disorders. The dry root of Asian ginseng (also called Panax ginseng) has a minimum of 13 different ginsenosides or panaxosides. The active constituents are triterpenoid saponin glycosides. Ginsenosides Ro, Rg1 and Rg2 and Panaxynol have anti-platelet activity. Panaxynol inhibits thromboxane formation.

The root of another variety of ginseng called Siberian ginseng contains lignans and coumarin derivatives.

Ginseng has been reported to reduce fasting blood glucose concentrations in patients with type 2 diabetes at a dose of 100 mg or 200 mg/d for 8 weeks. Statistically significant decreases in HbA_{1c} were also seen at a 200 mg/d dose [4]. The hypoglycemic effect of ginseng is apparently due to the glycans present in the herb that stimulate insulin secretion.

Interaction of ginseng with warfarin therapy has been reported. Consumption of ginseng for two weeks reduced the INR in a 47-year-old patient with a mechanical heart valve who was stabilized on warfarin therapy. His INR decreased from 3.1 prior to consumption of ginseng to 1.5 after two weeks of intake of the herb. The INR promptly increased to 3.3 after discontinuation of ginseng for two weeks [5]. Table 3.4.2 lists examples of other effects of Ginseng.

Table 3.4.1: Drug concentration in blood affected by the intake of St. John's wort (Modified from Narayanan and Young [1]).

Drug	Effect
Cyclosporine	Decrease due to induction of CYP3A4 and P-gp expression by the herb
Methadone	Decrease due to induction of CYP3A4 and P-gp expression by the herb
SN-38 (CPT-11) active metabolite of Irinotecan	Decrease due to induction of CYP3A4 and P-gp expression by the herb
Indinavir	Decrease
Theophylline	Decrease
Digoxin	Decrease due to induction of P-gp
Tacrolimus	Decrease
Amiodarone	Decrease
Amitryptyline, Nortryptyline	Decrease
Simvastatin	Decrease
Midazolam	Decrease
Imatinib (Gleevec)	Decrease
Verapamil	Decrease
Omeprazole	Decrease due to induction of CYP2C19
Vericonazole	Decrease due to induction of CYP2C19
Quazepam	Decrease
Warfarin	Decrease with decrease of prothrombin time and INR, although increased INR values have also been reported
Nelfinivir	Decrease
Phenobarbital	Decrease
Ritonavir	Decrease
Rosuvastatin (Crestor) and other statins	Decrease
Sirolimus and Tacrolimus	Decrease
Theophylline	Decrease
Vericonazole	Decrease by inducing CYP2C19

Note: Where the reason for the decrease is not listed it is assumed that it is due to the induction of CYP3A4 isoform by the herb.

Table 3.4.2: Other effects of Ginseng [1].

Drug or constituent affected	Effect
Digoxin	Increase
Nifedipine, a calcium-channel blocker	Increase due to inhibition of CYP3A4
Blood alcohol	Decrease due to enzyme induction (alcohol and aldehyde dehydrogenases)

Milk Thistle (Silybum marianum): An extract of milk thistle seeds contains a hepato-protective substance called Silymarin which is made of three isomers (Silibinin, Siliydianin and Silychristin). Silibinin is the most active component of Silymarin and formulations of Milk thistle are standardized on the basis of their Silybinin content.

In a study of 106 patients with liver disease due to alcohol abuse a 420 mg/d dose of Silymarin administered for four weeks resulted in a mean decrease of 57.5 % in the serum activity of aspartate aminotransferase (AST) and a mean decrease of 62.2 % in that of alanine aminotransferase (ALT) [6].

Saw Palmetto (Serenoa repens or Sabal serulata): This herb is also called Sabal fruit. The herb is dispensed as an extract standardized to contain 70–95 % free fatty acids and plant sterols such as β-sitosterol. The herb is used to treat urinary flow problems associated with benign prostatic hyperplasia (BPH).

Long-term treatment with Saw Plametto has resulted in a decrease in the plasma concentration of dihydrotestosterone (DHT) and an increase in the testosterone concentration due to the ability of the herb to inhibit the enzyme 5α-reductase [7]. Saw Palmetto, however, unlike the drug Finasteride, does not reduce the concentration of Prostate-specific antigen (PSA) and has a negligible effect upon prostate volume [8].

Grapefruit juice is a widely used breakfast beverage. Because of its effect in inhibiting intestinal CYP3A4 isoform and intestinal organic anion-transporting polypeptide (OATO1A2) it affects the plasma concentration of many drugs, some of which are listed in Table 3.4.3.

Table 3.4.3: Effect of grapefruit juice on plasma drug concentrations (Modified from Narayanan and Young [1]).

Drug	Effect
Felodipine	Increase*
Methadone	Increase*
Halofrantine	Increase*
Triazolam	Increase*
Simvastatin	Increase*
Amiodarone	Increase*
Atorvastatin and other statins	Increase*
Cyclosporine*	Increase*
Ripaglinide	Increase*
Fexofenadine	Decrease**
Celiprolol	Decrease**

Note: * Effect due to inhibition by grapefruit juice of intestinal CYP3A4 activity
** Effect due to inhibition by grapefruit juice of uptake of drug by OATO1A2

Ginkgo biloba: This herb is widely used to improve cognitive function especially in the elderly with mild to moderate memory loss. The flavanoids present in the herb have antioxidant properties. Ginkoglide B, one of the terpene lactones found in the herb, has anti-inflammatory properties due to its ability to inhibit platelet-activating factor (PAF). Gingko biloba, by its ability to induce CYP2C19 isoform, affects the pharmacokinetics of the proton pump inhibitor omeprazole which is used to reduce gastric acid production.

In one study a statistically significant decrease in plasma omeprazole concentration was observed in subjects consuming a 140 mg twice-daily dose of G. biloba [9]. Gingko biloba by inducing CYP2C9 isoform in the liver has been reported to elicit a statistically significant decrease in the plasma concentration of tolbutamide, ahypoglycemic agent. Both in healthy individuals, and in type 2 diabetics receiving oral hypoglycemic agents, plasma insulin and C-peptide concentrations are reduced by ginkgo biloba administration. The ability of the herb to inhibit CYP3A4 isoform activity in the liver and intestine results in a statistically significant increase in the plasma concentration of midazolam [10].

Fenugreek (Trigonella foenumgraceum): The ripe dried seeds of the plant have been used as both a hypoglycemic and antilipidemic agent in Ayurvedic (Indian) medicine. In one study 20 patients with mild type 2 diabetes were treated with 5 mg/d of fenugreek for a month. At the end of the month there was a statistically significant decrease of 18 % in their blood glucose concentrations.

In the same study when 30 patients with coronary artery disease and type 2 diabetes were treated with 5 mg/d of fenugreek for 3 months statistically significant decreases in serum cholesterol and triglyceride concentrations were seen. This statistically significant decrease was not seen after only one month's treatment [11].

Other herbs affecting Warfarin therapy: In addition to St. John's wort and Ginseng other herbs affect anticoagulant therapy with warfarin. Table 3.4.4 lists herbs that have an affect on INR which is used to monitor warfarin therapy. It should be noted that herbs rich in vitamin K, when coadministered with warfarin, are likely to reduce its anticoagulant effect.

Herbal mixtures: Often in oriental (Chinese, Korean and Japanese) herbal preparations we encounter mixtures of herbs. Examples of Chinese herbal mixtures are: Quillinggao, which contains nearly 25 herbs, and Bofu-Tsusho-San which has 18 herbs. The former is used as a restorative tonic and to improve skin complexion. The latter is used as a dietary supplement to promote weight loss. Japanese herbal medicine, also known as Kampo medicine, also offers Chinese herbal mixtures with a Japanese name. For instance a Chinese herbal mixture called Wen-Jing-Tang that contains 11 herbs is sold in Kampo medicine as "Unkei-To". One of the most widely used herbal preparations in Japan is called Hachimi-jio-gan. It is a mixture of 8 herbs.

Table 3.4.4: Other herbs affecting warfarin therapy (Modified from Narayanan and Young [1]).

Herb	Effect on INR
Angelica root	Increase
Cloves	Increase
Devil's claw	Increase
Danshen	Increase
Quilinggao	Increase
Dong quai	Increase
Lycium barbarum (Go-qi-zi)	Increase
Chamomile tea	Increase
Royal Jelly	Increase
Soy protein	Decrease
Ginger	Increase in patient on phenprocoumon therapy

It is used not only as a restorative tonic but also to treat prostate problems, diabetes, erectile dysfunction and other ailments that afflict the elderly. Among Korean herbal preparations mention should be made of Chunghyul-dan which has 6 components and is used to treat hyperlipidemia and Taeumjowi-Tang which is comprised of 9 herbs and is used to treat obesity [1].

Because of space constraints we will discuss the effect of just one herbal mixture, Hachimi-jio-gan. In a study where Hachimi-jio-gan was administered for 7 months to elderly subjects ranging in age from 66 to 82 years a statistically significant increase in serum high density lipoprotein (HDL) concentration was noted while at the same time the lipid peroxide (LPO) level decreased [12]. A dramatic decrease in serum prolactin concentration (from 259 ng/mL to 71 ng/mL) was noted in one female patient with a pituitary adenoma after four weeks of treatment with a 5 g/d dose of Hachimi-jio-gan. The patient was treated previously unsuccessfully with bromocriptine [13].

Herbs that have a toxic effect: Some herbs are toxic to the liver and kidney. Examples of herbs that are hepatotoxic are Black Cohosh, Chapparal and Germander.

Black Cohosh (Cimicifuga racemosa) is used to treat symptoms of menopause and premenstrual syndrome. However, acute hepatotoxicity was reported in a 47-year-old woman who took black cohosh for one week apparently to treat symptoms related to her menopause. Her serum aspartate aminotransferase (AST) activity increased to 91-times the upper limit of the reference interval (ULRI). The activity of other liver enzymes (alanine aminotransferase (ALT) and GGT (γ-glutamyltransferase)) and her bilirubin concentration also increased. The hepatotoxic effect was so severe that the patient required liver transplantation [14].

Chapparal (Larrea tridentate), a herb also available in tea form, which is used to treat a variety of conditions including weight loss and chronic skin disease was found to be hepatotoxic based on a review of 18 reports. These reports cited an increase in serum AST, ALT and alkaline phosphatase activity as well as an increase in the serum bilirubin concentration.

1–17 weeks after the herb was discontinued AST, ALT, alkaline phosphatase and bilirubin values returned to normal, thus implicating the herb for hepatotoxicity [15].

Germander (Teucrium chamaedrys): The blossoms of this plant are used to treat obesity. Seven cases of acute hepatitis occurring 3–18 weeks after healthy individuals consumed the recommended dose of the herb (600-1,620 mg/d) have been reported. In these patients there was a significant increase in the activity of liver enzymes (AST, ALT, GGT and alkaline phosphatase) and the serum bilirubin concentration. Upon discontinuation of the herb the patients recovered within 1.5–6 months by which time both the serum bilirubin concentration and the activity of the liver enzymes had returned to within the normal range [16].

An example of a herb that is nephrotoxic is **Aristolochia Fang Chi**. It is a component of a Chinese herbal preparation called Mu tong. Aristolochic acids I and II, found in the herb, are nephrotoxic and as such can lead to renal disease [17]. The herb is used to treat eczema and is also used in some slimming pills. Case reports describe two women who took this herbal preparation over a 2–6-year period to treat their eczema who developed end stage renal disease and became candidates for renal transplantation. Their serum creatinine concentrations were in the range of 7.5–9.5 mg/dL while their urea nitrogen concentrations were also significantly increased (58–100 mg/dL) [18].

Effect of contaminants in herbs and herbal preparations: Contaminants found in herbal preparations may have been added intentionally to boost the effect of the herb. These contaminants could be drugs or heavy metals. The latter may be present as a natural constituent of the herb having been absorbed from soil rich in such heavy metals. Substitution of one herb for another can occur due to misidentification. Some Chinese herbs may have the same Pin Yin name such as Fang Ji. This has led to the substitution of the herb Aristolochia Fangchi, which as we noted earlier contains nephrotoxins, for the herb Stephania tetranda, which is used as a slimming agent. This substitution resulted in patients developing severe nephropathy [19]. Substitution of one herb for another can also occur if the two plants at some stage in their life cycle appear similar in appearance. Substitution of a herb called Foxglove, (which describes plants of the digitalis family) for Comfrey (which belongs to the Symphytum species) resulted in adverse outcomes in two patients [20]. The substitution occurred since the plants from which both the herbs were isolated had a remarkably similar appearance prior to flowering.

An increasing number of Ayurvedic (Indian) and Chinese herbal medicines are known to be contaminated with lead, cadmium, arsenic and mercury. Lead poisoning from some Ayurvedic herbs is a result of such contamination [21]. Seaweed naturally contains arsenic, which being organically bound, has minimal toxicity. However, if arsenic is introduced as a contaminant it can lead to nephrotoxicity as was reported due to ingestion of bladderwrack (Fucus vesiculosus) [22].

Finally, bacterial contamination can occur in herb-based medicines due to improper storage and processing conditions [23]. Table 3.4.5 provides a list of herbs with known contaminants.

3.4.4 Quality control of Herb-based Medical Products (HMPs)

The quality of herb-based medicines can only be assured by strictly following good manufacturing practices (GMP). This includes not only establishing product specifications, but also setting up acceptable limits for heavy metals and documentation of every step in the manufacturing process. The purity of both the raw materials and the final product should be determined using well-established analytical procedures. Fortunately in Europe, Canada, Japan and Australia there are legal requirements for the manufacture of herb-based medical products. The European Union has a specific directive governing HMPs [24]. This European directive also applies to manufacturers from the United States and other nonEuropean countries if they wish to market their products in countries of the European Union.

Manufacturers of HMPs should report known adverse reactions associated with their products to regulatory agencies in their respective countries so that the public is alerted to the extent of the problem. Regulatory agencies should particularly look out for HMPs which are directed to treat specific conditions such as diabetes or erectile dysfunction to insure that these products are not adulterated with drugs. Case reports in peer-reviewed journals can be a vehicle for alerting consumers to the presence of extraneous drugs and contaminants in HMPs. To cite an example of the issue – the presence of analogues of a drug used to treat erectile dysfunction in herb-based products was communicated through a journal report [25].

Finally, the seriousness with which the quality of HMPs can be pursued by both the manufacturers and the user is only possible if these products are treated in the

Table 3.4.5: List of herbs with known contaminants [1].

Herb or Herbal preparation	Contaminant and comments
PC-SPES, has 3 herbs (Used for prostate problems)	Indomethacin, diethylstilbesterol, Coumadin (withdrawn from market due to bleeding)
Shen Loon	Betamethasone caused adrenal insufficiency
Tung Shueh	Indomethacin, diazepam, diclofenac, mefenamic acid
Artemisia vulgaris (mugwort)	Cocaine
Hai Ge Fen (36 ingredients)	225 mg/L (ppm) lead, Blood lead 76 µg/dL
Maha Yogran Guggul	65 mg lead/g. Blood lead 86 µg/dL (ref: <20 µg/dL)

same way as licensed drugs and practitioners who dispense HMPs are properly accredited and licensed.

Conclusion

The prevailing perception among the public is that just because herbs are natural products they must be safe and effective should be tempered with reality. The reality is what we discussed in this chapter. Herbs can in some instances interfere with therapy with possibly serious consequences. Some herbs may have beneficial effects in terms of treating conditions such as diabetes, hyperlipidemia and benign prostatic hyperplasia. But some herbs have toxic effects on the liver and kidney. Finally what you see on the product label may not be what you get. You may get more than what is stated on the product label in terms of extraneous drugs or heavy metal contaminants! With these caveats herbs do have a place in modern medicine and their use will continue to grow and enrich the already multi-billion dollar industry. It is up to the Regulators to be vigilant and police this burgeoning industry.

References

[1] Narayanan S, Young DS. Effects of Herbs and Natural Products on Clinical Laboratory Tests. AACC Press, Washington, DC 2007.

[2] Watkins RE, Wisely GB, Moore LB, Collins JL, Lambert MH, et al. The human nuclear xenobiotic receptor PXR: structural determinants of directed promiscuity. Science 2001; 292:2329–33.

[3] Jiang X, Williams KM, Liauw WS, Ammit AJ, Roufogalis BD, et al. Effect of St. John's wort and ginseng on the pharmacokinetics and pharmacodynamics of warfarin in healthy subjects. Br J Clin Pharmacol 2004; 57:592–9.

[4] Sotaniemi EA, Happakoski E, Rautio A. Ginseng therapy in non-insulin-dependent diabetic patients. Diabetes Care 1995; 18:1373–5.

[5] Janetzky K, Morreale AP. Probable interactions between warfarin and ginseng. Am J Health Syst Pharm 1997; 54:692–3.

[6] Salmi HA, Sarna S. Effect of Silymarin on chemical, functional and morphological alterations of the liver. A double-blind controlled study. Scand J Gastroenterol 1982; 17:517–21.

[7] Di Silverio F, Monti S, Sciarra A, Varasano PA, Martini C, et al. Effects of long-term treatment with Serenoa repens (Permixon) on the concentrations and regional distribution of androgens and epidermal growth factor in benign prostatic hyperplasia. Prostate 1998; 37:77–83.

[8] Carraro JC, Raynaud JP, Koch G, Chisholm GD, Di Silverio F, et al. Comparison of phytotherapy (Permixon) with finasteride in the treatment of benign prostatic hyperplasia: a randomized international study of 1,098 patients. Prostate 1996; 29:231–40.

[9] Yin OQP, Tomlinson B, Waye MMY, Chow AHL, Chow MSS. Pharmacogenetics and herb-drug interactions: experience with Ginkgo biloba and omeprazole. Pharmacogenetics 2004; 14:841–50.

[10] Uchida S, Yamada H, Li XD, Maruyama S, Ohmori Y, et al. Effects of Ginkgo Biloba extract on pharmacokinetics and pharmacodynamics of tolbutamide and midazolam in healthy volunteers. J Clin Pharmacol 2006; 46:1290–8.

[11] Bordia A, Verma SK, Srivastava KC. Effect of ginger (Zingiber officinale Rosc.) and fenugreek (Trigonella foenumgraceum L.) on blood lipids, blood sugar and platelet aggregation in patients with coronary artery disease. Prostaglandins Leukot Essent Fatty Acids. 1997; 56:379–84.
[12] Yoshida H, Kasukawa R, Watanabe N, Ohtsuki G, Sakamoto T, Haranaka R. The effects of Ba-wei-wan (hachimi-jio-gan) on plasma levels of high density lipoprotein cholesterol and lipoperoxide in aged individuals. Am J Chin Med 1985;13:71–6.
[13] Otani S, Usuki S, Iwasaki S, Inoue S, Yamashita K. Successful treatment of a hyperprolactinemic woman with a pituitary microadenoma using hachimi-jio-gan. Am J Chin Med 1991;19:145–54.
[14] Whiting PW, Clouston A, Kerlin P. Black cohosh and other herbal remedies associated with acute hepatitis. Med J Aust 2002; 177:432–5.
[15] Sheikh NM, Philen RM, Love LA. Chaparral-associated hepatotoxicity. Arch Intern Med 1997; 157:913–19.
[16] Larrey D, Vial T, Pauwels A, Castot A, Biour M, et al. Hepatitis after germander (Teucrium chamaedrys) administration: another instance of herbal medicine hepatotoxicity. Ann Intern Med 1992; 117:129–32.
[17] Vanherweghem JL. Misuse of herbal remedies: the case of an outbreak of terminal renal failure in Belgium (Chinese herbs nephropathy). J Altern Complement Med 1998; 4:9–13.
[18] Lord GM, Tagore R, Cook T, Gower P, Pusey CD. Nephropathy caused by Chinese herbs in the U.K. Lancet 1999; 354:481–2.
[19] Vanhaelen M, Vanhaelen-Fastre R, But P, Vanherweghem JL. Identification of ariatolochic acid in Chinese herbs. Lancet 1994; 343:174.
[20] Stillman AE, Huxtable RJ, Fox DW, Hart M, Bergeson P. Poisoning associated with herbal teas. Morb Mortal Wkly Rep 1997; 26:257–9.
[21] van Vonderen MGA, Klinkenberg-Knol EC, Craanen ME, Touw DJ, Meuwissen SG, et al. Severe gastrointestinal symptoms due to lead poisoning from Indian traditional medicine. Am J Gastroenterol 2000; 95:1591–2.
[22] Conz PA, LaGreca G, Benedetti P, Bevilacqua PA, Cima L. Fucus vesiculosus: a nephrotoxic alga? Nephrol Dial Transplant 1998; 13:526–7.
[23] Kneifel W, Czech E, Kopp B. Microbial contamination of medicinal plants - a review. Planta Med 2002; 68:5–15.
[24] European Directorate for the Quality of Medicines and Health Care: Directive 2004/24/EC of the European Parliament and of the Council of 31 March 2004 amending, as regards traditional herbal medicinal products, Directive 2001/83/EC on the community code relating to medicinal products for human use, 14th ed 2008; Brussels: Commission of the European Communities.
[25] Sze-Yin Oh S, Zou P, Low M-Y, Koh H-L. Detection of Sildenafil analogues in herbal products for erectile dysfunction [Part A]. J Toxicol Environmental Health 2006; 69:1951–8.

4. Source and Nature of Interferences of Analytical Procedures

Walter G. Guder, Friedrich da Fonseca-Wollheim, York M Schmitt, Gottfried Töpfer

4.1 The Hemolytic Sample

Hemolysis was described to be the most common cause of preanalytical errors [1]. It is therefore one of the most important variable when considering quality improvement in preanalytical phase [2, 3].

4.1.1 Definition and mechanisms of hemolysis

Hemolysis is defined as the release of intracellular components of erythrocytes and other blood cells into the extracellular space of blood [4, 5]. Hemolysis can occur in vivo (e. g. through a transfusion reaction or during malaria parasite infection affecting the invaded erythrocytes), and in vitro during all steps of the preanalytical phase (sampling, sample transport and storage). Transport time above 2 h, cooling versus room temperature during transport and traditional syringe extraction versus evacuated tube system were observed to result in higher rejection percentages due to hemolysis [6].

Hemolysis is caused by biochemical, immunological, physical and chemical mechanisms [7, 8, 9]. During blood transfusion, complement-dependent hemolysis may be caused by antibodies reacting with the major blood group antigens. Physical hemolysis is caused by destruction of erythrocytes by hypotonicity (e. g. dilution of blood with hypotonic solution), freezing, as well as decreased (vacuum) or increased pressure. Mechanical hemolysis may occur during the flow of blood through medical devices (e. g. catheters, heart valves) in vivo, during inadequate centrifugation as well as at elevated temperature in vitro. Contaminating substances may also cause in-vitro hemolysis. Finally, detergents (residual cleaning agents and disinfectants) and other contaminating substances may cause hemolysis.

After the separation of blood cells, hemolysis may be visible by the red color of serum or plasma. The sample may concomitantly be contaminated by constituents of other blood cells (leukocytes and platelets). For example, cell breakdown may result in changes in blood of patients with leukaemia; the disintegration of platelets during coagulation results in higher concentrations of intracellular platelet constituents in serum (10; see Chapter 2.6). On the other hand, the intracellular components of erythrocytes are also released into plasma without a concomitant increase in hemoglobin concentration during storage of whole blood in the refrigerator.

4.1.1.1 Hemoglobin based oxygen carriers used as blood substitutes

Therapeutic hemoglobin derivatives (so-called HbOC = hemoglobin-based oxygen carriers) were developed as blood substitutes. The substitutes occur at concentrations of up to 50 g/L in plasma of patients under blood substitute treatment [11]. Plasma or serum containing blood substitutes has a strong red color [12, 13]. Presently these therapeutic agents are of limited usefulness.

4.1.2 Detection and measurement of hemoglobin in serum or plasma

Visual detection

At extracellular hemoglobin concentrations above 0.3 g/L (0.0188 mmol/L), hemolysis is detectable by the red color of serum or plasma. This is slightly above the upper reference level of free hemoglobin, which was reported to be up to 50 mg/L in serum and 20 mg/L in plasma [14].

Spectrophotometric detection

Most analytical systems measure the extent of hemolysis by comparing the absorption of samples at two wavelengths [15]. This is reported as "hemolysis index" in most analytical systems [6, 16, 17]. The absorption spectrum of the hemoglobin derived oxygen carriers used as blood substitutes does not differ substantially from that of natural hemoglobin.

Analytical measurement

Hemoglobin in plasma or serum can be measured at concentrations that are below the concentration visible to the human eye [18, 19].

4.1.3 Distinction between in vivo hemolysis and in vitro hemolysis

In vivo hemolysis may be distinguished from in vitro hemolysis by comparing a hemolytic sample of a patient with other samples from the same patient, arriving at the same time.

In vivo hemolysis

Free hemoglobin in vivo rapidly binds to haptoglobin and the complex is eliminated from the circulating blood (as in hemolytic anaemia). Consequently, haptoglobin is reduced during intravasal hemolytic processes. The measurement of low concentration of haptoglobin thus permits in most cases a definitive assessment of in vivo hemolysis (exceptions are inborn haptoglobin deficiency and newborn children [20].

Likewise, the measurement of hemopexin and/or methemoglobin/albumin was used to characterize in vivo hemolysis [20].

A rise in concentration of indirect bilirubin and reticulocyte counts is a typical sign of in vivo hemolysis, which in turn leads to increased erythropoesis. Other consequences of in vivo hemolysis, such as a change in the LDH isoenzyme pattern, seem less suitable for the identification of hemolysis because of their low diagnostic sensitivity and specificity.

In vitro hemolysis

As a result of in vitro hemolysis the concentrations of all of erythrocyte constituents, including potassium, lactate dehydrogenase and aspartate aminotransferase activity, increase in addition to the concentration of free hemoglobin in plasma or serum [21]. In contrast, haptoglobin concentration in plasma/serum of hemolytic samples remains unchanged. Certain immunological methods differ in their ability to distinguish hemoglobin/haptoglobin complexes from free haptoglobin [20].

Identification of hemoglobin derived oxygen carriers

Therapeutic hemoglobin derivatives yield a visible hemoglobin concentration within the range of 10–50 g/L. The absorption spectrum of hemoglobin derived oxygen carriers is not distinguishable from that of hemoglobin [11, 12, 13]. However, hemoglobin concentrations of this magnitude rarely occur in vivo; therefore the use of therapeutic hemoglobin derivatives must be suspected at this plasma hemoglobin concentration. Haptoglobin cannot be used for discrimination, since the oxygen carriers form complexes with haptoglobin only slowly.

4.1.4 Mechanisms of interference by hemolysis

Hemolysis in-vivo or in-vitro can cause an apparent decrease or increase of results. A variety of mechanisms are contributing to these effects, some of which are summarized as follows.

Rise of intracellular constituents in the extra-cellular space

Cell constituents with an intracellular concentration 10 times higher than the extracellular concentration, will increase in plasma/serum during hemolysis (e. g. potassium, lactate dehydrogenase, aspartate aminotransferase). Differences of analyte concentrations between plasma and serum are also due to lysis of blood cells (essentially by platelets): thus, neurone-specific enolase, potassium and acid phosphatase are found to be higher in serum.

Interference with analytical procedure

Blood cell constituents can directly or indirectly interfere in the measurement of analytes. Adenylate kinase released from erythrocytes causes an increase of creatine kinase and CK-MB activity especially when inhibitors of adenylate kinase in the assay mixture are inadequate [22]. In contrast, adenylate kinase does not affect the immunochemical quantification of CK-MB. Pseudo-peroxidase activity of free hemoglobin interferes in the bilirubin procedure of Jendrassik and Groof by inhibiting the diazonium color formation [23]. Proteases released from blood cells reduce the activity of coagulation factors while fibrin split product formation may increase.

Optical interference by hemoglobin

The effect of hemolysis on various analytes measured in clinical chemistry has been thoroughly investigated [2, 8, 15]. Most often, the color of hemoglobin increases the absorption at a respective wavelength or changes the blank value. Dependent from the method applied and the analyte concentration, these changes lead to an apparently increased or decreased result. Likewise, the changes caused by therapeutic hemoglobin derivatives are primarily due to optical interference [8, 11, 12].

4.1.5 Means to avoid hemolysis and its interferences

Hemolysis in-vitro can almost always be avoided, when the mechanism of hemolysis is known. Therefore each hemolytic sample should be documented and the cause of hemolysis identified.

The most frequent causes of hemolysis, such as errors during sampling, are avoided using standardized materials and methods for the pre-analytical processes and by training and individual counselling.

Sometimes reliable results can only be obtained from a truly nonhemolytic sample. In some cases, the interference can be reduced or excluded using a method that is not sensitive to hemolysis or by pretreatment of the sample. Procedures, including ultrafiltration [24, 25] or molecular sieving and others have limited usefulness because of the work load involved. Today, a modification of the methodology, e. g. by using a blanking procedure by means of measurement at a second, appropriate wavelength [26], is preferred; although, this procedure may not be applicable for the analysis of blood from patients who received blood substitutes [27]. Likewise the ultrafiltration procedure, as applied in the multi-layer film technology, reduces the effect of interference by hemolysis [28]. Recently, so called hemolysis-resistant reagents have been reported [29]. It may however be noted that the degree of red color does not always directly correlate to the erroneously changed result due to the release of constituents from blood cells as has been observed in samples stored in refrigerator (potassium increased without visible hemolysis), in serum constituents from disintegrated platelets (increased potassium in acute phase in serum, but not plasma) and in some hematological diseases with increased degradation of leucemic cells (after cytotoxic treatment).

4.1.6 Recommendation regarding appropriate reaction upon the receipt of hemolytic samples

Each laboratory should document the procedures that are affected by hemolysis and to what extent they are affected. The procedures how to handle hemolytic samples should be described in the quality manual. This includes the criteria for rejecting the examination [30].

The hemolysis of each sample must be documented and reported to the requesting site who ordered the analysis.

When hemolysis occurs in all samples of a patient, hemolysis in vivo may be suspected. This must be immediately reported to the clinician to verify the possible causes of hemolysis or the possible use of synthetic hemoglobin derivatives.

After estimation of the degree of hemolysis the sample is treated for analysis according to the degree of interference. The results of measurement may be reported as follows:
- Method not impaired: Report results as with nonhemolysed samples.
- Method impaired, but eliminated by pretreatment: report results after pretreatment.
- Method impaired in a clinically relevant way: instead of providing a result, report: "Impaired by hemolysis".

It is not recommended to correct a measured result for hemolysis arithmetically using the hemoglobin concentration as an indicator.

References

[1] Bonini PA, Plebani M,Ceriotti F, Rubboli F. Errors in laboratory medicine. Clin Chem 2002; 48:691–8.
[2] Lippi G, Cervellin G, Favaloro J, Plebani M. In Vitro and in Vivo Hemolysis. Berlin, Boston: Walter de Gruyter 2012.
[3] Lippi G, McBride KB, Behulova D, Bowen RA, Church S, Delange J et al. Preanalytical quality improvement: in quality we trust. Clin Chem Lab Med 2013; 51:229–41.
[4] Guder WG. Haemolysis as an influence and interference factor in clinical chemistry. J Clin Chem Clin Biochem 1986; 24:125–6.
[5] Thomas L. Haemolysis as influence and interference factor. Biochem Clin 2002; 26:95–8.
[6] Fernandez P, Llopis MA, Perich C, Alsina MJ, Alvarez V, Biosca C et al. Harmonization in hemolysis detection and prevention. A working group of the Catalonian Health Institute (ICS) experience. Clin Chem Lab Med 2014; 52:1557–68.
[7] Blank DW, Kroll MH, Ruddel ME, Elin RJ. Hemoglobin interference from in vivo hemolysis. Clin Chem 1985; 31:1566–9.
[8] Sonntag O. Haemolysis as an interference factor in clinical chemistry. J Clin Chem Clin Biochem 1986; 24:127–39.

[9] Clinical and Laboratory Standards Institute. Hemolysis, icterus and lipemia/turbidity indices as indicators of interference in clinical laboratory analysis; approved guideline. CLSI document C56-A. Wayne PA:Clinical Laboratory Standards Institute 2012.

[10] Lutomski DM, Bower RH. The effect of thrombocytosis on serum potassium and phosphorus concentration. Am J Med Sci 1994; 307:255–8.

[11] Callas DD, Clark TL, Moreira PL, Lansden C, Gawryl MS, Kahn S, Bermes EW. In vitro effects of a novel haemoglobin-based oxygen carrier on routine chemistry, therapeutic drug, coagulation, hematology, and blood bank assays. Clin Chem 1997; 43:1741–8.

[12] Kazmierczak SC, Catrou PG, Best AE, Sullivan SW, Briley KP. Multiple regression analysis of interference effects from a haemoglobin-based-oxygen carrier solution. Clin Chem Lab Med 1999; 37; 453–64.

[13] Wolthuis A, Peek D, Scholten R, Moreira P, Gawryl M, Clark T, Westerhuis L. Effect of the haemoglobin-based oxygen carrier HBOC-201 on laboratory instrumentation: Cobas Integra, Chiron blood gas analyzer 840, Sysmex™ SE-9000 and BCT. Clin Chem Lab Med 1999; 37:71–6.

[14] Thomas L. Clinical Laboratory Diagnosis. Frankfurt/Main: TH-Books 1998.

[15] Grafmeyer D, Bondon M, Manchon M, Levillain P. The influence of bilirubin, haemolysis and turbidity on 20 analytical tests performed on automatic analysers. Eur J Clin Chem Clin Biochem 1995; 33:31–52.

[16] Lippi G, Salvagno G, Blanckaert, N et al. Multicenter evaluation of the hemolysis index in automated clinical chemistry systems. Clin Chem Lab Med 2009; 47:934–9.

[17] Dolci A, Panthegini M. Harmomization of automated hemolysis index assessment and use: is it possible? Clin Chim Acta 2013:doi:10.1016/j.cca.2013.10.012.

[18] Bauer K. Determination of free haemoglobin in serum by automated assay using 4-aminophenazone and the Cobas Bio system. Clin Chem Clin Biochem 1981; 19:971–6.

[19] Lammers M. Gressner AM. Immunonephelometric quantification of free haemoglobin. J Clin Chem Clin Biochem 1987; 25:363–7.

[20] Van Lente F, Marchand A, Galen RS. Evaluation of a nephelometric assay for haptoglobin and its clinical usefulness. Clin Chem 1979; 25:2007–10.

[21] Wisser H. Einflußgrößen und Störgrößen. In: Greiling H, Gressner A. (eds.) Lehrbuch der Klinischen Chemie und Pathobiochemie, 3rd ed. Stuttgart, New York: Schattauer, 1995, pp. 50–71.

[22] Szasz G, Gerhardt W, Gruber W, Bernt E. Creatinekinase in serum: 2. Interference of adenylate kinase with the assay. Clin Chem 1976; 22:1806–11.

[23] Van der Woerd - de Lange JA, Guder WG, Schleicher E, Paetzke I, Schleithoff M et al. Studies on the interference by haemoglobin in the determination of bilirubin. J Clin Chem Clin Biochem 1983; 21:437–43.

[24] da Fonseca-Wollheim F, Heinze KG, Lomsky K, Schreiner H. Serum ultrafiltration for the elimination of endogenous interfering substances in creatinine determination. J Clin Chem Clin Biochem 1988; 26:523–5.

[25] da Fonseca-Wollheim F da. Ultrafiltrate analysis confirms the specificity of the selected method for plasma ammonia determination. Eur J Clin Chem Clin Biochem 1992; 30:15

[26] da Fonseca-Wollheim F. Haemoglobin interference in the bichromatic spectrophotometry of NAD(P)H at 340/380 nm. Eur J Clin Chem Clin Biochem 1993; 31:595–601.

[27] Glick MR, Pieper J, Ryder KW. Interference-reduced methodologies for Boehringer Mannheim/Hitachi analyzers: validation using recombinant haemoglobin "blood substitute" product. Clin Chem 1998; 44, Suppl. A 140.

[28] Sonntag O, Glick MR. Serum-Index und Interferogramm – Ein neuer Weg zur Prüfung und Darstellung von Interferenzen durch Serumchromogene. Labmed 1989; 13:77–82.

[29] Plebani M, Lippi G. Hemolysis resistant reagent: another part of the puzzle for preventing errors in laboratory testing. Clin Chem Lab Med 2013; 51:1339–41.

[30] Plebani M, Lippi G. Hemolysis index: quality indicator or criterion for sample rejection? Clin Chem Lab Med 2009; 47:899–902.

Walter G. Guder, Nora Nikolac, York M. Schmitt, Gottfried Töpfer

4.2 The Lipemic Sample

Based on previous recommendations [1, 2, 3], this chapter summarizes the present state of knowledge of the diagnosis, handling and prevention of lipemia in samples. This chapter includes recommendations of a recently published review [4].

4.2.1 Definition

Lipemia manifests as turbidity of serum or plasma and is caused by elevated lipoprotein concentrations. Since lipemia is often visible to the eye, a sufficiently transparent sample container is a prerequisite to detection of lipemia. Visible detection of lipemia is also dependent on the type of plasma lipoproteins which are elevated in the sample. However, coagulation of serum samples after centrifugation from heparinized patients can also be the cause of turbidity.

4.2.2 Causes of lipemia (turbidity)

Most often, lipemia results from increased triglyceride concentration in plasma or serum. Increased trigycerides can be due to food intake, altered lipid metabolism or infusion of lipids. After intestinal absorption, triglycerides are present in plasma as chylomicrons and their metabolites (remnants) for 6–12 h. High levels of very low density lipoproteins (VLDLs) cause turbidity only at triglyceride concentrations exceeding the upper normal limit by a factor of 2. On the other hand administration of lipid emulsions such as Intralipid® (Fresenius Kabi) or Ivelip® (Baxter) used for parental nutrition causes turbidity for up to 6 h [5].

One to four hours after consumption of a "Continental" or "American" breakfast (containing bacon, eggs or sausage), plasma triglyceride concentrations increase substantially and cause turbidity in plasma or serum. The patient should be requested to fast before investigations are made which may be affected by lipemia.

Hypertriglyceridemia caused by lipid infusions can hardly be distinguished from metabolic disorders causing turbidity from cold agglutinins or monoclonal immunoglobulins.

4.2.3 Identification and quantification of lipemia

Optical and photometric methods for serum and plasma samples

In whole blood triglyceride concentrations above 1000 mg/dL (11.3 mmol/L) cause turbidity that is visually seen. Lipemia in plasma or serum can be observed at triglyceride concentrations above 300 mg/dL (>3.4 mmol/L) [6]. The extent of turbidity of serum or plasma samples is measured at wavelengths above 600 nm (e. g. 660/700 nm) [7]. These wavelengths are used on Roche Cobas and Architect platforms, whereas other wavelengths are used by other platforms (e. g. 660/800 by Beckman Coulter (formerly Olympus, or four different wavelength pairs by Abbott [4]. In addition monoclonal gammopathy may create falsely high lipemic indices [8]. Measurement of triglyceride concentration has likewise been applied in lipemic samples but is of limited concordance to the degree of turbidity. When the lipemic index is elevated, it may be useful to screen and diagnose different causes of hypertrigyceridemia [9].

Detection in EDTA blood

Hematologial tests are likewise influenced by lipemia. Hemoglobin concentration is increased by light scattering. However, because visual assessment of turbidity is only possible at triglyceride concentrations above 1000 mg/dL (11.3 mmol/L), turbidity is better assessed by spectrophotometric analysis. This analysis may be accomplished by measuring the lipemic index [10], resulting in 11 % of samples being lipemic. If spectrophotometric analysis is not possible, a centrifuged sample from the same patient taken at the same time can be used for comparison.

4.2.4 Mechanisms of the interference by lipemia on analytical methods

Interference in spectrophotometric analysis

Lipemia interferes in photometric measurement through light scattering and light absorption. The test result can be either increased or decreasd depending on the blanking procedure [11]. At high turbidity, measurement may not be possible due to the limits of the linearity of the method [12].

Volume depletion effect

Lipoproteins decrease the apparent concentration of the analyte by reducing available water, since the volume occupied by lipoproteins in plasma or serum is included in the calculation of the analyte concentration. This explains the reason behind lower sodium and potassium concentrations when measured in lipemic sera, when the plasma or serum is measured by flame photometry or by indirect measurement using ion-sensitive electrodes, in contrast to direct potentiometry [13]. The same observation of lower sodium and potassium measurements applies after centrifugation, when the lipoproteins are not homogeneously distributed in serum or plasma samples. In such instances, the concentration of an analyte dissolved in the aqueous phase is less in the upper layer than in the lower layers of the sample. The converse is true for concentration of lipids and lipid soluble constituents, including certain drugs that are taken up by lipoproteins.

Interference by physico-chemical mechanisms

A constituent such as an antibody that is extracted by lipoproteins may not be accessible to the reagent for detection. Similarly, electrophoretic and chromatographic procedures may be affected by lipoproteins present in the matrix.

4.2.5 Avoidance of lipemia and interferences caused by turbidity

In order to avoid interference of lipoproteins on measurements, the patient should fast at least 12 hours after oral intake of fat before blood samples are taken [4, 14, 15]. In patients receiving parenteral infusion of lipids, a period of 8 hours free of lipid infusion is necessary to avoid interference from turbidity [3]. If these approaches do not provide a non-turbid sample, other causes of turbidity should be suspected.

Several methods such as centrifugation have been recommended to remove lipids from serum or plasma, in order to produce a clear infranatant sample. Methods include the extraction of lipids with organic solvents or fluorine chlorinated hydrocarbons (e. g. Frigen®) and the precipitation of triglyceride rich lipoproteins by polyanion and cyclodextrin [16].

Centrifugation

Centrifugation at 1000 *g* is effective as a means of decreasing turbidity due to chylomicrons. In contrast, at least 10 min centrifugation at 10 000 *g* separates serum or plasma lipids by flotation [17]. The clear infranatant must be carefully separated for analysis. This approach is not recommended if lipophilic substances are to be measured like some hormones, drugs and hydrophobic metabolites. In this case falsely low test results will be obtained.

Ultra-centrifugation must be employed for the separation of intermediate density lipoproteins and high density lipoproteins, which seldomly cause turbidity of diagnostic importance. A centrifugation time of at least 30 min at a speed above 40 000 *g* is recommended. The separation of lipemic plasma from EDTA-blood in samples used in hematology can be performed by centrifugation and exchange of the cell-free supernatant with the same volume of isotonic NaCl solution.

Polyethylene glycol

The plasma or serum sample should be mixed 1 + 1 (v/v) with 8 % polyethylene glycol 6000, incubated for 30 min in a refrigerator at 4 °C and then centrifuged for 10 min at 4 °C at approximately 1000 *g*. The results obtained in the clear supernatant are multiplied by a dilution factor of 2 [18].

α-Cyclodextrin

200 g α-cyclodextrin should be dissolved in 1 L distilled water and kept in a refrigerator. Before use, the α-cyclodextrin solution must be brought to ambient temperature. Thoroughly mix one part of α-cylodextrin solution with two parts of serum, and centrifuge for 1 min at 10 000 *g*. The clear supernatant can be used for analysis. The dilution must be considered when calculating the concentration of the constituent in the original serum sample.

Experiments have revealed that the results of 20 serum constituents are not affected by the precipitation of lipoproteins using α-cyclodextrin [16].

Other methods for delipidation

Four different procedures for the extraction of lipids from serum samples were examined in one study [19], including Freon 113®, dextrane sulfate 500 S, Aerosil 300 and a butanol/diisopropylether mixture. Lipoclear® (StatSpin®, Norwood, USA) which contains these nontoxic polymers that bind lipids has been widely used. After centrifugation lipids are sedimented at the bottom of the tube. However, it was found that the delipidation methods may substantially alter the concentrations of certain analytes [20] or interfere with the analytical procedures resulting in higher results of troponin T [21]. Even the use of magnetic beads is not generally applicable [1].

Optical clearing systems and sample dilution

Commercial test kits may contain detergents such as triton X100, cholic and desoxycholic acid, lipase or cholesterol esterase to remove turbidity in plasma or serum samples. The assigned concentrations of these substances are method dependent and should not be changed by the user. Sample dilution up to 2–3 fold on the other hand may allow measurement of lipid soluble drugs and other critical analytes without artificial additives [4].

4.2.6 Recommendations

Visible turbidity of a sample must be documented and reported. Transparent sample containers should be used to detect turbidity. The methods used for the measurement of certain analytes that are affected by lipemia must be listed. Also the methods for delipidation and the criteria for their application must be documented in the quality manual.

The method of choice for removal of turbidity from serum and plasma is a 10 min centrifugation step in a micro-centrifuge with 10 000 *g*.

When chemicals are added (e. g. polyethylene glycol, α-cyclodextrin), the laboratory must prove that assigned method for measurement is not disturbed by the agent.

Samples submitted for the determination of lipids and other analytes may be delipidated only after measurement of the lipids. This criterion also applies to lipid soluble drugs.

4.2.7 Test of interference by lipemia

Various problems should be considered when assessing the influence of lipemia on analysis of samples by new methods. Unfortunately, there is no uniform human lipid standard available. Testing of interfering substances is mandatory for all reagent manufacturers. Results of such testing should be provided in the product insert. In general patient samples with high lipid concentrations should not be frozen, in order to prevent inhomogeneity after thawing out the specimens.

A 10 or 20 % emulsion of vegetable fat emulsion, similar to that applied in parenteral nutrition [22–28] is suitable to simulate lipemia. Significant differences between the "physiological lipemia" and the artificially produced lipemia were observed in one study, particularly in measurements of urea and potassium [24]. Therefore, the effect of lipemia should not be exclusively examined using a model that contains artificial fat emulsions, because the observations may not be biologically meaningful in vivo. It is therefore recommended to use standardized solutions containing lipids of known concentration and composition [29].

References

[1] Guder WG, da Fonseca-Wollheim F, Heil W, Schmitt Y, Töpfer G, Wisser H, et al. The haemolytic, icteric and lipemic sample. Recommendations regarding their recognition and prevention of clinically relevant interferences. J Lab Med 2000; 24:357–64.

[2] Guder WG, Fiedler M, da Fonseca-Wollheim F, Schmitt Y, Töpfer G, Wisser H, Zawta B. The Quality of Diagnostic Samples. 4th ed. BD-Diagnostics Oxford 2015.

[3] Guder WG, Narayanan S,Wisser H, Zawta B. Diagnostic Samples: From the Patient to the Laboratory, 4th ed. Weinheim: Wiley-Blackwell, 2009.

[4] Nikolac N. Lipemia: causes, interference mechanisms, detection and management. Biochem Med 2014; 24:57–67.

[5] Intralipid. http://www.drugs.com/pro/intralipid.htlm. Assessed March 2014.

[6] Kazmierczak SC. Hemolysis, lipemia and high bilirubin: Effects on laboratory tests. In: Dasgupta A, Sepulveda JL, ed. Accurate Results in the Clinical Laboratory: A Guide to Error Detection and Correction. Amsterdam: Elsevier 2013; pp, 53–62.

[7] Sonntag O, Glick MR. Serum-Index und Interferogramm – Ein neuer Weg zur Prüfung und Darstellung von Interferenzen durch Serumchromogene. Lab med 1989; 13:77–82.

[8] Fliser E, Jerkovic K, Vidovic T, Gorenjak M. Investigation of unusual high serum indices for lipemia in clear serum samples on Siemens analyzers Dimension. Biochem Med 2012; 22:352–63.

[9] De Haene H, Taes Y, Christophe A, Delange J. Comparison of triglyceride concentration with lipemic index in disorders of trigyceride and glycerol metabolism. Clin Chem Lab Med 2006; 44:220–2.

[10] Simundic AM, Nikolac N, Ivankovic V, Ferenec-Ruzic D, Magdric B, Kvaternik M, Topic E. Comparison of visual versus automated detection of lipemic, icteric and hemolyzed specimens: can we realy rely on a human eye? Clin Chem Lab Med 2009; 47:1361–5.
[11] Kroll MH. Evaluating interference caused by lipemia. Clin Chem 2004; 50:1968–9.
[12] Artiss JD, Zak B. Problems with measurements caused by high concentrations of serum lipids. CRC Crit Rev Clin Lab Sci 1987; 25:19–41.
[13] Külpmann WR. Determination of electrolytes in serum and serum water. Wiener Klin Wschr 1992; suppl:34–8.
[14] Guder WG, Nolte H. Das Laborbuch für Klinik und Praxis, 2nd ed. München: Elsevier Urban und Fischer 2009.
[15] Thomas L (ed) Clinical Laboratory Diagnosis. Frankfurt:TH Books.1998.
[16] Sharma A, Anderson K, Baker JW. Flocculation of serum lipoproteins with cyclodextrin: Application to assay of hyperlipidemic serum. Clin Chem 1990; 36:529–32.
[17] Dimeski G, Jones BW. Lipaemic samples: effective process for lipid reduction using high speed centrifugation compared with ultracentrifugations. Biochem Med 2011; 21:86–92.
[18] Rasbold K, Rendell MS, Goljan E. Simple removal of lipids from serum. Clin Chem 1985; 31:782.
[19] Agnese ST, Spierto FW, Hannon WH. Evaluation of four reagents for delipidation of serum. Clin Biochem 1983; 16:98–100.
[20] Vermeer HJ, Steen G, Naus AJ, Goeverts B, Agricola PT, Schoenmakers CH. Correction of patient results for Beckman Coulter LX-20 assays affected by interference due to hemoglobin, bilirubin or lipids: a practical approach. Clin Chem Lab Med 2007; 45:114–19.
[21] Saracevic A, Nikolac N, Simundic AM. The evaluation and comparison of consecutive high speed centrifugation and LipoClear® , reagent for lipemia removal. Clin Biochem 2013; 23:A50.
[22] Altura BT, Shirey TL, Young CC, Dell´Orfano K, Hiti J, Welsh R, Yeh Q, Barbour RL, Altura B. Characterization of a new ion selective electrode for ionized magnesium in whole blood, plasma, serum, and aqueous samples. Scand J Clin Lab Invest 1994; Suppl 217:21–36.
[23] Brady J, O´Leary N. Interference due to lipaemia in routine photometric analysis - survey of an underrated problem. Ann Clin Biochem 1994; 31:281–8.
[24] Cobbaert C, Tricarica A. Different effects of Intralipid™ and triacylglycerol rich lipoproteins on Kodak Ektachem serum cholesterol determination. Eur J Clin Chem Clin Biochem 1993; 31:107–9.
[25] Grafmeyer D, Bondon M, Manchon M, Levillain P. The influence of bilirubin, haemolysis and turbidity on 20 analytical tests performed on automatic analysers. Eur J Clin Chem Clin Biochem 1995; 33:31–52.
[26] Leary NO, Pembroke A, Duggan PF. Measuring albumin and calcium in serum in a dual test with the Hitachi 704. Clin Chem 1992; 38:1342–5.
[27] Nanji AA, Poon R, Hineberg I. Lipaemic interference: Effects of lipaemic serum and Intralipid. J Clin Pathol 1998; 41:1026–7.
[28] Roß RS, Eller T, Volbracht L, Paar D. Interferenzen durch Lipämie, Hämolyse und Hyperbilirubinämie am DAX 48-Analysator und ihre klinische Relevanz. Lab Med 1994; 18:233–9.
[29] Clinical and Laboratory Standards Institute (CLSI) Interference testing in clinical chemistry. Approved guideline- 3rd ed. CLSI document EP7-A3. Wayne, Pennsylvania, USA 2010.

Walter G. Guder, Friedrich da Fonseca-Wollheim, York M. Schmitt, Gottfried Töpfer

4.3 The Icteric Sample

4.3.1 Definitions

Bilirubin in plasma exists in a loosely or covalently bound form linked to albumin. In addition, water soluble conjugates exist as mono- and diglucuronides [1, 2]. Studies of bilirubin interference were based mainly on experiments in which free bilirubin or water-soluble di-taurobilirubin was added to serum [3]. Bilirubin molecules may interfere with patient specimens in a qualitative or a quantitative manner [4].

Conjugated bilirubin appears in urine when it is elevated in blood. In patients with proteinuria, bilirubin bound to albumin can also appear in urine. When the blood–brain barrier is compromised as permeable to substances, bilirubin bound to albumin may appear in the CSF. After intra-cerebral bleeds nonconjugated (free) bilirubin causes xanthochromia of the cerebrospinal fluid.

Plasma, serum, urine and cerebrospinal fluid samples are defined as icteric if bilirubin in either form appears in increased concentration. The presence of bilirubin may potentially interfere with measurements of other constituents in these samples.

4.3.2 Mechanisms of bilirubin interference

Spectral interference

Bilirubin has a high absorbance between 340 nm and 500 nm wavelengths. Therefore, the range of the linearity of a spectrophotometric procedure when these wavelengths are used for the measurement of an analyte can be a limiting factor because of the high background absorbance caused by bilirubin [5, 6]. In coagulation analyzers using the turbidimetric principle, a bilirubin concentration exceeding 25 µmol/L causes clinically relevant changes in the measured values of antithrombin III. Interference of bilirubin at higher concentrations may also be significant in other coagulation tests [7].

The decrease in absorption as a result of oxidation of bilirubin in alkaline solution is the main cause for bilirubin interference, when modifications of the Jaffe method for creatinine without deproteinization are used [8]. In a strongly acidic solution, the absorption of conjugated bilirubin shifts to the UV wavelengths. Therefore, in acidic solutions, bilirubin interferes in the determination of phosphate using the phosphomolybdate method through its reducing effect [4, 9].

Chemical interference

Bilirubin interferes in oxidase/peroxidase based test systems. Proportionally to its concentration bilirubin reacts with H_2O_2 formed in the test system which in turn causes systematically lower results in enzymatic procedures used for the measurement of glucose, cholesterol, triglycerides, urate and creatinine [4, 10]. Bilirubin competitively interferes with dyes binding to albumin [11]. However, di-taurobilirubin does not interfere in the procedure of dye binding to albumin [4].

4.3.3 Detection and documentation of increased bilirubin concentrations in clinical samples

The visual inspection of plasma or serum samples as an indicator of hyperbilirubinaemia is often insensitive. This is particularly true when samples are simultaneously pigmented by other substances (e. g. hemoglobin and its derivatives). Moreover, adhesive labels on primary containers can impair visual inspection.

Hyperbilirubinaemia is directly detected in diluted samples that are measured at 450 and 575 nm [12]. However, with the nutritional supply of carotines or carotinoids, bilirubin concentration by direct measurement is overestimated [13]. Therefore the direct procedure of bilirubin measurement is only applied for the determination of hyperbilirubinaemia in newborns. The common clinical chemical methods are applied to achieve quantitative data concerning the interference caused by bilirubin. To assess the mechanism of bilirubin interference, it is advisable to separate and measure the different bilirubin fractions [14].

4.3.4 Prevention of bilirubin interference

Method selection

The high prevalence of hyperbilirubinaemia in patients from intensive care, gastroenterological, surgical or paediatric departments makes it pertinent to select analytical methods that are less susceptible towards bilirubin interference.

Blanking procedures are useful in eliminating spectral bilirubin interferences [15]. Parallel sample blank values give better results than methods in which reagents are added successively into a cuvette [4]. Blanking procedures are often part of the analytical procedure, for example in the kinetic method for creatinine determination according to the Jaffé principle, in which automated analyzers are used [16].

The chemical interference of bilirubin in an analytical reaction is not eliminated by blanking procedures. K_4 [Fe $(CN)_6$] effectively eliminates bilirubin interference in H_2O_2-forming enzymatic methods based on the Trinder reaction [17]. Moreover, optimal concentrations of components of the Trinder reaction can reduce the interference by bilirubin. A mixture of nonionic tensides (detergents) may reduce bilirubin interference such as in the spectrophotometric determination of inorganic phosphate using phosphomolybdate [18].

Actions recommended for use in procedures sensitive to bilirubin

When procedures susceptible to bilirubin interference are used, the laboratory must know the limit of bilirubin concentration in which interference-free measurements are possible (e. g. the application limit). The limit depends on the maintenance status of the analytical system and other variables. Unfortunately, manufacturers' data are not always available. For the determination of the application limit, the following procedure should be performed in the dark: 2 mL of 20 mg free bilirubin should be dissolved in 0.1 mol/L NaOH; the mixture should be added to 20 mg di-taurobilirubin; and the preceeding mixture should be dissolved in 2 mL distilled water, resulting in the master solution. Next, 5 mL of nonicteric pool serum are added to 0.1 mL of the master solution to prepare a final bilirubin concentration of ~340 µmol/L (20 mg/dL). Serial dilutions are prepared by mixing a nonicteric pool serum with the master solution at different proportions. The test solution must be used on the same day [3].

Suitable alternative procedures must be applied for samples that have bilirubin concentrations beyond the application limit. The procedures may require pretreatment of samples to remove bilirubin. For the determination of serum creatinine using a bilirubin-sensitive susceptible enzymatic method the sample is preincubated with 4.4 kU/L bilirubin oxidase for 30 sec [17]. However, the low stability of bilirubin oxidase limits the practical application of this procedure. Ultrafiltration of serum was also used for the elimination of bilirubin interference in creatinine assays, since bilirubin binds to proteins [5]. In the ultrafiltration procedure, serum is centrifuged in a centrifugable ultrafilter (cutoff ≈ 20 kD) for 15 min at 2000 *g* to remove bilirubin and to obtain a completely protein-free ultrafiltrate. The volume depletion effect of proteins results in an approximately 4 % higher value for creatinine in the ultrafiltrate [8]. However, the distribution of ionised low-molecular weight analytes on the diaphragm may be pH dependent which has an effect on the measurement results [19].

If the above procedures for the elimination of bilirubin are not available, alternative analytical principles should be applied. Immunological procedures for the measurement of serum albumin can be used to replace dye binding methods such as those described above.

4.3.5 General recommendations

Each laboratory should specify in their quality manual which measurements are disturbed by increased bilirubin. The limits beyond which the analysis shall not be performed should be stated for each method that is subject to bilirubin interference. The European Directive for In Vitro-Diagnostics [20] 1998 states that providers of reagents must define the appropriate limiting conditions. The procedure for the detection of interfering properties as well as actions that should be taken in case of documented bilirubin interference, should be outlined in the laboratory's quality manual [21].

Each sample must be visually examined immediately after arrival or after centrifugation (in case of blood samples) and the potential or documented interference recorded in the laboratory internal system and clinical report. If no visible interfering property is observed, it should be registered in the list by the notation: "appearance unremarkable". If an interfering property, that is an increased concentration of bilirubin, appears to be present, it is recorded and quantified by measurement with a common clinical chemical method. The request should be reviewed to identify analytes that could be affected by the observed interfering sample property. Analytes that are not affected by bilirubin interference are measured in the same way as with samples that exhibit no interference. Analytes that may be expectedly affected by bilirubin interference before measurements are made; alternatively a measurement method may be used which is not subject to such interference. The analysis should not be made when a clinically relevant bias is expected, or if the interference cannot be eliminated or circumvented by an appropriate alternative method.

References

[1] Fuentes-Arderiu X, Fraser CG. Analytical goals for interference. Ann Clin Biochem 1991; 28:393–5.

[2] Guder WG, da Fonseca Wollheim F, Heil W, Schmitt YM, Töpfer G, Wisser H, Zawta B. The haemolytic, icteric and lipemic sample. Recommendations regarding their recognition and prevention of clinically relevant interferences. J Lab Med 2000; 24:357–64.

[3] Clincal and Laboratory Standards Institute (CLSI) Interference testing in clinical chemistry. Approved standard 6th edn CLSI Wayne, PA, 2010.

[4] Grafmeyer D, Bondon M, Manchon M, Levillain P. The influence of bilirubin, haemolysis and turbidity on 20 analytical tests performed on automatic analysers. Eur J Clin Chem Clin Biochem 1995; 33:31–52.

[5] da Fonseca-Wollheim F. Ringstudie zur Bilirubininterferenz bei der Bestimmung von Serum-Kreatinin. J Lab med 1988; 12:317–20.

[6] Roß RS, Eller T, Volbracht L, Paar D. Interferenzen durch Lipämie, Hämolyse und Hyperbilirubinämie am DAX 48-Analysator und ihre klinische Relevanz. Lab Med 1994; 18:233–9.

[7] Roß RS, Paar D. Analytisch und klinisch relevante Interferenzen in der Gerinnungsanalytik am Beispiel des MDA 180. J Lab Med 1998; 22:90–4.

[8] da Fonseca-Wollheim F, Heinze KG, Lomsky K, Schreiner H. Serum ultrafiltration for the elimination of endogenous interfering substances in creatinine determination. J Clin Chem Clin Biochem 1988; 26:523–5.

[9] Duncanson GO, Worth HGJ. Pseudohypophosphataemia as a result of bilirubin interference. Ann Clin Biochem 1990; 27:263–7.

[10] Spain MA, Wu AHB. Bilirubin interference with determination of uric acid, cholesterol, and triglycerides in commercial peroxidase-coupled assays. Clin Chem 1986; 32:518–21.

[11] Leonard PJ, Persaud J, Motwani R. The estimation of plasma albumin by BCG dye binding on the Technicon SMA 12/60. Clin Chim Acta 1971; 35:409.
[12] Sonntag O, Glick MR. Serum-Index und Interferogramm – Ein neuer Weg zur Prüfung und Darstellung von Interferenzen durch Serumchromogene. Laboratoriumsmedizin 1989; 13:77–82.
[13] Garb S. Clinical guide to undesirable drug interactions and interferences. New York: Springer, 1971.
[14] Balistreri WF, Shaw LM. Liver function. In: Tietz NW, ed. Textbook of Clinical Chemistry. Philadelphia: Saunders, 1986, 1373–433.
[15] Wolthuis A, Peek D, Scholten R et al. Effect of the haemoglobin-based oxygen carrier HBOC-201 on laboratory instrumentation: Cobas Integra, Chiron blood gas analyzer 840, Sysmex™ SE-9000 and BCT. Clin Chem Lab Med 1999; 37:71–6.
[16] Schoenmakers CHH, Kuller T, Lindemans J, Blijenberg BG. Automated enzymatic methods for creatinine measurement with special attention to bilirubin interference. Eur J Clin Chem Clin Biochem 1993; 31:861–8.
[17] Artiss JD, McEnroe RJ, Zak B. Bilirubin interference in a peroxidase coupled procedure for creatinine eliminated by bilirubin oxidase. Clin Chem 1984; 30:1389–92.
[18] Glick MR, Pieper J, Ryder KW. Interference-reduced methodologies for Boehringer Mannheim/ Hitachi analyzers: validation using recombinant haemoglobin "blood substitute" product. Clin Chem 1998; 44, Suppl. A 140.
[19] da Fonseca-Wollheim F. Ultrafiltrate analysis confirms the specificity of the selected method for plasma ammonia determination. Eur J Clin Chem Clin Biochem 1992; 30:15–9.
[20] European IVD Directive for in vitro diagnostics. Amtsblatt der Europäischen Gemeinschaften L331/1 vom 7.12.1998.
[21] ISO/EN/DIN 15189. Medical laboratories – Particular requirements for quality and competence Geneva/Bruxelles/Berlin, 2007.

Oswald Sonntag, Nils Tryding

4.4 Drug Interferences

4.4.1 Introduction

More often than not, if a laboratory result does not fit into the clinical picture it is called a laboratory error. In publications dealing with errors and patient safety in the field of laboratory medicine it is common found that the pre- and postanalytical phase are responsible for most of the errors. This was described first by Bonini et al [1] in 2002 and by Kalra in 2004 (Fig. 4.4.1) [2] followed by a high number of publications. If an error or problem is not detected we believe that there is no error or problem and therefore it was considered to be safe and reliable.

This chapter is dedicated to drug interferences which mainly occur in the analytical phase in a medical laboratory. In-vivo drug effects and drug–drug interactions will not be covered here. It is worthwhile to recognize that in the last 10 years only some individual publications adressed drug interferences. Several approaches to establish a database created by Tryding and his group in Sweden [3] and Young and team in USA [4] were not accompanied by the laboratorians and users of lab data. In August 2014 Wiley has cooperated with American Association for Clinical Chemistry (AACC) and presented now a new database named “Effects on Clinical Laboratory Tests” [5]. The assumption that industry hasn´t done much to adress and eliminate drug interferences is incorrect, since industry follows standards and recommendations given

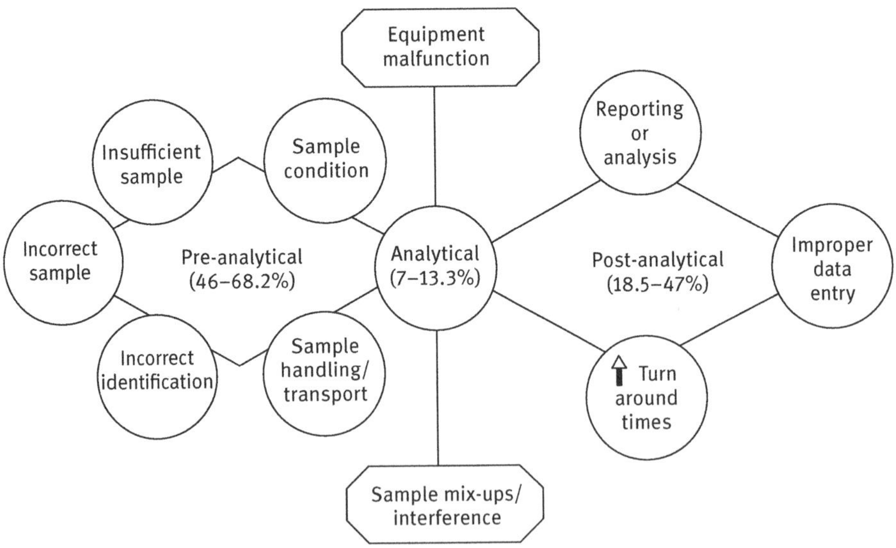

Fig 4.4.1: Percentage of errors in the pre-, analytical and postanalytical phase [2].

for drug interference testing [6]. Those testing procedures are performed in-vitro and do not reflect the real situation in native patient samples. Usually patients receive not only one drug but in addition may receive other medications. Rarely all interactions between the various medications are known and especially interactions occuring with the metabolites formed in the body of these patients remain often unclear. Drugs are one of the major source of interference in biochemical tests. A drug or its metabolite may interfere with the laboratory technique in several ways. It is known that drugs which interfere with one analytical procedure may not necessarily interfere with another method. Therefore the specific features of each analytical method should be understood. Sometimes a slight modification in the analytical procedure may affect the degree of interference caused by the interfering substance.
Drug influence or interfere with biochemical test results either
1. physiologically (In vivo effects) and/or
2. analytically (in vitro effects)

The In vivo effects are due to the intended (therapeutic) effects or side effects. The In vitro effects are caused due to
1. alterations of chemical reactions (enhancement or inhibition)
2. cause of turbidity in the reaction system
3. interference with enzyme reactions
4. cross-reaction with antibodies

4.4.2 Guidelines for interference testing

Drug interference studies are a necessary and integral part of method and instrument evaluation. A plethora of drugs are in clinical use; their blood and urine concentrations vary according to their clinical use and to pathological and physiological conditions. This presents a formidable problem to the diagnostic industry if it is to provide a systematic and clinically useful list of those drugs that interfere with laboratory tests. As part of the design of a new test or method a drug interference testing procedure has to be performed to evaluate possible limitations of the new procedure. Due to the difficulty to obtain real patient samples which may include a variety of commonly used drugs and their metabolites an in vitro study has to be performed. There are several guidelines or recommendations from various organizations or publications. In 1978 an expert group was convened within the European Community to study drug interferences and establish recommendations [7]. In 1984 the French Society of Clinical Chemistry (7) and the International Federation of Clinical Chemistry (IFCC) also addressed this issue and published, in French and in English, recommendations for the evaluation of drug interferences [8]. The National Committee for Laboratory Standards (now named the Clinical Laboratory Standards Institute (CLSI) published proposed guidelines for interference testing in 1986 resulting in approved guidelines in 2010 [6].

International Federation of Clinical Chemistry (IFCC) [8]

The IFCC guidelines suggested a 2-step approach in their guideline which was published in 1989 [8]. The first step for testing for possible drug interferences is the performance of a screening procedure in which the potential interfering drug is added to a sample to achieve a final concentration that is 10 times the normally expected therapeutic concentration. A second step will become necessary if the difference between the specimen which contains the drug and a corresponding control free of drug is statistically different.

In the second step the specimen matrix used in step 1 will be used and one has to mix this with the control pool to produce a series of specimens containing the drug at five or more different concentrations. At least two of the samples in this procedure should have concentrations of the drug within the therapeutic range. Each of the specimens should be analyzed at a minimum of five times. The resulting dose-response curve should be analyzed with regression methods using a third-degree polynomial equation.

CLSI (former NCCLS) [6]

The two step approach is also recommended by the Clinical and Laboratory Standards Institute (CLSI guidelines for interference testing in clinical chemistry in the Evaluation Protocol (EP 07). The first step is similar to the IFCC recommendation. In the second approach, samples from a selected population, for example patients taking the drug, are analyzed by using the method under investigation and an alternate method known not to be affected by the drug being evaluated. If either of these two screening methods indicates the presence of interference, CLSI guidelines suggest that a dose-response experimental evaluation be performed to assess the quantitative effects of the interferent. Statistical analysis of the dose-response data, if linear, is analyzed with linear regression analysis. If the dose-response data do not appear to be linearly related, nonlinear regression or graphic methods are recommended. Currently a new updated and modified guideline is in preparation and will be published in 2015.

Sonntag and Scholer 2001 [9]

A list of drugs recommended for interference study was provided by International experts. Each expert prepared a list of candidate drugs for interference testing. From a list of 101 drugs the experts identified 18 in serum (see Table 4.4.1) and 13 in urine (see Table 4.4.2) which was likely to present frequent, clinically significant interference. This exercise drew support from Swedish, French, English and American databases, and from other published reports. From a review of the literature, the group identified toxic and therapeutic concentrations for each of the 24 drugs.

Inclusion criteria used to select drugs for the in vitro study were:
- known interferent
- frequent clinical use

- clinical relevance
- absorbs light at a wavelength close to that used for test measurement

For the urine studies, if metabolites are excreted in place of the parent drug the major metabolite should be used; however, this is subject to the availability of the metabolite. Two different concentrations of each drug were adopted. To reveal potential interference in in vitro screening, a very high drug concentration (C1) was used. C1 represents a typical toxic concentration, and where interference was identified a validation experiment was performed using the therapeutic drug concentration (C2).

Two tables were prepared, one for blood (see Table 4.4.1) and the other for urine (see Table 4.4.2). These include the international nonproprietary name (INN) of the drug, its clinical use, drug concentrations C1 and C2, and the highest blood and

Table 4.4.1: Recommended drug list for in vitro drug interference studies for clinical chemical methods using serum or plasma as sample [9].

Active agents soluble in blank serum or plasma	Clinical use	Test concentration C1 (mg/L)	Test concentration C2 (mg/L)	Peak (mg/L)	Therapeutic (mg/L)	Toxic (mg/L)
Acetaminophen[1]	Analgesic	200	20	50	10–20	150–300
Acetylcysteine	Mucolytic agent	150	30	33		
Acetylsalicylic acid	Analgesic	1000	300	300	150–300	300–500
Ampicillin-Na	Antibiotic	1000	200		2–20	>320
Ascorbic acid	Vitamin	300	30	18	4–15	
Ca-dobesilate	Vasotherapeutic	200	20	400	6–18	
Cefoxitin	Antibiotic	2500	250	2230	up to 150	
Ciclosporine	Immunosuppressant	5	1	4	0.05–0.6	>0.6[2]
Doxycycline	Chemotherapeutic	50	10	9		
Heparin[3]	Anticoagulant	5000 U	10 U		0.5–1	>400
Ibuprofen[1]	Analgesic	500	50	50	10–30	>200
Intralipid[4]	Supplement	10000	2000			
Levodopa	Antiparkinson's Agent	20	4	18	10	>200
Methyldopa	Antihypertonic	20	4	7	up to 2	>9
Metronidazole	Chemotherapeutic	200	10		2–5	200
Phenylbutazone	Analgesic	400	100	120	50–110	200–400
Rifampicin	Chemotherapeutic	60	20		4±10	>10
Theophylline[1]	Bronchodilator, pulmonary vasodilator100	100	10		10–20	20–80

1 For these substances a separate reference serum is necessary: 100µL ethanol (70 %) + 9.9 mL separate 100 from microL and 70 from % and 0,9 from % in line 4 (WG) agent free serum/plasma

2 Depends on individuals, drug interactions and other influences

3 Liquemin Na (Hoffmann-La Roche) 5000 IE (= 5000 U) 0.5mL

4 For Intralipid a separate reference serum is necessary: 9.5 mL serum pool+0,5 mL NaCl (0.9 %)

urine drug concentrations found in the literature. Therapeutic peak and toxic concentrations have been included where they are available.

When studying drug interference the analyte concentration used should be close to a clinical decision level, perhaps the upper or lower normal reference limit.

This recommendation was adopted in a validation experiment and was published.

Procedure

The study covered two phases: a screen for effects at elevated drug levels (C1) and a validation phase to check for interference at more clinically relevant levels (C2).

1. Step: Screening (C1)

In the screening experiment five replicates of a drug-free pool were analyzed, followed by analysis of five replicates of a pool containing the drug under investigation. A simple automatic wash step as typically programmed in the instrument was performed between each sample to minimize sample carry-over.

2. Step: Validation (C2)

In the validation experiment, analysis of 10 replicates of the drug-free pool was followed by analysis of 10 replicates of the spiked drug pool, with a wash step between each sample.

Table 4.4.2: Recommended drug list for in vitro drug interference studies for clinical chemical methods using urine as sample [9, 18].

Active agents soluble in urine	Clinical use	Test concentration C1 (mg/L)	Test concentration C2 (mg/L)	Peak (mg/L)	Toxic (mg/L)
Acetaminophen	Analgesic	3000	500		140–830
Acetylcysteine	Mucolytic	10	1		
Salicyluric acid[1]	Analgesic	6000	100		19–1350
Ascorbic acid	Vitamin	4000	400	4000	
Ca-dobesilate	Vasotherapeutic	1000	200		1000
Na_2-cefoxitin	Antibiotic	12 000	2000	24 000	3000
Gentamicin sulphate	Antibiotic	400	80		16–125
Ibuprofen	Analgesic	4000	500		1000
Levodopa	Antiparkinson's agent	1000	250		
Methyldopa	Antibiotic	2000	200		1400
Ofloxacin	Analgesic	900	100		250–350
Phenazopyridine	Antibiotic	300	50		
Tetracycline	Antibiotic	300	100		

1 Metabolite of 4-aminosalicylic acid and salicylic acid

Results

In vitro studies cannot accurately reflect the in vivo interference effects of drugs and especially their metabolites. In the study those drugs that caused interference under in vitro conditions had been identified. It cannot be considered either definitive or comprehensive. Rapid developments in new analytical methods, technologies and the introduction of new and more complex drug preparations makes it essential that the clinical chemist remains vigilant to the potential for drug interference.

4.4.3 Collection of data

There exists a number of compilations on drug interferences in different countries. The first critical review of the original literature was published by Caraway and Kammeyer in 1972 [10]. In 1975 Young et al. published "Effects of Drugs on Clinical Laboratory Tests" and in 2007 he published the final edition [11]. In Sweden Tryding started in 1977 [3] with his first compilation and printed the last version (7th edition) in 1996 [12]. In France it was Siest et al in 1985 [13], in England Salway, who published a summary in 1990 [14] and Sonntag from Germany in 1985 [15]. In 1994 Kroll and Elin wrote a critical review about interference with clinical laboratory analyses [16]. Several more publications and compilations appeared between 1975 and 2000. Of late only a few cases of drug interferences have been published.

Information of drug interference

Two tables have been prepared extracted from two publications. The first is from Hagemann in 1998 who made a short version of available information at that time (Table 4.4.3) [17]. The other literature is from Young who compiled the information in 2000 (Table 4.4.4) [11].

Database

First a database project named "DEEC" was initiated in Sweden under the leadership of Tryding. Later a joint effort by Trying and Young under the auspice of the American Association for Clinical Chemistry (AACC) resulted in new database project. After about two years this database project was stopped due to low interest.

Since August 2014 a new database program is available [5]. A cooperation between AACC and Wiley finalized this database which is named "Effects on Laboratory Tests". Using the link: http://clinfx.wiley.com/aaccweb/aacc/

Drug concentration

Since 1975 Baselt et al is publishing frequently updates of drug concentrations in therapeutic and toxic clinical cases (situations), which is therefore a helpful source of information when interpreting drug interferences (18).

Table 4.4.3: Drug interferences published by Hagemann in 1998 [17].

Drug	Analytes in serum increased	Analytes in serum decreased	Analytes in urine increased	Analytes in urine decreased
Acetyl salicylic acid	CO_2, free thyroxine, uric acid	Triglycerides		Glucose
Aminoglycosides			Total protein	
Ascorbate	Creatinine, glucose	CO_2, cholesterol, creatine kinase, creatinine, glucose, HDL-cholesterol, uric acid	Oxalate	Uric acid
Bromides	CO_2, chloride			
Ca dobesilate		Creatinine, glucose, triglycerides, uric acid		
Caffeine	Theopylline			
Cephalosporins	Creatinine		Creatinine	
Chlorpropamide	Calcium			
Dextran	Cholesterol, total protein	Triglycerides		
Diflunisal	Salicylate			
Fat emulsion	Alanin amino transferase, albumin, aspartate amino transferase, bilirubine, c-reactive protein, calcium, cholesterol, creatinine, glucose, blood-hemoglobin, iron, phosphate, total protein, theophylline, uric acid, urea	Bilirubine, cholesterol, creatine kinase, creatinine, magnesium, phosphate, total protein, theopylline, uric acid		
Fluorescein	Amylase, cortisol, digoxin, quinidine, thyroxine	Total protein		
Heparin	Antithrombin, free thyroxine	Albumin, gamma glutamyl transferase, sodium, free thyroxine		
Ibuprofen			Barbiturates, benzodiazepines, cannabinoids	
Indomethacine	Glucose			
Iodine	CO_2, chloride			
Isoniazid	Uric acid			
Labetalol	Catecholamines		Amphetamine, phenothiazine, catecholamines	
Levodopa	Bilirubin	Glucose, triglycerides, uric acid		Glucose, vanillyl mandelic acid
Metamizole		Creatinine, triglycerides		

Table 4.4.3: (continued)

Drug	Analytes in serum increased	Analytes in serum decreased	Analytes in urine increased	Analytes in urine decreased
Methotrexate	Bilirubin			
Methyldopa	Bilirubin, catecholamines, creatinine, glucose	Cholesterol, triglycerides, uric acid	Catecholamines, creatinine, vanillyl mandelic acid	
Naproxen	5-hydroxy indoleacetic acid			
Nifipipine			Vanillyl mandelic acid	
Nitrofurantoin	Bilirubin			
Paracetamol	Glucose		Catecholamines	
Phenolphthalein			Total protein	
Spironolactone	Digoxin			
Sufasalazine	Bilirubin	Bilirubin		
Sufonamides		Total protein	Creatinine	
Theophylline		Alkaline phosphatase		
X-ray contrast agent		Glucose	Total protein	Glucose

Table 4.4.4: Drug interferences published by Young [11].

Drug	Analyte in serum increased	Analyte in serum decreased	Analyte in urine increased	Analyte in urine decreased
4-Aminosalicylic acid	Bilirubin			
6-α-methyl-prednisolone	Cortisol			
Acetaminophen	ALT, ALP, AST, creatinine, glucose, LDH, uric acid	P-amylase, AST, glucose	Catecholamines, uric acid	
Acetazolamide			Total protein	
Acetohexamide	Creatinine, urea			
Acetylcysteine	Chloride			
Acetylsalicylic acid	AST, CO_2, cortisol, glucose, total protein, triglycerides, uric acid	CO_2, cholesterol, CK, LDH, total protein, triglycerides		
Aldrin		CHE		
Allopurinol		Chloride, cholesterol,		
Amikacin		Bilirubin, cholesterol, CK, creatinine, glucose, LDH, urea		

Table 4.4.4: (continued)

Drug	Analyte in serum increased	Analyte in serum decreased	Analyte in urine increased	Analyte in urine decreased
Amino acids	Bilirubin, blood Hb, total protein, urea			
Aminoantipyrine		Cholesterol		
Aminohippurate	Creatinine			
Aminophenazone	ALT, AST	AST, bilirubin, glucose		
Aminosalicylic acid	ALT, AST, bilirubin, glucose, urea		Total protein	
Amiodarone	Thyroxine			
Ammonium chloride	Creatinine			
Ammonium heparinate	Creatinine, phosphate	Creatinine		
Ammonium hydroxide		Total protein		
Amphotericin B	Bilirubin, cholesterol			
Ampicillin	Total protein, sodium, theophylline, uric acid	Cholesterol	Glucose	Glucose
Antipyrine			Acetoacetate	
Arsenicals		ALP		
Ascorbic acid	P-amylase, AST, bilirubin, creatinine, glucose, phosphate, potassium, sodium, uric acid	AST, CO_2, bilirubin, chloride, cholesterol, CK, creatinine, glucose, HDLC, LDH, triglycerides, urea, uric acid	Creatinine, glucose, total protein	Acetaminophen, glucose
Aspirin	Acetaminophen, chloride, uric acid	Albumin	Acetoacetate, glucose, total protein, uric acid	Glucose
Azlocillin	Total protein, creatinine		Total protein, Glucose	
Azopropazone	Thyroxine			
Benzylsulfonic acid	CO_2			
Beryllium salts		ALP		
Bicarbonate	Magnesium, urea	Chloride, sodium	Total protein	
Bisacodyl				Glucose
Bromide	CO_2, chloride	Cholesterol	Chloride	
Bromisovalum	Chloride			
Bunamiodyl			Total protein,	
Ca dobesilate		CK, creatinine, glucose, HDLC, triglycerides, uric acid		
Ca gluconate		Magnesium		Magnesium

Table 4.4.4: (continued)

Drug	Analyte in serum increased	Analyte in serum decreased	Analyte in urine increased	Analyte in urine decreased
Calcium bromo-galactogluconate	Chloride			
Caproxamine	Albumin	Urea, total protein		
Carbamazepine	Chloride, lithium, sodium	Uric acid		
Carbenicillin	Bilirubin, total protein, triglycerides	Albumin	Glucose	
Carbimazole		Cortisol		
Carbromal	Chloride			
Carinamide			Total protein	
Cefaclor	Creatinine	Creatinine	Glucose	
Cefamandole	Creatinine	Creatinine	Glucose, total protein	
Cefazolin	Creatinine, theophylline	Creatinine	Glucose	
Cefdinir			Glucose	
Cefipime			Glucose	
Cefixime	Creatinine		Glucose	
Cefoperazone	Creatinine		Glucose	
Cefotaxime	Albumin, ALP, calcium, chloride, cholesterol, CK, creatinine, glucose, iron, magnesium, phosphate, potassium, sodium	Ammonia, amylase, chloride, GGT, LDH, magnesium, phosphate, total protein, urea, uric acid		
Cefotiam	Bilirubin, creatinine			
Cefoxitin	Creatinine		Creatinine, glucose	
Cefpirome	Creatinine			
Cefsoludine	Creatinine			
Ceftazidime			Glucose	
Cefuroxime	Creatinine, glucose	Creatinine, glucose	Glucose	
Cephalexine	Creatinine		Glucose	
Cephaloridine	Creatinine		Total protein	
Cephalosporines			Glucose	
Cephalothin	Creatinine, total protein, theophylline	Creatinine,	Creatinine, total protein	
Cephradine	Creatinine,	Creatinine,		
Chloral hydrate	Urea, uric acid		Catecholamines	Glucose
Chloramphenicol	Total protein, urea	Urea, uric acid		
Chlordiaze-poxide	ALT, AST			
Chlorhexidine			Total protein	
Chloride salts	Amylase			
Chlorine		Uric acid		

Table 4.4.4: (continued)

Drug	Analyte in serum increased	Analyte in serum decreased	Analyte in urine increased	Analyte in urine decreased
Chlorothiazide	Theophylline			
Chlorpromazine	Cholesterol		Total protein	
Chlorpropamide	Calcium	Glucose		
Chlortetracycline			Catecholamines	
Cibenzoline			Total protein	
Citruplexina	Urea			
Clindamycin	Theophylline			
Clothiapine		AST, CK		
Co-trimoxazole	Urea			
Contraceptives, oral	Cortisol			
Corticosteroids	Cholesterol			
Cyclacillin			Glucose	
Cyclopropane		Catecholamines		
Danazole	Cortisol	Cortisol, thyroxine		
Deferoxamine		Iron		
Delalande 69276	Thyroxine			
Demeclocycline			Catecholamines	
Deoxyglucose		Glucose		
Dextran	Bilirubin, cholesterol, glucose, total protein, urea, uric acid	Total protein		
Dextran 40	Glucose	Total protein		
Dextran 70	Total protein			
Dextrothyroxine	Thyroxine			
Diazepam				Glucose
Dicofenac	AST, cortisol, glucose			
Dieldrin		CHE		
Diflunisal		Thyroxine		
Digitoxin	Cholesterol			
Digoxin				Glucose
Dihydrotachy-sterol			Total protein	
Dihydroxyaceton	Triglycerides			
Dimenhydrinate	Theophylline			
Dimethadione		CO_2		
Dimpylate		CHE		
Dipyrone	Cortisol	AST, cholesterol, CK, creatinine, LDH, trigly-cerides, urea, uric acid		Glucose
Dithiazanine			Total protein	
Dobutamine		Cholesterol, creatinine, triglycerides, uric acid		

Table 4.4.4: (continued)

Drug	Analyte in serum increased	Analyte in serum decreased	Analyte in urine increased	Analyte in urine decreased
Doxorubicin	Total protein			
Doxycycline			Catecholamines	
DTNB		Cholesterol		
Erythromycin	AST			
Estrogens	Cortisol			
Ethamsylate		Creatinine		
Ether		Catecholamines		
Ethionamide			ALP	
Ethoxazene	Bilirubin			
Etoposide	Uric acid			
Fenfluramine	Cortisol			
Fenoprofen	Cortisol			
Ferrous sulfate				Glucose
Flucytosine	Creatinine			
Fluosol-DA	Albumin, bilirubin, chloride, cholesterol, glucose, LDH, phosphate, potassium, total protein, sodium, triglycerides	Albumin, AST, CO_2, chloride, potassium, total protein, sodium		
Flurazepam				Glucose
Flurbiprofen	Cortisol			
Furazolidone			Glucose	
Furosemide	Cortisol			Glucose
Gabapentin			Total protein	
Gallium nitrate	Creatinine			
Gentamicin			Total protein	
Glyburide			Total protein	
Glycocyamidine	Creatinine		Creatinine	
Guanethidine	Urea			
Guanoclor		Glucose		
HBOD	Bilirubin	ALP, AST		
Hydantoin derivates	Urea			
Hydralazine	AST, calcium, glucose, uric acid	Glucose		
Hydroquinone				Glucose
Hydroxyurea		Triglycerides		
Hyoscine-N-butylbromide		Sodium		
Ibuprofen	ALP	AST		
Indocyanine green	Bilirubin			
Indomethacin	Thyroxine			

Table 4.4.4: (continued)

Drug	Analyte in serum increased	Analyte in serum decreased	Analyte in urine increased	Analyte in urine decreased
Iodate	Cholesterol			
Iodide	CO_2, chloride, cholesterol, potassium			
Iodine	Thyroxine			
Iodoalphionic acid	Thyroxine		Total protein	
Iodopyracet			Total protein	
Iopanoic acid	Thyroxine		Total protein	
Iophenoxic acid			Total protein	
Iothalamate			Total protein	
Iothiouracil	Thyroxine			
Ioxaglate			Total protein	
Iproniazid		Glucose		
Iron	Calcium			
Iron dextran	Bilirubin, glucose, iron	Calcium		
Iron sorbitex	Glucose			
Isocarboxazid		Glucose		
Isoniazid	AST, uric acid	Glucose		
Isoproterenol	AST, bilirubin, glucose			
Isosorbide dinitrate		Cholesterol		
Ketoprofen		AST, LDH		
Labetalol			Catecholamines	Catecholamines
Levarterenol	Glucose			
Levodopa	AST, bilirubin, catecholamines, CHE, creatinine, glucose, uric acid	Glucose, triglycerides, urea, uric acid	Creatinine, uric acid	Glucose
Levoglutamide		Ammonia		
Lidocain	Creatinine			
Lithium	Creatinine			
Magnesium salts	ALP, calcium			
Mannitol	Phosphate	Phosphate		Phosphate
Meralluride				Glucose
Mercaptopurine	Glucose, uric acid			
Mercurial diuretics				Glucose
Metahexamide			Total protein	
Methapyrilene			Glucose	
Methenamine	Catecholamines			
Methicillin	Phos, total protein, triglycerides			
Methimazole	Glucose, thyroxine	Triglycerides		
Methotrexate	ALP, bilirubin, cholesterol, phosphate	LDH, triglycerides, uric acid		

Table 4.4.4: (continued)

Drug	Analyte in serum increased	Analyte in serum decreased	Analyte in urine increased	Analyte in urine decreased
Methylaminoantipyrine		Cholesterol		
Methyldopa	AST, bilirubin, catecholamines, creatinine, glucose, uric acid	Cholesterol, creatinine, glucose, triglycerides, uric acid	Catecholamines, creatinine, uric acid	
Methylparaben	Sodium	Lithium		
Metronidazole	Glucose	AST, glucose, LDH, triglycerides		
Metyrosine			Catecholamines	
Mezlocillin			Glucose, total protein	
Moxalactam	Creatinine			
N-acetylprocainamide	Lithium			
Nafcillin	Total protein			
Nalidixic acid	Glucose		Glucose	
Naproxen	ALP, CO_2, phosphate, uric acid	Triglycerides		
Neostigmine	Chloride			
Niacin	Catecholamines		Catecholamines, glucose	
Nitrazepam		Glucose		
Nitrofurans			Creatinine	
Nitrofurantoin	ALP, AST, bilirubin, creatinine, sodium		ALP, glucose	
Nitrofurazone			Creatinine	
Nitroglycerin	Triglycerides	Cholesterol		
Nitromethane	Creatinine			
Norfenefrin	Sodium	Creatinine		
Novaminsulfon	Creatinine	Cholesterol, glucose, triglycerides, uric acid		
Novobiocin	Bilirubin			
Osazepam	Glucose			
Oxacillin	Total protein			
Oxyphenbutazone	Uric acid	Glucose, thyroxine		
Oxypurinol		Chloride		
Oxytetracycline	Bilirubin, catecholamines, uric acid			Glucose
P-aminophenol	AST, bilirubin, calcium, glucose, urea, uric acid	Glucose		
P-aminosalicylic acid	Acid phosphatase, albumin		Total protein	
Pancreozymin	Amylase			

Table 4.4.4: (continued)

Drug	Analyte in serum increased	Analyte in serum decreased	Analyte in urine increased	Analyte in urine decreased
Paraldehyde			17-Hydroxycorticosteroids, ketones	
Paraquat	Creatinine			
Penicillamine		Cholesterol		
Penicillin		Albumin	Glucose, total protein	
penicillin G	Total protein		Glucose, total protein	
Phenacemide		Creatinine		
Phenazopyridine	Albumin, bilirubin, total protein	Glucose	Glucose, total protein	Glucose
Phenelzine	AST, bilirubin, uric acid			
Phenobarbital	Calcium, LDH, total protein, theophylline			Glucose
Phenolphthalein			Total protein	
Phenolsulfonphthalein	Creatinine		Creatinine, total protein	
Phenothiazine		Phosphate	Acetoacetate	
Phenytoin	Cholesterol, theophylline			
Pindolol	ALP	AST, bilirubin, CK		
Piperacillin			Glucose, total protein	
Piperazine		Uric acid		
Prednisolone	Cortisol			
Prednisone	Cortisol, theophylline			Glucose
Primidone	Total protein	Chloride, phosphate		
Probenecid	Theophylline		Glucose	
Procainamide	Lithium, potassium	CHE, lithium		
Promazine			Total protein	
Promethazine		Phosphate		
Propantheline	Chloride, CO2			
Propoxyphene				Glucose
Propranolol	Bilirubin			
Propylidone	Thyroxine			
Propylthiouracil	Glucose, thyroxine, uric acid			
Pseudoephedrine	Theophylline			
Pyrazinamide		Iron		
Pyridostigmine	CO_2, chloride			
Quinidine	Lithium	Lithium		

Table 4.4.4: (continued)

Drug	Analyte in serum increased	Analyte in serum decreased	Analyte in urine increased	Analyte in urine decreased
Radiographic agents	Total protein, thyroxine		Total protein	Glucose
Ranitidine			Total protein	
Reserpin	Bilirubin			
Rifampicin	Bilirubin, glucose, iron, LDH, phosphate, total protein, uric acid	AST, cholesterol, triglycerides		
Salicylate	CO_2, chloride, glucose, theophylline	Glucose		
Secobarbital				Glucose
Sodium bromide	CO_2, chloride			
Spironolactone	Cortisol			
Streptomycin		Urea		
Sulbactam	Creatinine			
Sulfamethoxazole	Creatinine	CK	Total protein	
Sulfanilamide	Uric acid			
Sulfasalazine	Creatinine, total protein	Potassium, total protein		
Sulfathiazole	AST			
Sulfhydryl compounds		ALP		
Sulfinpyrazone		Cyclosporine		
Sulfisoxazole			Total protein	
Sulfobromophthalein	ALP, calcium, creatinine, total protein			
Sulfonamides			Glucose	
Sulfonylureas	Urea			
Sulpiride		Cholesterol, glucose		
Suramin	Amylase	AST, calcium, LDH		
Terbutaline	Theophylline			
Tetracycline	Cholesterol, glucose, urea, uric acid	Acetaminophen, glucose, uric acid	Catecholamines	Glucose
Tetraiodoacetic acid	Thyroxine			
Theobromine	Theophylline			
Theopylline	Catecholamines, uric acid	ALP, bilirubin, LDH	Uric acid	
Thioneine	Uric acid			
Thiouracil	Cholesterol	Cholesterol		
Thiouric acid	Uric acid			
Ticarcillin			Glucose, total protein	
Tolazamide		Glucose		
Tolbutamide	AST, glucose	Glucose	Total protein	

Table 4.4.4: (continued)

Drug	Analyte in serum increased	Analyte in serum decreased	Analyte in urine increased	Analyte in urine decreased
Tolmetin			Total protein	
Triamterene				Catecholamines
Trifluoperazine			Total protein	
Trimethoprim	Creatinine			
Trimetozine			Glucose	
Triple bromide	CO_2, chloride			
Tromethamine		Ammonia, creatinine		
Trypan Blue		Glucose		
Valproic acid	Sodium	Total protein		
Vitamin B complex	Catecholamines			
Vitamin D	Cholesterol			
Vitamin preparations				Glucose
Vlomycin	Cholesterol			

Abbreviations:
ALP: alkaline phospatase
ALT: alanine aminotransferase
AST: aspartate aminotransferase
CHE: choline esterase
CK: creatine kinase
CO_2:carbon dioxide
GGT: gamma-glutamyl transferase
HDLC: high density lipoprotein cholesterol
LDH: lactate dehydrogenase

References

[1] Bonini PA, Plebani M, Ceriotti F, Rubboli F. Errors in laboratory medicine. Clin Chem 2002; 48:691–8.
[2] Kalra J. Medical errors: introduction to concepts. Clin Biochem 2004; 37:1043–51.
[3] Tryding Drug effects in Clinical Chemistry Stockholm: Apoteksbolaget 1977.
[4] Young DS, Pestaner lC, Gibberman V. Effects of drugs on clinical laboratory tests. Clin Chem 1975; 21:1D–423D.
[5] http://clinfx.wiley.com/aaccweb/aacc/
[6] Clinical Laboratory Standards Institute (CLSI) Interference testing in clinical chemistry. Approved guidelines. CLSI publication EP7-A3; Wayne, PA 2010.
[7] Galteau MM, Siest G. Drug effects in clinical chemistry. Part 2: Guidelines for evaluation of analytical interference. J Clin Chem Clin Biochem 1984; 22:275–9.
[8] Kallner A, Tryding N. International Federation of Clinical Chemistry (IFCC) guidelines to the evaluation of drug effects in clinical chemistry. Scand J Clin Lab Invest 1989; 49 (suppl 195):1–28.
[9] Sonntag 0, Scholer A. Drug interferences in Clinical Chemistry: recommendation of drugs and their concentrations to be used in drug interference studies. Ann Clin Biochem 2001; 38:376–85.

[10] Caraway WT, Kammeyer CW. Chemical interference by drugs and other substances with clinical laboratory test procedures. Clin Chim Acta 1972; 41:395–434.
[11] Young D S. Effects of Drugs on Clinical Laboratory Tests. Vol 1 + 2. 5th ed. Washington: AACC press, 2000.
[12] Tryding N, Tufvesson C, Sonntag O. Drug Effects in Clinical Chemistry. Vol. 1 + 2. 7th edition, Stockholm: Apoteksbolaget 1996.
[13] Siest G, Galteau MM, Schiele F, Henny J. Examens de laboratoire et médicaments. Paris: Expansion Scientifique Française 1985.
[14] Salway JG. Drug Test Interactions Handbook. London: Chapman and Hill 1990.
[15] Sonntag O Arzneimittel-Interferenzen, Stuttgart-New York: G Thieme Verlag 1985.
[16] Kroll MH, Elin RJ. Interference with clinical laboratory analyses. Clin Chem 1994; 40:1996–2005.
[17] Hagemann P. Drug Interference in Laboratory Tests, Basel: Editiones Roche 1998.
[18] Baselt RC, Disposition of Toxic Drugs and Chemicals in Man. 9th ed.,Foster City: Biomed Publ 2011.

Raffick A. R. Bowen, Dorothy M. Adcock-Funk

4.5 Interferences from Blood Sampling Device Materials on Clinical Assays: I Blood Collection Devices and Their Constituents and Additives

4.5.1 Introduction

The role of the laboratory in-patient care is significant, because laboratory data informs 60–70 % of critical decisions related to admission, discharge, and the administration of medications [1, 2, 3]. Consequently, in important practices related to specimen collection, transport and storage (preanalytical phase), testing (analytical phase), and reporting (postanalytical phase), errors that affect patient safety may occur and unnecessarily burden hospital budgets. We argue that by minimizing errors in the preanalytical phase through a better understanding of blood collection tubes (BCTs) and their components; laboratories can improve the quality of blood test results, reduce the number of specimens requiring re-collection, increase efficiency with respect to turnaround time (TAT), and improve patient management. On average, errors in the preanalytical phase represent between 0.23 and 1.2 % of total hospital operating costs [3]. For a US hospital with approximately 650 beds, such errors translate to an estimated cost of $1,199,122 per annum – a significant draw on any institution's operating budget [3]. Similarly, the cost of preanalytical errors with respect to time efficiency is evident in the additional hours dedicated to specimen redraws, retesting, and delayed patient care. For example, in one year alone, an 850-bed US hospital paid costs associated with an additional 24,027 h of patient care due to the need for redraws and additional patient treatment [3]. Of these hours, 2,507 (~10 %) represented spending associated with laboratory-redraws and processing time [3].

Preanalytical errors can result from complications related to the devices used for blood collection. The ways in which these devices affect blood samples are not fully understood, and they may have a greater influence on test results than many health professionals are aware. These devices have complex interactions with blood and the potential to alter composition, serum, and plasma fractions. In some cases, such alterations can adversely affect laboratory test results. For example, components of BCTs, including stoppers, stopper lubricants, tube walls, surfactants (SFs), clot activators, tube additives, and separator gels, may add constituents to blood, adsorb elements, or interact with protein and cellular components – potentially leading to inaccuracies in test results [4]. Additives and chemicals associated with the manufacture of BCTs can significantly alter the stability of analytes in blood specimens. Because collection devices can introduce a major source of error in the preanalytical phase of laboratory testing, manufacturers of collection devices, vendors of laboratory tests, and clinical laboratories could all benefit from knowledge of the interactional effects devices may have on blood sample test results.

This chapter aims to improve the understanding of sampling device materials in order to:
1. ensure blood specimen tests are more accurate by providing information about how specific chemicals can affect results;
2. reduce additional costs related to redrawing blood by suggesting methods for anticipating and avoiding preanalytical errors; and
3. accelerate medical interventions and prevent delayed patient care by reducing laboratory processing times.

Accordingly, we offer a detailed discussion of how blood collection materials and devices can alter chemistry, hematology, and coagulation test results, with an emphasis on BCT additives. We begin this chapter with a brief history of blood collection devices. Then, in an effort to increase awareness of and reduce device-related errors during the preanalytical phase, we identify current practices and areas of concern related to blood collection components, BCTs, order of draw, and the impact of protease inhibitors. The chapter concludes with recommendations for clinical laboratories and sampling device manufacturers.

4.5.2 Blood collection device history

Reusable glass syringes with steel hypodermic needles and hard rubber hubs were among the first devices used to collect blood [5]. Glass evacuated tubes containing anticoagulants were the device of choice from the 1950s to the 1990s [6]. Glass tubes were used because it was impossible to preserve a vacuum seal for long periods of time using other available materials. Early modifications to glass tubes included a refined needle, replacement of the rubber hub with glass, and the Luer-Lok syringe, which modified the needle tip for more secure attachment to the syringe [5]. Glass syringes were eventually abandoned as a result of three drawbacks: their high manufacturing costs, their susceptibility to breakage [7], and more significantly, the multiple hepatitis outbreaks that resulted from their use [5]. The latter, coupled with modern chemical and radiation sterilization techniques, ultimately inspired the full replacement of glass syringes with disposable plastic syringes [5].

In the 1940s, Joseph Kleiner invented the first evacuated blood collection tube (BCT), calling it the "Evacutainer," and it has become the most common blood collection device in use today [8]. Evacuated BCTs allow for the draw of a predetermined volume of blood and facilitate switching between tubes to obtain additional samples without spillage and needle-stick injury [9]. This innovation in blood collection brought with it increased specimen quality, greater workflow efficiency, and enhanced safety for patients and health-care professionals with respect to bloodborne pathogens.

There are currently numerous manufacturers of plastic evacuated and nonevacuated BCTs. Becton Dickinson (BD) (Franklin Lakes, NJ, USA), Greiner Bio-One (Kremsmuenster, Austria and Monroe, NC, USA), Sekisui Chemical Co. (Tokyo, Japan), Sarstedt (Nuembrecht, Germany), and Terumo (Tokyo, Japan) are well-known examples. BCTs are differentiated by color-coded stoppers that indicate the presence of additives and other constituents. These plastic tubes usually contain polymer gels and clot activators [10]. If plasma rather than serum is preferred for testing, several types of anticoagulants that differ in their mechanism of action may be added [11]. Despite their similarities, evacuated tubes supplied by different manufacturers vary in composition and in the additives used, which can potentially affect specimen test performance [12]. BCTs tend to function properly under most circumstances; consequently, many laboratorians are unaware of the implications of their use and of their potential limitations. For instance, a widespread surfactant problem reported by the author revealed how these devices can adversely affect laboratory test results [13, 14]. This incident illustrates how important it is to understand the benefits, limitations, and disparities within and among BCTs utilized by our laboratories.

4.5.3 Blood collection device components

The following section will elucidate blood collection procedures and devices, discuss instances in which blood collection device components interferences can occur, and provide some general recommendations.

4.5.3.1 Alcohol and other disinfectants

Immediately before blood specimen collection, the skin is typically disinfected with 70 % isopropyl alcohol or chlorhexidene gluconate/alcohol [15]. If the alcohol does not dry completely before venipuncture, it may be inadvertently introduced into the blood sample. This contamination can cause hemolysis or interfere with blood ethanol level measurements [2, 16]. Stronger antiseptics such as betadine (povidone-iodine solution) may be used when stringent infection control is needed, as with the collection of blood cultures or arterial punctures [2]. When disinfecting with betadine, contamination can falsely elevate phosphorus, uric acid, and potassium levels [17]. This is due to the oxidative effects of betadine when using guaiac or toluidine tests, which have been identified as being responsible for false-positive results for hemoglobin in stool samples and for glucose in urine samples [18]. To minimize the interference of these antiseptics, the skin should be completely dry before obtaining blood specimens [3]. For patients with iodine allergies, chlorhexidine gluconate or benzalkonium chloride is available [3]; however, benzalkonium compounds have been observed to affect electrolyte test results [19].

4.5.3.2 Needles

Needles used with evacuation tubes, syringes, catheters, and butterfly systems are composed of various materials, including stainless steel, aluminum, titanium, chromium, iron, manganese, nickel, and alloys [20, 21]. Typically, needles have long, sharp ends (covered by a protective sheath) for puncturing the skin and blood vessels and shorter ends for piercing the rubber stopper of the BCT [20]. Needles are calibrated by gauge, which is inversely related to the needle size [22, 23]. More specifically, needles range from a 13-gauge (1.80 mm internal diameter) to a 27-gauge (0.190 mm internal diameter), with lengths from 3.5 inches (8 cm) for the 13-gauge to 0.25 inches (0.6 cm) for the 27-gauge [23]. In clinical settings, venipuncture is usually performed with needles ranging from 19-gauge (0.686 mm internal diameter) to 25-gauge (0.241 mm internal diameter); 21-gauge needles are the standard for routine venipuncture among adults [22, 24].

One problem encountered with needles is hemolysis, which causes the release of hemoglobin and other intracellular constituents (e. g., potassium, lactate dehydrogenase, alanine aminotransferase, inorganic phosphorus, and magnesium) into the serum or plasma [22, 23, 25]. These analytes can therefore be falsely increased in hemolyzed specimens, whereas albumin, alkaline phosphatase, and sodium can be falsely decreased [26, 27]. Release of phospholipid from fragmented erythrocytes may activate clotting, resulting in shortened clotting times [27]. Free hemoglobin in serum or plasma can interfere with several clinical assays, leading to inaccurate results or necessitating repeat blood draws [28]. Lippi et al [27] found that small-bore needles (25-gauge or smaller) were associated with statistically significant increases in serum potassium and other constituents due to hemolysis [22, 27]. They recommended that small-bore needles be reserved for neonates and patients with poor venous access [23]. Slower flow rates in smaller-bore needles are also associated with increased clotting, occlusion, and test-result variations [23, 28–31]. Large-bore needles (greater than 19-gauge) may cause hemolysis as a result of turbulence from increased non-laminar blood flow [23, 28, 32, 33]. Therefore, it is important to match the needle to the vein size. Under most collection conditions, 21-gauge needles are preferred [2, 34].

Another issue related to needle use is lubricant coating. These coatings serve to reduce
1. Penetration force (the force measured prior to the needle puncture through the tissue),
2. Drag force (the force required to continue tissue penetration), and
3. Pain associated with venipuncture [35, 36].

The most widely used lubricants are silicones, [35, 36] specifically polydimethylsiloxane and curable amino-functional silicone dispersions [36, 37]. Silicone lubricants impart hydrophobicity to the needle to minimize blood and metal contact [36, 37]. Silicone lubricants may prevent drugs from binding with proteins and interfering with chemical reactions or antigen–antibody reactions in immunoassays [34]. Furthermore, agents used to prevent the release of lubricants into blood specimens include gelling agents, [38] polar lubricants (which strongly adhere to surfaces) [39], plasma treatments [40], graft polymerization [41], and ultraviolet photopolymerization [42].

Additionally, there are concerns about metal-based needle components. Chromium, iron, manganese, and nickel can contaminate blood specimens and interfere with subsequent chemical reactions or falsely elevate metal levels in blood [34, 43]. We suggest that metals and alloys used in the construction of needles be tested more thoroughly to evaluate their potential role in the contamination of whole blood, serum, and plasma.

4.5.3.3 Butterfly collection devices

Because of their petite size (21-gauge or 23-gauge), butterfly needles are preferred to conventional venipuncture needles for pediatric patients and when accessing small or fragile veins. US butterfly-device manufacturers include BD (Vacutainer Safety-Lok™) (see also chapter 5.3), Kendall Co. (Angel Wing™), and Wingfield (Shamrock Safety Winged Needle™). The butterfly collection set consists of a stainless-steel needle with a protective shield and plastic wings, which are connected to plastic tubing on one end and, on the opposite end, to a Luer adapter that is inserted into the BCT [22]. The design facilitates the collection of multiple samples, but the short needle length (0.5–0.75 inches) limits the butterfly needle's utility when collecting from surface veins.

Problems arising from the use of butterfly collection devices include increased risks of hemolysis, exposure to blood-borne pathogens, needle-stick injuries, and the incomplete filling of BCTs [23]. It should be noted that no clinically significant differences have been found between test results obtained from specimens collected with butterfly devices and those obtained from specimens collected with straight needles [22].

Hypodermic syringes are preferred when obtaining venous blood specimens from small or fragile veins that may collapse under the forces associated with withdrawing blood into evacuated tubes [44, 45]. This is often the case with elderly patients and neonates. Syringes are also used to collect arterial blood specimens for blood gas analyses, because leaving a vacuum or air space in the collection device affects gas pressure [44, 45]. Syringes may be used to draw specimens from intravenous lines or catheters and are often used for the transfer of blood to collection tubes Although CLSI H3-A6 [46] does not recommend the routine use of needles and syringes for the collection of arterial blood due to increased risk of needle-stick injury and poorer blood specimen quality, [47, 48] they are still commonly used in this capacity.

Syringes for blood collection are typically composed of polypropylene (PP) or polyethylene and include additives and modifiers (e. g. antioxidants, antistatic agents, heat stabilizers, ultraviolet stabilizers, lubricants, and plasticizers) to meet industry standards for production and to improve the sterilization process for plastics [34].

Materials used for stoppers in plastic syringe plungers (the plasticizer di(2-ethylhexl) phthalate) have been shown to contaminate blood specimens and interfere with drug assays [34, 49]. When present in syringe stoppers, 2-mercaptobenzothiazole can be transformed during sterilization into 2-(2-hydroxyethylmercapto)benzothiazole, which interferes with toxicological analyses [34, 50]. Phthalates, including diethyl phthalate, can cause co-migration of gas chromatography peaks, thereby mimicking drugs with similar retention times [34]. To address this issue, some syringe manufacturers have developed barrier films (fluoropolymers) to lubricate syringe components and prevent leaching of vulcanizing agents from the rubber stopper of the syringe plunger [51]. Medical-grade silicone oil, which is applied to the plungers and the inside walls of the syringe barrels to smooth the stopper action, is a component of the device that may affect co-oximetry measurements of different hemoglobin species [52]. To limit potential contamination by syringe lubricants, some manufacturers bake the silicone onto the inside wall of the syringe barrel [51].

A cause for concern regarding blood specimen collection that relates to both human error and the devices themselves is excessive suctioning and forceful plunger depression during blood collection or transfer, which results in shear forces, the breakage of red blood cells (RBCs), and the activation of platelets [16, 32, 53]. Several studies have shown increased in vitro hemolysis when using syringes rather than evacuated tubes for collection [16, 32]. One study indicated that 19 % of syringe-collected specimens were hemolyzed, compared to 3 % of tube-collected specimens. Stankovic and Smith [2] and Ashavaid et al [54] reported that the incidence of hemolysis was 200 times greater in specimens collected with needles and syringes compared to those collected with evacuated tubes. To minimize hemolysis, the syringe plunger can be moved up and down gently to reduce stress on RBC membranes. We recommend the use of a syringe no greater than 20 mL in size when collecting blood for hemostasis testing, which will help minimize clot formation prior to the addition of liquid sodium citrate.

Historically, arterial blood gas specimens were collected in glass syringes (which were impermeable to atmospheric gases) and then placed in an ice slurry for transport to the clinical laboratory [55, 56]. Today, high-density plastic syringes (usually PP) have almost entirely replaced glass syringes in clinical and research laboratories because of their low cost, convenience (single use, disposable, preheparinized), easy-to-use plunger function, and resistance to breakage [57, 58]. Nonetheless, the use of plastics is not without drawbacks. With respect to blood gas measurement, a major drawback of plastic syringes is oxygen (and to a lesser extent, carbon dioxide) permeation [56, 59]. Oxygen can permeate the barrel walls and plunger tips [56, 59]. This gas permeability is influenced by the type of plastic materials used, syringe size (surface-to-volume ratio), and barrel-wall thickness [60, 61]. Numerous studies have reported clinically significant changes in the partial pressure of oxygen (pO_2) in blood gas specimens obtained from glass versus plastic syringes, especially when blood pO_2 is high [55, 56, 58, 62, 63]. A study that compared pore size and density determined that plastic syringes have 4 to 150 times the oxygen diffusion area of glass syringes [64]. Additional studies have shown that initial oxygen levels, oxygen–hemoglobin dissociation, total hemoglobin, length of time between blood draw and analysis, and temperature during storage may also affect oxygen measurements in specimens from plastic syringes [58, 63, 65–67]. CLSI C46-A2 [53] currently recommends that blood gas specimens collected via plastic syringes be kept at room temperature and analyzed within 30 minutes, whereas glass syringes may still be used when analyses will be delayed longer than 30 minutes [68].

A safePICO™ syringe, which has a safe tip cap for removing air bubbles and a soft magnetic steel ball for dissolving anticoagulants, was developed to standardize the mixing of whole-blood specimens for blood gas, electrolyte, metabolite, and hemoglobin measurements on the ABL FLEX™ blood gas analyzer (Radiometer America, Westlake, OH, USA) [69]. Despite concerns that automatic magnetic mixing in the safePICO syringes may hemolyze RBCs and falsely elevate potassium concentrations measured by direct potentiometry, a recent study showed that the results were comparable to those obtained with a hematology analyzer (LH 750™) and chemistry analyzer (LX-20™) [69].

The direct transfer of blood specimens from syringes to BCTs by piercing of the rubber stopper of the tube is a practice that should be avoided. This practice may cause hemolysis when cells under pressure from the plunger collide with the tube wall [28]. It will also activate platelets during transfer and is especially problematic when large-bore needles are used [2, 25]. Other devices for the safe transfer of blood from syringes to BCTs are now commercially available, and we recommend their use [22]. In cases where a syringe must be used to transfer blood to a collection tube, the blood should be added to the indicated volume level (being careful not to overfill or underfill the BCT) to avoid an incorrect blood-to-anticoagulant ratio, which will generate unreliable assay results [22]. Interestingly, in a recent study that evaluated four different brands/types of heparin-coated syringes, Lima-Oliveira et al [70] found statistically and clinically significant differences among four syringes commonly used for blood gas analyte concentrations. These findings show that different brands/types of blood gas syringes can be a significant source of preanalytical variability in blood gas test results.

4.5.3.4 Catheters

Catheters are manufactured from many polymers, including polytetrafluoroethylene (PTFE, Teflon™), polyethylene, PP, polyurethane, silicone, polyether urethane, polyvinylchloride, polyimide, and fluoropolymer [71, 72]. Catheters are used to administer fluids and medications and to withdraw blood and other body fluids [71, 72]. Lubricants are used with catheters to minimize pain, facilitate the catheter's insertion or removal, and permit easy removal of the needle from the catheter [71, 72]. Catheter lubricants include polydimethylsiloxanes, curable and noncurable silicones, silicone surfactants, and lecithin [35].

Blood flowing through catheters is exposed to shear forces, which may modify cell shape, activate cells, damage cells, and cause the efflux of intracellular constituents into the serum [28]; these changes affect hematological, electrolyte, and enzymatic determinations [2, 73, 74]. Unequal diameters of the catheter, adapter device, and stopper-piercing needle may cause varying levels of pressure on RBCs, which can lead to hemolysis during blood collection via conventional evacuated tubes [2]. Specimens obtained by intravenous catheters are up to three times more likely to be hemolyzed than those obtained by venipuncture [75–77]. Hemolysis is most common in smaller (24-gauge to 20-gauge) catheters [75]. Furthermore, air entering evacuated tubes from loose catheter connections or assemblies may also cause hemolysis [75]. If clinicians are unaware of the method of blood collection and its effects on laboratory results, patient care may be complicated by inconsistent values and difficulty determining correct results. For example, benzalkonium heparin, a catheter coating used to prevent blood clot formation and decrease infections, can be released into blood specimens and interfere with ion-sensitive electrodes, thereby falsely elevating sodium and potassium levels [74, 78]. Falsely elevated potassium levels likely result from the interaction of benzalkonium (a monovalent cation) with electrodes [74, 78]. Extensive flushing of the catheter reduces and eventually eliminates this interference, but catheter coatings may leach into blood specimens during initial blood draws immediately after catheter placement [74, 78]. Under these circumstances, we propose that direct venipuncture be used for accurate electrolyte measurements [2].

Many studies have suggested that intravenously administered drugs may be adsorbed into catheter surfaces. This adsorption is of primary concern when blood specimens are drawn soon after a catheter is used for a drug infusion. For example, polyvinyl chloride catheters adsorb drugs such as glycerol trinitrate, hydralazine hydrochloride, thiopental, benzodiazepines [79, 80], and phenothiazines [81], and polyurethane catheters adsorb a variety of drugs [82]. For example, tacrolimus infusion through polyurethane catheters can lead to falsely elevated tacrolimus levels (~ eight times higher than those samples drawn from a peripheral vein) when subsequently drawn through a saline-flushed catheter [83]. Ciclosporine, another drug that interacts with test results, does so by binding to silicone, polyurethane, and silastic material in catheters, even in lines not used for the infusion [71, 72, 84, 85]. In general, specimens for therapeutic drug monitoring should not be obtained from a catheter or lumen previously used for drug infusion, even if the catheter has been flushed.

In this section, we have summarized blood collection devices and components that may lead to preanalytical errors and suggested ways to reduce contamination and assay interference. We now take a look at BCT-associated materials and their history in order to discuss how they have been known to interact with blood specimens and to make recommendations for future improvements.

4.5.4 Blood collection tubes

BCTs consist of tube walls, rubber stoppers, lubricants, anticoagulants, separator gels, clot activators, and SFs, each of which can affect the quality of specimens and the accuracy of laboratory tests (Fig. 4.5.1). Each of these components will be explored in detail, and recommendations for reducing test-result inaccuracies will be presented.

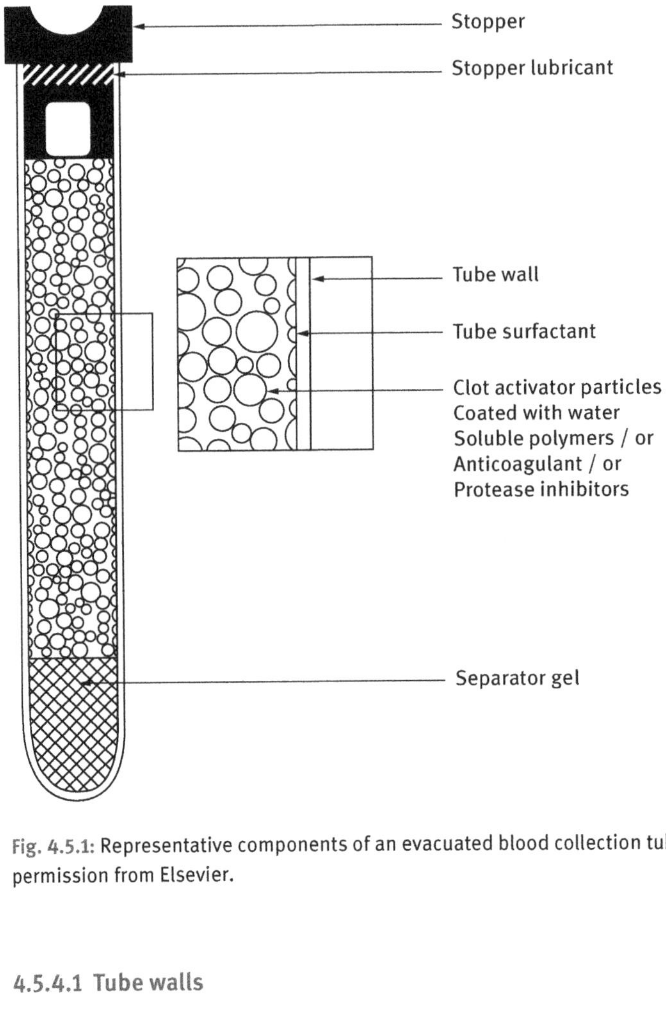

Fig. 4.5.1: Representative components of an evacuated blood collection tube. Reprinted from [4] with permission from Elsevier.

4.5.4.1 Tube walls

Evacuated BCTs are generally cylindrical, measuring 50–150 mm in length and 10–20 mm in diameter [86]. Most tubes for adult clinical specimens are 75–100 mm in length and 13 mm in diameter and are designed to collect 2–10 mL of whole blood [86, 87]. Microcollection tubes for pediatric specimens are 40–50 mm in length and 5–10 mm in diameter [88, 89]. Evacuated tubes were originally made from soda-lime or borosilicate glass; the former were found to release calcium and magnesium into blood specimens [34]. Glass evacuated tubes are manufactured to be airtight, waterproof, and thermally resistant, which allows for vacuum preservation and long shelf life [90]. Contact between certain blood coagulation factors, such as Factor XII (Hageman factor), and hydrophilic glass surfaces activates the clotting cascade [91]. More specifically, hydrophilic glass surfaces cause activation of the contact factors (Factor XII, prekallikrein, and high-molecular-weight kininogen) of the coagulation cascade as well as proteolytic activation of Factor VII, potentially leading to shortened clotting times [92]. Using glass that is siliconized will prevent such activation and is strongly recommended for use in the collection of samples for blood coagulation testing.

With respect to plastic tubes, in the mid-1960s, Sarstedt invented the first plastic blood sampling system (see ref. 10 in Chapter 1.1 and chapter 5.2) and Greiner Bio-One the first plastic evacuated BCTs, which are still in use today (Greiner Bio-One website and chapter 5.1). These plastic tubes replaced most glass tubes following the establishment of the Occupational Safety and Health Administration guidelines for improving safety for laboratorians and reducing their exposure to blood-borne pathogens [93]. Plastic tubes are manufactured through injection-molding using polyethylene terephthalate (PET) and polyolefins (e. g. polyethylene, PP, polyesters, polyacrylic, polytetrafluoroethylene, polysiloxane, polyvinyl chloride, polyacrylonitrile, and polystyrene) [86, 87]. When compared with glass, plastic minimizes exposure to biohazardous materials following breakage, has greater shock resistance, tolerates higher centrifugation speeds, weighs less, has excellent dimensional precision, and is more easily disposed of through low-cost incineration [94, 95]. Plastic does, however, have greater gas permeability compared with glass tubes [96]. There have been numerous studies comparing glass and plastic tubes for use in chemistry [94, 97], endocrinology [97], molecular testing [98], serology [99], and coagulation testing [87, 97]. Although there are small statistically significant differences between plastic and glass tube analyte determinations, none are considered clinically significant.

Plastic collection or aliquot tubes for coagulation testing should be polypropylene (PP) and not polystyrene in composition. Polystyrene may cause contact activation leading to erroneous results in clot-based assays [100]. Polyethylene (PET), a plastic commonly used in the manufacture of BCTs, is unbreakable and maintains a vacuum for a prolonged period of time [101]. PP, another plastic routinely used for BCTs, has lower water permeability than PET, allowing it to retain liquid anticoagulant volume and concentration [101]. Combined PET tubes have double walls to minimize evaporation, especially for coagulation-based tests and the internal PP layer protects against citrate solution evaporation, whereas the outer PET layer is more transparent, allowing easier visualization of tube fill levels [101]. The PP–PET duo improves shelf life and anticoagulant volume retention [101].

Plastic tubes generally have a hydrophobic surface and do not efficiently activate the coagulation process; clots formed on the plastic surfaces of tubes are more gelatinous compared to clots formed in glass tubes [102]. Furthermore, blood does not flow smoothly over hydrophobic plastic surfaces, which can result in the adherence of platelets, fibrin, or clotted blood on the tube walls [102]. This can make it difficult to cleanly separate serum from the blood clot by centrifugation, especially for microcollection tubes or during centrifugation of vacuum tubes. The hydrophilicity of plastic surfaces can be increased by using plasma-enhanced chemical vapor deposition to introduce polar functional groups [103]. Alternatively, the interior plastic surfaces can be coated with SFs – water-soluble polymers or hydrophilic-hydrophobic copolymers [102] – but surfactants may dissolve in blood and interfere with clinical tests [89]. There are ongoing efforts to incorporate SFs into plastic tubes to prevent exudation into blood specimens [86, 87]. For instance, efforts are underway to cross-link the interior surfaces of plastic BCTs with a hydrogel via electron beam or gamma irradiation to permanently bind a hydrophilic coating to the walls of plastic tubes. This technology will improve the integrity of blood specimens; however, the process will be time-consuming and expensive [86, 87]. Increased manufacturing costs will make hydrophilic-coated BCTs more expensive, and production companies may lose their competitive edge in the marketplace if they chose to produce them.

4.5.4.2 Rubber stoppers

Rubber stoppers are routinely color-coded according to anticoagulant type and the presence of a separator gel. The stopper should be readily penetrable by a needle and should self-seal upon needle removal, [102] maintaining the internal pressure differential [102]. Suitable materials include polychloroprene, silicone, styrene butadiene, isobutylene-isopropene, chlorinated ethylene–propylene copolymers, and isobutylene–isoprene rubber [102, 104, 105]. Butyl rubber, a copolymer of isobutylene and isoprene, and halogenated butyl rubber are commonly used materials; [102, 104, 105] butyl rubber exhibits superior air and moisture impermeability, superior resistance to chemical attack, heat resistance, and good process ability [104, 105].

Unfortunately, splatter may occur when rubber stoppers are removed from collection tubes, posing an infectious risk. A stopper shield can be used (e. g. Hemogard™) to prevent this from happening. Ideally, laboratories would use stopper shields made from thermoplastic materials, such as polyethylene, PP, and polyvinylchloride [89, 105].

Discrepancies in the bioavailability and bioequivalence of tests for blood specimens collected via tubes with rubber stoppers containing the plasticizer tris-(2-butoxyethyl)-phosphate (TBEP) have been reported [106]. TBEP, which is used to make stoppers soft, displaces certain drugs from plasma–protein binding sites, such as the α_1-acid glycoprotein, resulting in increased drug uptake by RBCs [107]. This artificially lowers serum or plasma levels. TBEP has also been reported to alter the drug distribution of quinidine, propranolol, lidocaine, tricyclic antidepressants, and several phenothiazine drugs, including fluphenazine and chlorpromazine [108]. In light of this, tube manufacturers have decreased or eliminated production of rubber stoppers containing TBEP. Shah et al [106] and Janknegt et al [109] demonstrated that rubber stoppers made without TBEP do not interfere with therapeutic drug monitoring; however, other stopper components can pose problems. Curry et al [110] reviewed how materials from elastomeric closures, including butyl rubber stoppers, can contaminate specimens with these container closures. Additionally, metals such as calcium, aluminum, magnesium, and zinc are used to manufacture rubber stoppers; it is essential that these metals are not extracted upon contact with blood [94]. Specially formulated rubber stoppers have been developed to limit divalent cation leaching [111]. Sulfur, sulfur-containing vulcanization accelerators, fatty acids, and peroxides in stoppers may also affect lab tests; consequently, most stoppers are manufactured with low-extractable rubber or have been modified to minimize leaching into the blood specimens [108]. The complete filling of BCTs dilutes any leached material and helps reduce these effects [112]. Furthermore, it is recommended that specimens in tubes with rubber stoppers be stored in an upright position and at low temperatures (2–8°C) to minimize leaching [112]. A recent publication by Lippi et al [113] reinforced the importance of maintaining BCTs in a vertical, closure-up position after centrifugation in order to reduce bias in clinical chemistry test results.

Magnesium is another element that can leach from rubber stoppers into sodium citrate BCTs, and it may have a statistically significant influence on the prothrombin time (PT) and international sensitivity index (ISI) of the thromboplastin reagent [94]. The magnitude of the effect depends on the thromboplastin reagent and may result in up to an 8.8 % effect on the ISI, leading to variation in the International Normalized Ratio (INR) result.

Recently, Van den Besselaar et al [114] compared these lower-magnesium-content rubber stopper tubes to conventional tubes with respect to PT and international normalized ratio (INR) results. It was found that BCTs with lower-magnesium-content stoppers resulted in longer PT times and higher INR values than conventional tubes with a higher-magnesium-content rubber stopper [114]. The mechanism for the longer PT and higher INR with lower-magnesium-content rubber stopper tubes is reduced acceleration of tissue factor–induced coagulation, because there is less magnesium available to bind to the γ-carboxyglutamic acid domain of Factor VII/Factor VIIa and Factor X [115, 116]. Thus, tube manufacturers should standardize their BCTs to provide low magnesium content in order to provide accurate and consistent coagulation test results.

4.5.4.3 Stopper lubricants

Lubricants, such as silicone oils, fluids, and glycerol, facilitate the insertion and removal of stoppers [86, 87, 117]. Lubricants minimize RBC and clot adherence to stoppers in order to prevent serum or plasma contamination [86, 87, 117]. It is important to state that glycerol should not be chosen to lubricate stoppers for specimens measuring glycerol or triglyceride when a nonglycerol blank assay is used [118]. Siliconized stoppers are generally preferred because they are less likely to interfere with assays, although silicone may falsely elevate ionized magnesium and total triiodothyronine levels and may confound peaks during mass spectrometry (MS) analysis and peak interpretation [13, 89]. A binding agent like tridodecylmethylammonium chloride (TDMAC) may be applied to the stoppers along with anticoagulants to reduce adhesion of cellular components to the stopper and prevent contamination of the plasma/serum layer in the tube [119]. Moreover, TDMAC does not leach into the blood specimen [119].

4.5.4.4 Anticoagulants

Although serum specimens are used for many assays, plasma is a worthwhile alternative, because plasma samples are required for clot-based hemostasis assays and plasma affords rapid processing times. Containing fibrinogen and other clotting factors, plasma has a higher viscosity and total protein content than serum [120]. Serum has a higher concentration of potassium, activation peptides for coagulation factors, platelet factor 4 (PF4), and platelet components released during platelet activation [120]. Anticoagulants used to preserve analytes may interfere with other analyte determinations [121]. Ethylenediaminetetraacetic acid (EDTA), heparin, and citrate are the most commonly used anticoagulants [46]. The merits of each anticoagulant will be reviewed below.

4.5.4.4.1 Ethylenediaminetetraacetic acid (EDTA)

Potassium EDTA (Table 4.6.1), an anticoagulant and chelating agent, interferes with calcium assays and clot generation [122], but it is preferred for hematology testing [123, 124]. EDTA binds the metallic ions europium (immunoassay reagent), zinc, calcium, and magnesium (enzyme cofactors for immunoassay reagents such as alkaline phosphatase) [125, 126]. Insufficient sample volumes produce relatively elevated EDTA levels, which can increase the chelation of magnesium, calcium, and zinc and consequently affect reagent enzymes used for signal generation, such as alkaline phosphatase [125]. Reagent antibodies recognize divalent cation complex binding sites on proteins; thus, decreased calcium and magnesium levels may induce conformational changes that decrease antibody binding [125].

With respect to the hematology laboratory, EDTA is the anticoagulant of choice, because it preserves the cellular components and morphology of the blood cells. The EDTA (di-potassium, tripotassium) salts must be used as the anticoagulant in BCTs, as opposed to the free acid, which is not soluble in an aqueous media. Dipotassium EDTA is the anticoagulant recommended for hematology by the International Council for Standardization in Hematology and [123, 124]. Dipotassium EDTA is recommended due to its good solubility and stable micro-hematocrit results. K_2EDTA in a concentration of 1.5–2.0 mg/mL (4.1–6.8 mmol/L) of blood does not have any significant effect on the blood count parameter [126].

Table 4.5.1: Evacuated blood collection tube stopper color and additives.

Stopper Color* (Tube Wall Material)	Additive(s)	Amount/Common Applications
Red (glass)	Clot activator Uncoated interior	Serum for chemistry, serology, and blood bank tests
Gold (plastic) Red/black (plastic) Red/gray (plastic)	Clot activator with separator gel Clot activator with separator gel Clot activator with separator gel	Serum for chemistry, serology, and blood bank tests
Orange (plastic)	Thrombin with separator gel	Serum for STAT chemistry tests 10–15 National Institute of Health units per tube[10]
Light blue (plastic)	Sodium citrate (liquid additive) (1 part additive to 9 parts blood)	0.109 mol/L (3.2%) or 0.129 mol/L (3.8%) Plasma for coagulation tests
Dark green (plastic)	Heparin, sodium (dry additive)	10–30 USP units/mL blood plasma for chemistry tests
Light green (plastic)	Heparin, lithium (dry additive) with separator gel	10–30 USP units/mL blood
Green/gray	Heparin, lithium (dry additive) with separator gel	10–30 USP units/mL blood plasma for chemistry tests
Lavender (plastic)	EDTA, dipotassium (dry additive)	1.5–2.2 g/L blood Whole-blood hematology, immunohematology and chemistry testing, molecular viral load
Lavender (plastic)	EDTA, tripotassium (liquid additive)	1.5–2.2 g/L blood
Lavender (plastic)	EDTA, disodium (dry additive)	1.4–2.0 g/L blood
Gray (plastic)	Sodium fluoride/potassium oxalate (dry additive)	Sodium fluoride: 2.5 g/L blood; potassium oxalate: 2.0 g/L blood Plasma or serum for test requiring inhibition of glycolysis
Gray (plastic)	Sodium fluoride/sodium EDTA (dry additive)	Sodium fluoride: 2.5 g/L blood; sodium EDTA: 1.5 g/L blood
Gray (plastic)	Lithium iodoacetate	Iodoacetate: ~2 g/L blood
Yellow (glass)	Acid citrate dextrose – solution A (1 part additive to 5.67 parts blood)	Citrate, disodium, 22.0 g/L; citric acid, 8.0 g/L; dextrose, 24.5 g/L Blood bank, DNA testing
Yellow (glass)	Acid citrate dextrose – solution B (1 part additive to 3 parts blood)	Citrate, disodium, 13.2 g/L; citric acid, 4.8 g/L; dextrose, 14.7 g/L
Royal blue (glass) (with red band on label)	None	Serum for trace element, toxicology, and nutrition tests
Royal blue (glass) (with lavender band on label)	EDTA, dipotassium (dry additive)	~1.8 g/L blood

Table 4.5.1: (continued)

Stopper Color* (Tube Wall Material)	Additive(s)	Amount/Common Applications
Tan (plastic)	EDTA, dipotassium (dry additive)	1.8 g/L blood Whole-blood lead testing
Black (glass)	Sodium citrate	0.105 M (~3.2%) Westegren sedimentation rate
Clear (plastic) Red/light gray (plastic)	None None	Serum for chemistry, serology, and blood bank tests

* Single or multiple stopper color combinations may vary among different tube manufacturers and different countries.
Table modified from [124] and information from Young et al [198] and in the BD website [122].
Reprinted from [4] with permission from Elsevier.

EDTA is an excellent preservative of blood cells and morphology parameters, providing 48-h stability for hemoglobin and 24-h stability for erythrocytes. Because EDTA's hypertonic activity can alter erythrocyte indices and hematocrit, peripheral smears should be made within two or three hours of the blood draw. The white blood cell count remains stable for at least three days in EDTA-anticoagulated blood stored at room temperature. EDTA is also adequate for platelet enumeration; however, morphological changes occur over time [127].

In some individuals, EDTA may cause inaccurate quantitative platelet results, significantly underestimating the results of platelet count. The underestimation of the platelet counts is due to platelet clumping and satellitism, that may occur when the calcium ion is removed by EDTA, allowing for the binding of preformed antibodies. The prevalence of EDTA-induced pseudothrombocytopenia is nearly 0.1 %, and its frequency appears to be higher in thrombocytopenic patients, ranging from 1.25 % to 15.3 % [128]. This form of spurious thrombocytopenia is caused by IgM autoantibodies directed against the glycoprotein IIa and IIIb on the platelet surface. The incidence of autoantibodies causing clumping of platelets due to the presence of EDTA is 1/1000 in humans [129]. EDTA may induce structural morphology and externalization of the glycoproteins IIa and IIIb, triggering the immunological reaction with the IgM autoantibodies [130]. Such problems can be eliminated by re-collecting samples in sodium citrate tubes, ensuring that the proper ratio of 9 parts blood to 1 part anticoagulant is practiced. These new specimens can be analyzed in the usual way by automated instrumentation. Platelet count and white blood cell counts from sodium citrate specimens must be corrected for the dilution of blood with the anticoagulant, using the appropriate dilution factor. All other complete blood count parameters should be reported from the original EDTA tube and slide.

EDTA tubes for hematology testing need to be checked for proper fill volumes, and appropriate action based on hospital protocol should be taken if tubes are underfilled. Gros [131] found significant differences and variations in the draw volume and the EDTA anticoagulant concentration among tubes produced by different blood collection manufacturers, which may contribute to the preanalytical variability in test results. Underfilling the EDTA BCT can lead to erroneously low blood cell counts and hematocrits, morphologic changes to RBCs, and staining alteration. Excess EDTA can shrink cells. Conversely, overfilling the BCT will not allow the tube to be properly mixed and may lead to platelet clumping and clotting. CLSI H01-A6 [124] recommends filling the tube to ± 10 % of the stated draw volume. Thus, BCTs must be able to maintain, for a predetermined minimum life span, a well-defined blood volume capacity by ensuring that there is a certain degree of vacuum still present in the tube compared to the vacuum pre-set at the time of manufacturing. Lima-Oliveira et al [132] reported clinically significant differences in mean platelet volume and platelet distribution width among different brands of BCTs with EDTA. We hypothesize that the differences observed among the tubes are due to preparation, quality, quantity of EDTA, and other tube components. We suggest that laboratories determine normal reference intervals using the same BCT manufacturer that are used to collect patient specimens.

Anticoagulants other than sodium citrate, such as EDTA and heparin, are not acceptable for hemostasis testing. Samples collected in EDTA or heparin tubes and the use of serum for clot-based coagulation testing will lead to aberrant results. In serum samples, Factors VII and IX are activated, resulting in supranormal values, whereas Factors V and VIII are consumed such that PT and activated partial thromboplastin time (aPTT) assays result in no clot detected [11, 133]. EDTA plasma, on the other hand, demonstrates moderate prolongation of the PT and aPTT with significant spurious reduction of Factor VIII and Factor V activities [11, 133]. Importantly, in mixing studies, plasma collected in EDTA shows a factitious inhibitor effect and may lead to spurious identification of a Factor V or Factor VIII inhibitor [90]. The receipt of serum or plasma other than that collected in sodium citrate for the performance of clot-based assays must result in specimen rejection.

4.5.4.4.2 Heparin

Heparin was first discovered in 1916 as an inhibitor of coagulation [134]. Heparin salts (typically from porcine intestinal mucosa) are also extensively used as anticoagulants in BCTs (Table 4.6.1) [135]. The recommended concentration of heparin in blood collection is 10–30 USP units per milliliter of blood. Plastic microcollection tubes may contain less than 15 USP units of heparin per milliliter of blood. Heparin complexes with and induces a conformational change of antithrombin that accelerates the inhibition of thrombin, which inhibits thrombin activation and the generation of fibrin from fibrinogen [125]. Because heparin binds electrolytes and changes the concentration of bound and free ions, [44] manufacturers have created electrolyte-balanced formulations [136]. However, heparin can still interfere with a variety of clinical assays. For example, specimens assayed with Dimension™ Vista 1500 (Siemens Healthcare Diagnostic, Newark, DE, USA) may produce negative anion gaps due to heparin interference with chloride electrode membranes (unpublished observation by author). Heparin also slows some antibody–antigen reaction rates [137], particularly during the precipitation step in second-antibody systems, but this problem can be avoided by using solid-phase systems [138]. Exogenously administered heparin alters serum thyroid hormone levels, and it should be avoided in cryoprotein investigations because it precipitates cryofibrinogen [125, 138]. Furthermore, falsely low albumin levels have been observed when heparinized tubes have been used on hemodialysis patients [139, 140]. It has been proposed that heparin inhibits the binding of bromocresol green to albumin, leading to less colorimetric complex formation [139]. Use of bichromatic procedures eliminates this problem [140]. Proteomic studies have shown that heparinized plasma causes nonspecific protein binding, which influences the separation and mass spectrometry of peptides [141]. Recently, Lippi et al [52] demonstrated that incomplete filling of lithium heparin tubes produced significantly higher creatine kinase and γ-glutamyltransferase activities on a Unicel DxC™ 800 analyzer. Interestingly, Lima-Oliveira et al [142] found clinically significant variations in glucose, creatinine, amylase, aspartate aminotransferase, lactate dehydrogenase, calcium, magnesium, and potassium concentrations in different brands of serum and lithium heparin BCTs. This demonstrates that different brands of lithium heparin BCTs may not be used interchangeably and are a source of preanalytical variability.

Results of hematology tests are often influenced by a number of preanalytical variables, such as the type of anticoagulant used in the BCTs. Heparin is not recommended for the preparation of blood smears when using Wright-Giemsa stain, because it causes a blue background to form on the peripheral smear, although it does not affect cell size or shape [143]. Moreover, when hematology samples are collected, there is frequent platelet clumping that interferes with the morphological interpretation of platelets and platelet count estimates [143].

4.5.4.4.3 Sodium citrate

Used as a liquid anticoagulant, sodium citrate is not suitable for hematological analysis unless the dilution factor is taken into account to avoid reporting erroneously low values. Most samples referred for coagulation testing are drawn into sodium citrate. Trisodium citrate in a 3.2 % (105–109 mmol/L) or 3.8 % (129 mmol/L) solution is the preferred anticoagulant for coagulation testing (Table 4.5.1) [144].

The anticoagulant effect of sodium citrate is attributed to its ability to bind calcium, making the calcium unavailable to promote clot formation. BCTs may contain either 105–109 mM or 129 mM sodium citrate, also referred to as 3.2 % or 3.8 %, respectively [80, 145, 146]. The CLSI H21-A5 [100, 145] guideline favors the use of 3.2 % sodium citrate, [145, 147]. A higher citrate concentration leads to greater calcium binding and longer clotting times [147]. Specimens collected in 129 mM (3.8 %) buffered sodium citrate may overestimate the PT and aPTT and underestimate fibrinogen if the normal range is based on 3.2 % citrated samples [145, 146]. Due to the variation in clotting times and sodium citrate concentration, it is imperative that laboratories standardize to one citrate concentration and develop reference intervals appropriate to it. Four variables peculiar to sodium citrate BCTs are appropriate fill volume, the need for a discard tube, the need to promptly mix the sample, and the need to prevent in vitro clot formation. We will briefly discuss each of these below.

Sodium citrate tubes must be adequately filled (to the mark noted on the tube if provided) or to no less than 90 % of total volume [80, 145–147]. The required ratio of sodium citrate to whole blood is 1:9. Underfilling may cause significant sample dilution due to the volume of liquid anticoagulant, and may also result in falsely prolonged clotting times due to the excess calcium-binding citrate present [80, 145–147]. The degree of this effect depends on the citrate concentration, the tube size, and the test performed. This effect is more pronounced with 3.8 % citrate tubes and in small volume (pediatric) collection tubes [148, 149]. Unless local studies have been performed to demonstrate the acceptability of reduced fill volumes or package inserts provide for such an allowance, sodium citrate tubes filled to less than 90 % of total volume are considered unacceptable for testing [80, 146–149]. Conversely, overfilling of evacuated tubes may occur if the rubber stopper is removed and additional sample is added [145, 148, 149]. This should be avoided, because it may lead to inadequate anticoagulant volume and limited sample mixing potential, with resultant in vitro clot formation. Blood should never be transferred from one collection tube to another in an effort to provide the required complete fill volume [148, 149]. The restriction applies even to the combination of two sodium citrate tubes, because this practice may lead to doubling up of anticoagulant citrate levels and further dilution of the plasma sample [80, 145–149] .

Historically, a discard tube was required prior to collecting a sodium citrate tube for coagulation testing in order to avoid tissue factor contamination and sample activation that may be present in the first tube, but not in subsequent tubes [148, 150, 151]. Studies have demonstrated no effect of BCT with sodium citrate on both routine and special coagulation testing, except in select circumstances, such as when butterfly devices are used or when samples will be subject to platelet function analysis [147, 150, 151]. Butterfly devices cause underfilling of the first-drawn tube because the air space in the tubing partially fills the BCT [152–154].

Blood samples should be procured in a relatively atraumatic fashion, and during collection the blood should flow freely into the collection container. When obtaining plasma for coagulation testing, it is imperative that clotting of the sample in the BCT is avoided. In vitro clots may develop in samples for which the blood is slow to fill the collection container, where there is prolonged use of a tourniquet, or when considerable manipulation of the vein by the needle has occurred [155]. These situations must be avoided. The presence of a clot in the BCT is cause for specimen rejection. The integrity of the sample may be affected even if the clots are not visible to the naked eye [155]. Clot development may result in in vitro consumption of clotting factors, activation of clotting factors, activation of platelets, and platelet granule release; any of these conditions may alter results of hemostasis assays.

Another important way to prevent in vitro clot formation is to adequately and promptly mix the sample following collection to ensure complete distribution of anticoagulant CLSI H21-A5 (2008). When using evacuated collection tubes, three to six complete end-over-end inversions are recommended CLSI H21-A5 [100]. Vigorous shaking is to be avoided in order to prevent inducing hemolysis or spurious platelet and factor activation that may result in shortened clotting times or false elevation of clotting factor activity in specimen tests (e. g. Factor VII) [155].

When sodium citrate is used for chemistry testing, it can inhibit both aspartate aminotransferase and alkaline phosphatase by the chelation of cations [146].

4.5.4.4.4 Potassium oxalate

Potassium oxalate, another calcium-chelating anticoagulant (Table 4.6.1) often combined with antiglycolytic agents (sodium fluoride and sodium iodoacetate), can actually decrease hematocrits by as much as 10 % by drawing water from cells into plasma [145]. Oxalate can also inhibit several enzymes, such as amylase, lactate dehydrogenase, and acid and alkaline phosphatase [145].

Potassium oxalate also shrinks RBCs and is therefore not recommended for packed cell volume (PCV), erythrocyte sedimentation rate (ESR), or generation of peripheral smears. Ammonium oxalate, on the other hand, causes swelling of RBCs and should not be used for PCV, ESR, or peripheral smears. To balance the swelling effects of ammonium oxalate and the shrinking effects of potassium oxalate, the two anticoagulants are combined in a mixture in the ratio of 3 parts ammonium oxalate to 2 parts potassium oxalate (double oxalate). The morphology of the blood cells, however, is not well preserved in this mixture.

4.5.4.4.5 Sodium fluoride

Sodium fluoride (Table 4.5.1) inhibits the glycolytic enzyme enolase and is used to limit the ex vivo consumption of glucose by cells in a collected blood specimen [145]. However, in fluoridated, nonseparated blood samples, glucose is still metabolized at approximately 5–7 % per hour at room temperature (ex vivo glycolysis), because upstream enzymes continue to convert it to glucose-6-phosphate [156]. Hence, complete inhibition of glycolysis in fluoride-containing tubes can take up to 4 h at room temperature with a normal blood cell count [157]. Fluoridated tubes can affect diabetes diagnosis, which uses fixed plasma glucose levels established using blood that was iced and centrifuged and had the plasma removed [158]. In fact, the American Diabetes Association no longer recommends using only sodium fluoride to inhibit in vitro glycolysis [158]. A BCT with EDTA and fluoride in a citrate buffer (pH <5.9) has been proposed to preserve glucose concentrations due to its immediate inhibition of glycolysis if plasma cannot be separated from cells within 30 min [157]. Recently, Norman and Jones [159] showed an increase in glucose results of 14 % (0.80 mmol/L) from fasting patients after switching from fluoride/oxalate to citrate/fluoride/EDTA tubes manufactured by Terumo™. This increase in glucose results was due to the inhibition of glycolysis in erythrocytes by lowering the pH of the blood sample to about 5.5 using a citrate buffer in the tubes. A previous study showed, using extracted data from a laboratory information system, higher glucose results and significantly more patients above the decision limit for diabetes when blood was collected in citrate compared to sodium fluoride tubes [160, 161]. A recent investigation by del Pino et al [160] demonstrated that implementation of citrate buffered tubes during an oral glucose tolerance test caused an increase in positive diagnostic tests, although it was statistically significant only when screening for gestational diabetes mellitus, not for diabetes in general.

Sodium fluoride may be unsuitable for enzymatic immunoassays because of its enzyme inhibitory activity [145]. Fluoride may also interfere with electrolyte measurements by altering cell membrane permeability and promoting hemolysis by RBC ATP with subsequent potassium efflux [145]. Iodoacetate preserves glucose concentrations by inhibiting glyceraldehyde-3-phosphosphate dehydrogenase, but it can cause hemolysis and interfere with the measurement of glucose, sodium, potassium, chloride, and lactate dehydrogenase measurements [145].

Although anticoagulants and antiglycolytics can be unsuitable for certain assays, tube manufacturers do not always specify the plasma sources used to validate their tests. Consequently, it is important that clinical laboratories assess tube performance with their particular assays, instruments, and platforms. Tube manufacturers' fill-volume recommendations should be followed to ensure proper additive-to-blood ratios and to minimize assay interference and resultant laboratory errors, repeat testing, and unnecessary troubleshooting.

4.5.4.4.6 Citrate, theophylline, adenosine, and dypyridamole (CTAD)

Citrate, theophylline, adenosine, and dypyridamole (CTAD) is a cocktail of additives that prevent in vitro platelet activation. These tubes afford more reliable measurement of unfractionated heparin (UFH) levels by diminishing the effect of platelet factor 4, a potent neutralizer of unfractionated heparin (UFH) that is released from platelets [91, 93]. Whole-blood samples containing UFH must be processed within 1 h of collection to allow separation of the cellular fraction from the plasma. Specimens have been reported to remain stable over a 4-h period if they are collected into CTAD and stored at room temperature unprocessed [91]. In general, CTAD tubes are recommended for measurement of platelet-activation markers such as β-thromboglobulin or platelet factor 4 [162]. These tubes may also be useful when determining plasma levels of analytes that have significant platelet stores, such as plasminogen activator inhibitor-1 [163]. In this situation, CTAD prevents release of the analyte from the platelet store during collection and processing, allowing for more accurate measurement of plasma concentration.

4.5.4.5 Separator gels

Separator gels are used to separate serum from clotted whole blood or to separate plasma from cells [164]. In general, serum separator tubes are easy to use, require short processing times, yield higher serum levels, and limit hazardous aerosolization. Separator tubes require only one centrifugation step, allow primary tube sampling, and require a single label [164].

During centrifugation, the thixotropic gel used in these tubes lodges between packed cells and the top serum layer [165]. The position of the gel after centrifugation is influenced by many tube characteristics, such as specific gravity, yield stress, viscosity, density, and tube material [165, 166]. Gel position can also be affected by temperature, centrifugation speed, acceleration and deceleration, storage, low hematocrit, elevated plasma protein, serum/plasma specific gravity, and patient factors, such as heparin therapy [167]. Polymeric gels affect viscosity, density, and other physical properties [102, 167, 168]. Separator gels are typically made from viscous liquids, fillers, or tackifiers with substances like dibenzylidene sorbitol as a gelling agent [102]. The inner tube surface may have a hydrophobic coating to ensure separator gel adherence and a complete barrier to prevent mixing between RBCs and serum/plasma [86, 87]. Because the serum/plasma specific gravity ranges from 1.026 to 1.031 g/cm^3 and the clot specific gravity ranges from 1.092 to 1.095 g/cm^3, the separator gel specific gravity should ideally be between 1.03 and 1.09 g/cm^3 [169]. If the serum/plasma specific gravity is elevated due to hyperproteinemia or radio-contrast dye, the serum may not float above the gel [167]. Fatas et al [170] demonstrated that specific gravity has a more important effect than viscosity on improper BCT gel separation. In addition, Faught et al [166] found differences in separator gel specific gravity in different BCTs and between some tube lots.

Several reports of gels affecting analyte concentrations have been published. Hydrophobic drugs, such as phenytoin, phenobarbital, carbamazepine, quinidine, and lidocaine, can adsorb onto hydrophobic separator gels and lead to a decrease in serum drug concentrations by as much as 20–50 % after 24 h at 4°C [74, 170]. Organochlorine, polychlorinated biphenyl, and progesterone levels may also be significantly reduced [138]. For example, Streete and Flanagan et al [171] reported interference in volatile assays from ethylbenzene and xylene originating in Sarstedt Monovette™ serum gel tubes. Furthermore, Ritter and Mayo [172] identified an interference that was estimated to have caused erroneous claims of hypercalcemia in 5 % of all total calcium laboratory reports. It was speculated that the interference came from the separator gel in Vacuette™ plasma separator tubes using the Arsenazo III dye method on the Architect c8000 (Abbott Laboratories, Abbott Park, IL, USA). Various lots of Vacuette plasma tubes resulted in this interference [173]. Wang et al [173] reported that the separator gel clot activator tubes (KeHua, China) can cause variable increased troponin I results from healthy volunteers on the Vitros ECiQ™ system. They did not determine the mechanism of interference from the tube components. A small but statistically significant difference in myoglobin and CK-MB levels has been reported between tubes with and without separator gels [10]. Interestingly, newer separator gels (e. g. polydimethylsiloxane-polyethylene oxide copolymers) that minimize drug and analyte adsorption have been developed (e. g. the BD SST II™ tube) [86, 87, 175]. RBCs have been observed to surpass the separator gel barrier in plasma, and serum tubes increased plasma/serum potassium levels [175]. Separator gels may also release materials (e. g. gel pieces and silicone oil) into the specimens and spuriously interfere with assays, sample probes, tubes and cuvettes, solid-phase immunoassay systems, and electrode surfaces; the rate of degradation and release may be increased by improper storage or extreme temperatures [138, 176]. Shi et al [177] demonstrated that the separator gel components in some types of BCTs (i. e. SST™ and lithium heparin plasma separator tubes) from a specific tube manufacturer were the source of interference in the quantitation of serum testosterone levels using liquid chromatography-tandem MS. The interference increased according to the length of storage of serum in the tubes and was more pronounced with specimens containing low testosterone levels [177]. Modifications of the assay and liquid chromatography-tandem MS parameters did not resolve the problem of tube interference with the quantitation of serum testosterone levels [177]. Recently, Gounden and Soldin [178] showed that liquid chromatography-tandem mass spectrometry using electrospray ionization for total triiodothyronine and thyroxine was affected by tube type, which may be reflected by the type of separator gel in the BCTs. Thus, new technologies applied in the clinical laboratory to determine analyte concentrations can be significantly affected by BCT components like separator gel. Because no commercially available separator gel is completely chemically inert to all analytes, we suggest that ideal separator gels in BCTs:

1. Not be affected by temperature changes during transportation and storage;
2. Be stable during sterilization (e. g., chemical or irradiation) of the BCT;
3. Not contaminate the blood specimen with separator gel material; and
4. Be truly inert to the blood specimen.

4.5.4.6 Clot activators and water-soluble agents

Plastic tubes require clot activators that use either intrinsic or extrinsic coagulation pathways to ensure rapid and dense clot formation [87]. We will describe each process. Clot activation by the intrinsic pathway is surface dependent, and a greater density of activating surface sites speeds clotting time. Siliceous substances (e. g. glass, silica, kaolin, bentonite, diatomaceous earth) accelerate clot formation through contact activation [102], but particulate clot activators work relatively slowly (30–60 min) [87]. The amount of clot activator provided in BCTs varies by manufacturer [102]. Clot activators also diminish latent fibrin formation in the separated serum [139]. Silica, which is commonly used in BCTs, is preferably made hydrophobic, because it is highly dispersible and prevents hemolysis as a result of its decreased solubility in blood, thus allowing blood to coagulate in a shorter time [179].

Clot activation by the extrinsic pathway occurs when coagulation is initiated by the addition of substances extrinsic to blood, such as thrombin, snake venoms, thromboplastin, thiol proteases such as cathepsin B and ficin, and hydrolases such as metal proteases, including kininase I [87]. Although these clot activators produce rapid clotting (10–20 min) compared to those discussed above, the clots formed are gelatinous and do not easily separate from serum [87]. Clot activators can be added to tubes by inserting small beads or paper-coated discs, or they can be sprayed on interior tube surfaces with a carrier (e. g. polyvinylpyrrolidone [PVP], carboxymethyl cellulose, polyvinyl alcohol, and polyethylene oxide) [87, 102]. These carriers allow for rapid clot activator suspension into blood so that the carriers dissolve into both serum and clots as the clotting is initiated [102]. PVP and water-soluble SFs also release clot activators into blood specimens to reduce the need for mixing [102]. BD has recently released a serum tube containing thrombin (RST™; Table 4.6.1, orange stopper) for rapid clot activation (within five minutes). Dimeski et al [180] demonstrated that the use of RST™ tubes would not be appropriate for patients on high-dose heparin or warfarin therapy because latent clot formation in the tube may clog instrument probes and produce erroneous test results. Based on these findings, it is clear that additional studies are needed to ensure that RST™ tubes give results clinically equivalent to those given by other commercially available serum tubes, especially when tubes are only partially filled. Snake venom prothrombin activators may be used to resolve some of the aforementioned issues [181]. Some tube manufacturers may add stabilizers, such as β-alanine, glycine, L-serine, protamine, benzamidine, and L-tryptophan, for the clot-promoting enzymes (e. g. thrombin) [182].

Some clot activators are problematic in that they must be thoroughly mixed to allow complete pelleting with the clot. If soluble fibrin clots form, they can interfere with pipetting-device accuracy or in solid-phase binding in immunoassays [183]. To minimize these problems, plasma gas treatment (e. g. plasma-enhanced chemical vapor deposition) may be used to modify the tube wall surface chemistry by adding polar functional groups that are clot activating [103]. Through this method, clotting is accelerated but no particulate or soluble clotting activators or binders are present to contaminate the serum specimen.

Various studies have revealed the impact of clot activators on laboratory test performance. Sampson et al [184] found that silica and silicone surfactant (SF) are associated with elevated lithium concentrations when using the Lytening™ 2Z (Lytening, West Peabody, MA, USA) ion-specific electrode analyzer. The clot activators or silicone

SFs can interact with ion-specific analyzer membranes, which increase the measured voltage and falsely elevate serum lithium ion concentration. Clot activators can also falsely elevate serum testosterone measurements, but changing the ion pair eliminates this problem [185]. Proteome analysis by mass spectrometry may also be altered by clot activators [186]. Silica and silicate clot activators, when sprayed onto plastic tubes, induce the release of pro-, active, and complexed matrix metalloproteinases [187]. Recently, it was shown that levels of ficolin-1 and ficolin-2, a group of proteins that can activate the complement pathway and binding capacities, were significantly affected, presumably by the silicate material in SST™ tubes [188]. Hence, it is critical that the optimal amounts and composition of clot activators and water-soluble agents be determined and consistently added to different types and sizes of BCTs in order for these substances to function properly without adversely affecting the quality of the blood specimens and test results. Ultimately, there is a need to develop plastic BCTs with enhanced rates of clot formation that do not leave any soluble or particulate material in the serum layer or in the clot during centrifugation, thereby avoiding potential interference with clinical tests.

4.5.4.7 Surfactants

Surfactants (SFs) are commonly used to decrease nonspecific adsorption, and they must be carefully selected and optimized for immunoassays because, in high concentrations, they may cause the loss of antibodies through their passive adsorption onto the solid support beads used in immunoassays [189]. Commercially available tubes contain a variety of SFs that can improve blood flow, distribute clot activator, and prevent proteins, RBCs, and platelets from adsorbing to tube walls [13, 14, 87]. For example, Terumo™ has coated the inner wall of their BCTs with a water-soluble silicone material (SF8421™, manufactured by Torei Silicone Kabushiki Kaisha, Japan) to prevent the adhesion of RBCs, blood, clots, and other cellular material to the inner tube wall [190]. Some other tube manufacturers may use water-soluble silicone oil to prevent clots from attaching to the inner walls of tubes and stoppers [179]. The silicone oil can be made water-soluble by the introduction of polar groups to the molecule [179]. Examples of such polar groups include hydroxyl, amino, carboxyl, epoxy, and ether groups [179].

Unfortunately, silicone SF-coated tubes have been shown to interfere with ion-specific electrode measurement of ionized magnesium and lithium [13, 184]. Silicone SFs seem to interact with ion-specific electrode membranes to increase the measured voltage during magnesium and lithium determinations [13, 184]. In addition, water-soluble silicone polymer coatings in separator tubes can physically mask antibodies and alter avidin-biotin binding reactions in immunoradiometric assays [191].

Bowen et al [14] demonstrated that the nonionic polydimethylsiloxane – polyethylene oxide and polypropylene oxide graft copolymer SF, Silwet™ L-720 (Figure 2; OSI Specialties, Danbury, CT, USA) in BD SST™ tubes falsely elevates

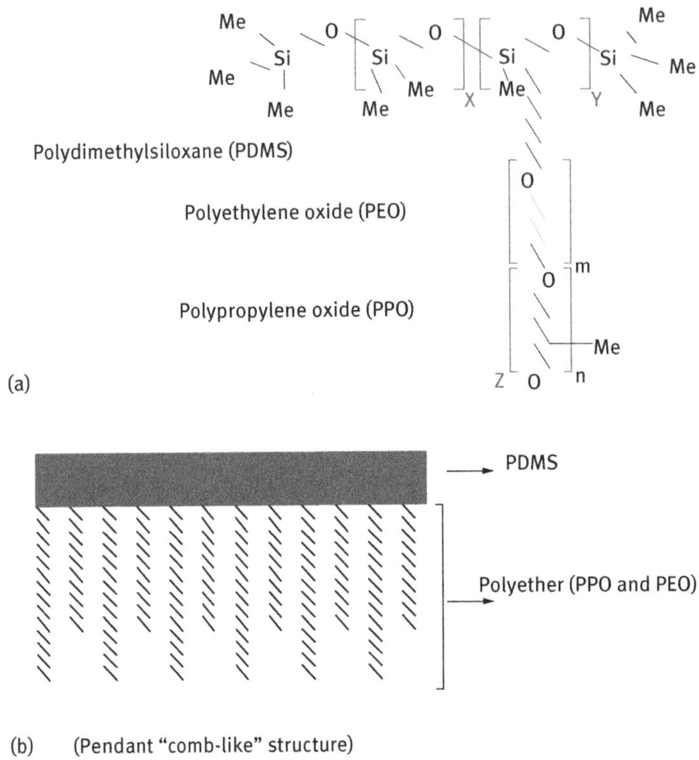

Fig. 4.5.2: Silwet™ silicone surfactant. (a) general molecular formula and (b) schematic structure with polyether (polyethylene oxide and polypropylene oxide) attached (via hydrosilation reaction) to the polydimethylsiloxane backbone. x, y, m, n are integers independently greater than zero; z can be hydrogen or alkyl radical [86].

triiodothyronine in a dose-dependent manner by causing the desorption of captured antibodies from the solid phase used in immunoassays [13, 14]. Competitive immunoassays (e. g. vitamin B_{12}) and noncompetitive immunoassays (e. g. cancer antigen 15–3] are also affected by Silwet L-720™, but the mechanism is unclear [13, 14].

Due to immunoassay interference, BD reformulated their tubes to reduce SF levels [14]. Morovat et al [192] showed statistically significant but clinically insignificant biases in immunoassay results using these reformulated tubes; yet, in this study, the control tubes were coated with the problematic SF. Wang et al [105] reported that the reformulated tubes produced clinically significant biased results for free triiodothyronine and free thyroxine.

Further studies have examined whether additive supply molecules may interfere with MS components, such as SF and polyvinylpyrrolidone, which produced multiple MS signals in the range of 1,000–3,000 m/z. These tube additive peaks complicate and compromise mass spectra interpretation in the low molecular mass range, particularly for matrix-assisted laser desorption ionization (MALDI) or surface enhanced laser desorption ionization (SELDI) techniques [13]. Tube additives may also affect the ionization process during liquid chromatography-MS analysis, thereby suppressing metabolite ionization. Yin et al [193] recently reported that five different S-Monovettes™ (Sarstedt, Nümbrecht, Germany and Newton, NC, USA) BCT additives produced chemical noise in the mass spectra that interfered with metabolic profiling. Thus, an initial step in MS investigations should be the examination of BCTs.

SF detergent properties can also alter cell membrane permeability and lipophilic structures [194]. One study showed that SFs in tubes affected free fatty acid concentrations in specimens rather than interfering with their analytical detection [195]. There is a need to develop BCTs with SFs covalently bonded to the interior wall to prevent leaching and possible interference with diagnostic tests. For example, a BCT manufactured to minimize assay interference could have an internal tube wall surface modified via graft polymerization of various monomers, induced by UV, plasma, ozone, and radiation, physical or chemical treatments [196]. However, manufacturing BCTs like this would be very laborious, expensive, and time-consuming.

Interestingly, Ottinger [197] reported that reformulated SST™ tubes with decreased concentrations of SF resulted in significantly elevated serum potassium levels ranging from 0.1 to 0.4 mmol/L compared to glass red-top tubes. The data used in Ottinger's study was analyzed using different statistical methods. In fact, the author showed that the reference interval for potassium would have to be changed depending on the tube type used by the laboratory [197]. This is not surprising, because decreasing the concentration of SF in the BCTs increases the possibility of RBC, protein, and/or platelet adherence to the tube walls with subsequent release of intracellular constituents like potassium into the serum/plasma layer. Therefore, producing BCTs with SFs that do not contaminate the blood specimens and cause assay interferences would be warranted.

4.5.5 Blood Collection Tube Storage Conditions

Evacuated BCTs are not completely empty; they contain a small amount of gas (air) at low pressure [165]. The pressure of the gas within the evacuated tube determines the draw volume based on the ideal gas law [165]. The state of the gas within the tube is determined by ambient pressure, temperature, and volume [165]. The blood volume capacity of a BCT is therefore affected by these conditions. For instance, lower fill volumes can occur when tubes are maintained and drawn at high altitude (≥5,000 ft) and when they are maintained at high temperature [165]. High temperature may also negatively influence tube additives such as the biochemicals and gel, leading to degradation [165]. Moreover, very high humidity may lead to the accumulation of water vapor inside the tube; conversely, low humidity may result in the escape of liquid additive [165]. Some tube additives, such as CTAD, are photosensitive and degraded by exposure to light [165]. Additionally, evacuated plastic tubes lose vacuum over time, and this will affect draw volume. Given the effects of environmental conditions and time, BCTs should be stored under appropriate conditions and used as specified [165]. At the very least, BCTs should be used within their expiration dates, because additives within the tubes may have limited shelf lives.

Acknowledgements: The authors would like to thank Ms. Krista Tanquary and Ms. Raven Bowen for editing and reviewing the manuscript.

References

[1] Forsman RW. Why is the laboratory an afterthought for managed care organizations? Clin Chem 1996; 42:813–6.
[2] Stankovic AK, Smith S. Elevated serum potassium values: the role of preanalytic variables. Am J Clin Pathol 2004; 121 Suppl:S105–12.
[3] Green SF. The cost of poor blood specimen quality and errors in preanalytical processes. Clin Biochem 2013; 46:1175–9.
[4] Bowen RA, Hortin GL, Csako G, Otanez OH, Remaley AT. Impact of blood collection devices on clinical chemistry assays. Clin Biochem 2010; 43:4–25.
[5] Lilienfeld AM, Bross ID, Sartwell PE. Observations on an outbreak of infectious hepatitis in Baltimore during 1951. Am J Public Health 1953; 43:1085–96.
[6] Boeynaems JM, De Leener A, Dessars B, add up to three authors before et al. Evaluation of a new generation of plastic evacuated blood-collection tubes in clinical chemistry, therapeutic drug monitoring, hormone and trace metal analysis. Clin Chem Lab Med 2004; 42:67–71.
[7] Funderburk JV, inventor "System for lubricating a syringe barrel", US patent 5,456,940. October 1995.
[8] Assembly and method to improve vacuum retention in evacuated specimen containers. 2009 (Accessed April, 2013, at www.freshpatents.com/-dt20090625ptan20090162587.php).
[9] MacNutt MJ, Sheel AW. Performance of evacuated blood collection tubes at high altitude. High Alt Med Biol 2008; 9:235–7. is Josef Kleiner named here?
[10] Daves M, Trevisan D, Cemin R. Different collection tubes in cardiac biomarkers detection. J Clin Lab Anal 2008; 22:391–4.
[11] Banfi G, Salvagno GL, Lippi G. The role of ethylenediamine tetraacetic acid (EDTA) as in vitro anticoagulant for diagnostic purposes. Clin Chem Lab Med 2007; 45:565–76.
[12] Bush Vj, Leonard L, Szamosi DI. Advancements in blood collection devices. Lab Med 2008; 29:616–22.
[13] Drake SK, Bowen RA, Remaley AT, Hortin GL. Potential interferences from blood collection tubes in mass spectrometric analyses of serum polypeptides. Clin Chem 2004; 50:2398–401.
[14] Bowen RA, Chan Y, Ruddel ME,add 3 authors before et al. Immunoassay interference by a commonly used blood collection tube additive, the organosilicone surfactant silwet L-720. Clin Chem 2005; 51:1874–82.
[15] Kiechel F, Funk AD, Rosse R, Sesok-Pizzin iD. So you're going to collect a blood specimen. An introduction to phlebotomy College of American Pathologists press; 2013.
[16] Burns ER, Yoshikawa N. Hemolysis in serum samples drawn by emergency department personnel versus laboratory phlebotomists. Lab Med 2002; 33:377–80.
[17] Meites S. Skin-puncture and blood-collecting technique for infants: update and problems. Clin Chem 1988; 34:1890–4.
[18] Blebea J, McPherson RA. False-positive guaiac testing with iodine. Arch Pathol Lab Med 1985; 109:437–40.

[19] Bach PR. When a sodium is not a sodium. Pediatr Lab Med 1991; 9:5.
[20] Higaki Y, Uchida Y, inventors; "Blood collection needle". US-patent 7,267,653. 2007.
[21] Sunderman FW, Jr., Crisostomo MC, Reid MC, Hopfer SM, Nomoto S. Rapid analysis of nickel in serum and whole blood by electrothermal atomic absorption spectrophotometry. Ann Clin Lab Sci 1984; 14:232–41.
[22] Lippi G, Salvagno GL, Guidi GC. No influence of a butterfly device on routine coagulation assays and D-dimer measurement. J Thrombos Haemostas 2005; 3:389–91.
[23] Lippi G, Salvagno GL, Montagnana M, Brocco G, Cesare Guidi G. Influence of the needle bore size used for collecting venous blood samples on routine clinical chemistry testing. Clin Chem Lab Med 2006; 44:1009–14.
[24] Ernst DJ, Ernst C. Phlebotomy tools of the trade. Home Healthcare Nurse 2002; 20:151–3.
[25] Sharp MK, Mohammad SF. Hemolysis in needleless connectors for phlebotomy. Am Soc Artificial Internal Organs 2003; 49:128–30.
[26] Yucel D, Dalva K. Effect of in vitro hemolysis on 25 common biochemical tests. Clin Chem 1992; 38:575–7.
[27] Lippi G, Montagnana M, Salvagno GL, Guidi GC. Interference of blood cell lysis on routine coagulation testing. Arch Pathol Lab Med 2006; 130:181–4.
[28] Sharp MK, Mohammad SF. Scaling of hemolysis in needles and catheters. Ann Biomed Engineering 1998; 26:788–97.
[29] Wilcox GJ, Barnes A, Modanlou H. Does transfusion using a syringe infusion pump and small-gauge needle cause hemolysis? Transfusion 1981; 21:750–1.
[30] Laessig RH, Hassemer DJ, Paskey TA, Schwartz TH. The effects of 0.1 and 1.0 per cent erythrocytes and hemolysis on serum chemistry values. Am J Clin Pathol 1976; 66:639–44.
[31] Prabhu S, Kazarian T, Hakobyan N, Jabbar A, Dunham T, Valentino LA. Needles and needleless devices for infusion of anti-haemophilic factor concentrate: impact on protein structure and function. Haemophilia 2006; 12:58–61.
[32] Carraro P, Servidio G, Plebani M. Hemolyzed specimens: a reason for rejection or a clinical challenge? Clin Chem 2000; 46:306–7.
[33] Savory J, Bill JG. Hemolysis of specimens drawn in the ER? [Q&A?]. Lab Med 1996; 27:802.
[34] Narayanan S, Lin FC. Sampling technique. In: Wong SHY, ed. Therapeutic drug monitoring and toxicology by liquid chromatography 1985:79–88.
[35] Khan AJ, Khan MA, inventors; "Medical device lubricant containing lecithin" US patent 5,824,359. 1998.
[36] Steube GA, inventor Medical device lubricant comprising radiation curable silicon material US-patent 7,354,628. 2008.
[37] Pelkey BJ, inventor Lubricant system for hypodermic needles and method for its application US-patent 5,911,711. June 1999.
[38] Williams JL, Shepard TA, inventors; Method for lubricating articles US-patent 5,186,972. 1993.
[39] Williamitis VA, Spielvogel DE, inventors; Lubricated metal articles and assembly containing same US-patent 5,185,006. February 1993.
[40] Hu CM, Solom on DD, Williamitis VA, inventors; Method for preparing lubricated surfaces US-patent 4,842,889. 1989.
[41] Uyama Y, Tadokoro H, Ikada Y. Surface lubrication of polymer films by photo induced graft polymerization J Appl Polym Sci 1990; 39:489–98.
[42] Kim BS, Hrkach JS, Langer R. Biodegradable photo-crosslinked poly(ether-ester) networks for lubricious coatings. Biomaterials 2000; 21:259–65.
[43] Versieck J, Cornelis R. Trace elements in human plasma or serum. Boca Raton: CRC Press; 1989.
[44] Young DS, Bermes EW, Haverstick DM. Specimen collection and other preanalytical variables. In Fundamentals of Clinical Chemistry. Philadelphia: Elsevier Saunders; 2008.

[45] Lippi G, Blanckaert N, Bonini P, Green S, Kitchen S, Palicka V et al. Haemolysis: an overview of the leading cause of unsuitable specimens in clinical laboratories. Clin Chem Lab Med 2008; 46:764–72.
[46] Clinical Laboratory Standrads Institute (CLSI). Procedures for the collection of diagnostic blood specimen by venipuncture. Approved standard. 6th ed. Wayne, PA: CLSI Document H3-A6 2007.
[47] McGuff J, Popovsky MA. Needlestick injuries in blood collection staff. A retrospective analysis. Transfusion 1989; 29:693–5.
[48] Bilski B. Needlestick injuries in nurses – the Poznan study. Int J Occup Med Environm Health 2005; 18:251–4.
[49] Pike E, Skuterud B, Kierulf P, Fremstad D, Abdel Sayed SM, Lunde PK. Binding and displacement of basic, acidic and neutral drugs in normal and orosomucoid-deficient plasma. Clin Pharmacokin 1981; 6:367–74.
[50] Petersen MC, Vine J, Ashley JJ, Nation RL. Leaching of 2-(2-hydroxyethylmercapto) benzothiazole into contents of disposable syringes. J Pharmaceut Sci 1981; 70:1139–43.
[51] Lehandro B. The next generation of prefillable syringes: specialized plastics lead the way. On Drug Delivery 2007:7.p?
[52] Clinical Laboratory Standards Institute (CLSI). Blood gas and pH analysis and related measurements; approved guideline – 2nd ed. Wayne PA: Document C 46-A2, 2007.
[53] Verresen L, Lins RL, Neels H, De Broe ME. Effects of needle size and storage temperature on measurements of serum potassium. Clin Chem 1986; 32:698–9.
[54] Ashavaid TF, Dandekar SP, Keny B, Bhambhwani VR. Influence of blood specimen collection method on various preanalytical sample quality indicators. Ind J Clin Biochem 2008; 23:144–9.
[55] Eldrige F, Fretwell LK. Change in oxygen tension of shed blood at various temperature. J Appl Physiol 1965; 20:790–2.
[56] Scott PV, Horton JN, Mapleson WW. Leakage of oxygen from blood and water samples stored in plastic and glass syringes. Brit Med J 1971; 3:512–6.
[57] Winkler JB, Huntington CG, Wells DE, Befeler B. Influence of syringe materials on arterial blood gas determinations. Chest 1974; 66:518–21.
[58] Smeenk FW, Janssen JD, Arends BJ, et al. Effects of four different methods of sampling arterial blood and storage time on gas tensions and shunt calculation in the 100 % oxygen test. Europ Respiratory J 1997; 10:910–3.
[59] Rosenberg E, Price N. Diffusion of CO_2 and O_2 through the walls of plastic syringes. Can J Physiol Pharmacol 1973; 51:608–11.
[60] Müller-Plathe O, Heyduck S. Stability of blood gases, electrolytes and haemoglobin in heparinized whole blood samples: influence of the type of syringe. Europ J Clin Chem Clin Biochem 1992; 30:349–55.
[61] Wu EY, Barazanji KW, Johnson RL, Jr. Source of error on A-aDO2? (aCO_2?) calculated from blood stored in plastic and glass syringes. J Appl Physiol 1997; 82:196–202.
[62] Evers W, Racz GB, Levy AA. A comparative study of plastic (polypropylene) and glass syringes in blood-gas analysis. Anesthes Analgesia 1972; 51:92–7.
[63] Pretto JJ, Rochford PD. Effects of sample storage time, temperature and syringe type on blood gas tensions in samples with high oxygen partial pressures. Thorax 1994; 49:610–2.
[64] Wiwanitkit V. Glass syringes are better than plastic for preserving arterial blood gas for oxygen partial pressure determination: an explanation based on nanomaterial composition. Int J Nanomed 2006; 1:223–4.
[65] Schmidt C, Mueller-Plathe O. Stability of pO_2, pCO_2 and pH in heparinized whole blood samples: influence of storage temperature with regard to leukocyte count and syringe material. Eur J Clin Chem Clin Biochem 1992; 30:767–73.

[66] Mahoney JJ, Harvey JA, Wong RJ, Van Kessel AL. Changes in oxygen measurements when whole blood is stored in iced plastic or glass syringes. Clin Chem 1991; 37:1244–8.
[67] Knowles TP, Mullin RA, Hunter JA, Douce FH. Effects of syringe material, sample storage time, and temperature on blood gases and oxygen saturation in arterialized human blood samples. Respiratory care 2006; 51:732–6.
[68] Clinical Laboratory Standards Institute (CLSI). Procedures for the handling and processing of blood specimens; approved guideline. 3rd ed. Wayne PA: Document 2004.
[69] Benoit MO, Paul JL. Evaluation and advantages of an automatic magnetic mixing of syringes integrated to a whole blood gas analyser. Scand J Clin Lab Invest 2009; 69:628–32.
[70] Lima-Oliveira G, Lippi G, Salvagno GL, Montagnana M, Picheth G, Guidi GC. Different manufacturers of syringes: a new source of variability in blood gas, acid-base balance and related laboratory test? Clin Biochem 2012; 45:683–7.
[71] Busca A, Miniero R, Vassallo E, Leone L, Oddenino O, Madon E. Monitoring of cyclosporine blood levels from central venous lines: a misleading assay? Therap Drug Monitor 1994; 16:71–4.
[72] Duffner U, Bergstraesser E, Sauter S, Bertz H, Niemeyer C. Spuriously raised cyclosporin concentrations drawn through polyurethane central venous catheter. Lancet 1998; 352:1442.
[73] Raisky F, Gauthier C, Marchal A, Blum D. Haemolyzed samples: responsibility of short catheters. Ann Biol Clin 1994; 52:523–7.
[74] Koch TR, Platoff G. Suitability of collection tubes with separator gels for therapeutic drug monitoring. Therap Drug Monit 1990; 12:277–80.
[75] Kennedy C, Angermuller S, King R, et al. A comparison of hemolysis rates using intravenous catheters versus venipuncture tubes for obtaining blood samples. J Emerg Nurs 1996; 22:566–9.
[76] Grant MS. The effect of blood drawing techniques and equipment on the hemolysis of ED laboratory blood samples. J Emerg Nurs 2003; 29:116–21.
[77] Tanabe P, Kyriacou DN, Garland F. Factors affecting the risk of blood bank specimen hemolysis. Acad Emerg Med 2003; 10:897–900.
[78] Gaylord MS, Pittman PA, Bartness J, Tuinman AA, Lorch V. Release of benzalkonium chloride from a heparin-bonded umbilical catheter with resultant factitious hypernatremia and hyperkalemia. Pediatrics 1991; 87:631–5.
[79] Parker WA, Morris ME, Shearer CA. Incompatibility of diazepam injection in plastic intravenous bags. Am J Hospit Pharm 1979; 36:505–7.
[80] Cloyd JC, Vezeau C, Miller KW. Availability of diazepam from plastic containers. Am J Hospit Pharm 1980; 37:492–6.
[81] Jostell KG. Interaction between clomethiazole and pediatric intravenous administration sets. Acta Pharmac Sueica 1983; 3:61–7.
[82] Smith JC, Davies MC, Melia CD, Denyer SP, Derrick MR. Uptake of drugs by catheters: the influence of the drug molecule on sorption by polyurethane catheters. Biomaterials 1996; 17:1469–72.
[83] Grouzmann E, Buclin T, Biollaz J. Misleading tacrolimus concentration value in blood taken from a catheter used for tacrolimus administration. Am J Health-Sys Pharm 2008; 65:226–8.
[84] Blifeld C, Ettenger RB. Measurement of cyclosporine levels in samples obtained from peripheral sites and indwelling lines. New Engl J Med 1987; 317:509.
[85] Claviez A, Glass B, Dreger P, Suttorp M. Elevated blood drug levels obtained from indwelling silicon catheters during oral cyclosporine A administration. Bone marrow transplant 2002; 29:535–6.
[86] Vogler EA, Shepard TA, Graper JC, inventors; Blood collection assembly including mechanical phase separating USpatent 5,53,518. 1996.
[87] Dubrowny NE, Harrop AJ, inventors; "Collection device", US-patent 6,686,204. 2004.
[88] Burns JA, inventor Blood microcollection tube assembly. US-patent 5,288,466. 1994.
[89] Bowen RA, Chan Y, Cohen J, et al. Effect of blood collection tubes on total triiodothyronine and other laboratory assays. Clin Chem 2005; 51:424–33.

[90] Hulon W, inventor Blood collection tube assembly. US- patent 5,297,561. 1994.
[91] Gorbet M, Sefron M. Biomaterial-associated thrombosis: roles of coagulation factors, complement platelets and leukocytes. Biomaterials 2004; 25:5681–703.
[92] van den Besselaar AM, Hoekstra MM, Witteveen E, Didden JH, van der Meer FJ. Influence of blood collection systems on the prothrombin time and international sensitivity index determined with human and rabbit thromboplastin reagents. Am J Clin Pathol 2007; 127:724–9.
[93] Ernst DJ. Plastic collection tubes decrease risk of employee injury. Med Lab Obs 2001; 33:44–6, 8, 50.
[94] van den Besselaar AM, van Dam W, Sturk A, Bertina RM. Prothrombin time ratio is reduced by magnesium contamination in evacuated blood collection tubes. Thrombos Haemostas 2001; 85:647–50.
[95] Anderson DR, Wiseman J, MacLeod J, Burton E, Zayed E. Evaluation of polyethylene terephthalate for ABO and Rh typing and alloantibody screening. Transfusion 2000; 40:669–72.
[96] Kratz A, Stanganelli N, Van Cott EM. A comparison of glass and plastic blood collection tubes for routine and specialized coagulation assays: a comprehensive study. Arch Pathol Lab Med 2006; 130:39–44.
[97] Smets EM, Dijkstra-Lagemaat JE, Blankenstein MA. Influence of blood collection in plastic vs. glass evacuated serum-separator tubes on hormone and tumour marker levels. Clin Chem Lab Med 2004; 42:435–9.
[98] Landry ML, Garner R, Ferguson D. Use of plastic vacutainer tubes for quantification of human immunodeficiency virus type 1 in blood specimens. J Clin Microbiol 2001; 39:354–6.
[99] Feng CS, Wan CP. The plastic tube compared with the glass tube in blood bank tests. Pathology 1988; 20:309–10.
[100] Clinical and Laboratory Standards Institute (CLSI). Collection, transport and processing of blood specimens for testing plasma-based coagulation assays and molecular hemostasis assays; approved guideline, 5th ed. Wayne, PA: Clinical and Laboratory Standards Institute; Document H21-A5,2008.
[101] Flanders MM, Crist RA, Rodgers GM. A comparison of blood collection in glass versus plastic vacutainers on results of esoteric coagulation assays. Lab Med 2003; 34:732–5.
[102] Anraku H, Shoji Y, inventors; Vacuum blood collection tubes. US-patent 8,565,334. 1989.
[103] Montgomery DB, Vogler EA, inventors; Tube having regions of different surface chemistry and method. US-patent 5,318,806. 1994.
[104] Percarpio EP, Holderith WJ, Wyckoff RW, inventors; Vacuum indicator closure for a blood collection tube. US-patent 4,293,078. 1979.
[105] Wang S, Ho V, Roquemore-Goins A, Smith FA. Effects of blood collection tubes, including pediatric devices, on 16 common immunoassays. Clin Chem 2006; 52:892–3.
[106] Shah VP, Knapp G, Skelly JP, Cabana BE. Interference with measurements of certain drugs in plasma by a plasticizer in vacutainer tubes. Clin Chem 1982; 28:2327–8.
[107] Borga O, Piafsky KM, Nilsen OG. Plasma protein binding of basic drugs. I. Selective displacement from alpha$_1$-acid glycoprotein by tris (2-butoxyethyl) phosphate. Clin Pharmacol Ther 1977; 22:539–44.
[108] Cotham RH, Shand D. Spuriously low plasma propranolol concentrations resulting from blood collection methods. Clin Pharmacol Ther 1975; 18:535–8.
[109] Janknegt R, Lohman JJ, Hooymans PM, Merkus FW. Do evacuated blood collection tubes interfere with therapeutic drug monitoring? Pharm Weekbl Sci 1983; 5:287–90.
[110] Curry W, Conway S, Goodfield C, Miller K, Mueller RL, Polini E. Reducing the risk of contamination of sterile parenteral products via ready-to-use closure components. Am Assoc Pharm SciTech 2010; 11:1572–9.
[111] Cummings J. Evacuated tubes for monitoring heparin treatment. J Clin Pathol 1981; 34:939–40.

[112] Cohen RR, inventor Evacuated sample collection tube with aqueous additive. US-patent 5,860,937. 1999.
[113] Lippi G, Salvagno GL, Danese E, Lima-Oliveira G, Brocco G, Guidi GC. Inversion of lithium heparin gel tubes after centrifugation is a significant source of bias in clinical chemistry testing. Clin Chim Acta 2014; 436c:183–7.
[114] van den Besselaar AM, van Vlodrop IJ, Berendes PB, Cobbaert CM. A comparative study of conventional versus new, magnesium-poor Vacutainer(R) Sodium Citrate blood collection tubes for determination of prothrombin time and INR. Thromb Res 2014; 134:187–91.
[115] Vadivel K, Agah S, Messer AS et al. Structural and functional studies of gamma-carboxyglutamic acid domains of factor VIIa and activated Protein C: role of magnesium at physiological calcium. J Mol Biol 2013; 425:1961–81.
[116] Persson E, Ostergaard A. Mg(2+) binding to the Gla domain of factor X influences the interaction with tissue factor. J Thromb Haemostas 2007; 5:1977–8.
[117] Bush VJ, Janu MR, Bathur F, Wells A, Dasgupta A. Comparison of BD Vacutainer SST Plus Tubes with BD SST II Plus Tubes for common analytes. Clin Chim Acta 2001; 306:139–43.
[118] Chowdhury FR, Rodman H, Bleicher S. Glycerol-like contamination of commercial blood sampling tubes. J Lipid Res 1971; 12:116.
[119] Crouther R, inventor Sherwood Medical INdustries INc., assignee. Anticoagulant Stopper Coating. US-patent 4,308,232. 1981.
[120] Ladenson JH, Tsai LM, Michael JM, Kessler G, Joist JH. Serum versus heparinized plasma for eighteen common chemistry tests: is serum the appropriate specimen? Am J Clin Pathol 1974; 62:545–52.
[121] Sevastos N, Theodossiades G, Efstathiou S, Papatheodoridis GV, Manesis E, Archimandritis AJ. Pseudohyperkalemia in serum: the phenomenon and its clinical magnitude. J Lab Clin Med 2006; 147:139–44.
[122] Becton Dickinson www.bd.com May 2013.
[123] International Council for Standardization in Haematology. Recommendations of the ICSH for ethylenediamine tetraacetic acid anticoagulants of blood for blood cell counting and syzing. Expert panel of cytometry. Am J Clin Pathol 1993; 100:371–2.
[124] Clinical and Laboratory Standards Institute (CLSI). Tubes and Additives for Venous and Capillary Blood Specimen Collection; Approved Standard – 6th ed. Wayne, PA: Clinical and Laboratory Standards Institute; Document H 01-A6.2010.
[125] Tate J, Ward G. Interferences in immuno assay. Clin Biochem Rev 2004; 25:105–20.
[126] Lewis SM, Stoddart CT. Effects of anticoagulants and containers (glass and plastic) on the blood count. Lab Pract 1971; 20:787–92.
[127] Green D, McMahon B, Foiles N, Tian L. Measurement of hemostatic factors in EDTA plasma. Am J Clin Pathol 2008; 130:811–5.
[128] Silvestri F, Virgolini L, Savignano C, Zaja F, Velisig M, Baccarani M. Incidence and diagnosis of EDTA-dependent pseudothrombocytopenia in a consecutive outpatient population referred for isolated thrombocytopenia. Vox Sanguinis 1995; 68:35–9.
[129] Chiurazzi F, Villa MR, Rotoli B. Transplacental transmission of EDTA-dependent pseudothrombocytopenia. Haematologica 1999; 84:664.
[130] Banfi G, Germagnoli L. Preanalytical phase in haematology. J Med Biochem 2008; 27:348–53.
[131] Gros N. Evacuated blood-collection tubes for haematological tests – a quality evaluation prior to their intended use for specimen collection. Clin Chem Lab Med 2013; 51:1043–51.
[132] Lima-Oliveira G, Lippi G, Salvagno GL, Montagnana M, Poli G, Solero GP et al. K_3 EDTA vacuum tubes validation for routine hematological testing. ISRN Hematology 2012; 2012:875357.http://dx.doi.org/10.5402/2012/875357.

[133] Crist RA, Gibbs K, Rodgers GM, Smock KJ. Effects of EDTA on routine and specialized coagulation testing and an easy method to distinguish EDTA-treated from citrated plasma samples. Lab Hematol 2009; 15:45–8.
[134] Barrowcliffe TW. History of heparin. Handbook of Experimental Pharmacology 2012:3–22.
[135] CuhadarS. Preanalytical variables and factors that interfere with the biochemical parameters: a review. OA Biotechnology 2013; 2:19.
[136] Toffaletti JG, Wildermann RF. The effects of heparin anticoagulants and fill volume in blood gas syringes on ionized calcium and magnesium measurements. Clin Chim Acta 2001; 304:147–51.
[137] Zaninotto M, Mion M, Altinier S, Forni M, Plebani M. Quality specifications for biochemical markers of myocardial injury. Clin Chim Acta 2004; 346:65–72.
[138] Wild D JR, Sheehan C. The Immunoassay Handbook. 4th ed. Oxford, UK: Elsevier, 2013.
[139] Meng QH, Krahn J. Lithium heparinised blood-collection tubes give falsely low albumin results with an automated bromcresol green method in haemodialysis patients. Clin Chem Lab Med 2008; 46:396–400.
[140] Hallbach J, Hoffmann GE, Guder WG. Overestimation of albumin in heparinized plasma. Clin Chem 1991; 37:566–8.
[141] Tammen H, Schulte I, Hess R, et al. Peptidomic analysis of human blood specimens: comparison between plasma specimens and serum by differential peptide display. Proteomics 2005; 5:3414–22.
[142] Lima-Oliveira G, Lippi G, Salvagno GL, Montagnana M, Picheth G, Guidi GC. Preanalytical management: serum vacuum tubes validation for routine clinical chemistry. Biochem Med (Zagreb) 2012; 22:180–6.
[143] Houwen B. Blood film preparation and staining procedures. Clin Lab Med 2002; 22:1–14, v.
[144] Adcock DM, Hoefner DM, Kottke-Merchant K, Marlar RA, Szamosi DI, Warunek DJ. Collection, Transport, and Processing of Blood Specimens for Testing Plasma-Based Coagulation Assays and Molecular Hemostasis Assays: Approved Guideline.5th ed. Wayne, PA: Clinical Laboratory Standards Institute; 2008.
[145] Narayanan S. The preanalytic phase. An important component of laboratory medicine. Am J Clin Pathol 2000; 113:429–52.
[146] Narayanan S. Preanalytical aspects of coagulation testing. Haematologica 1995; 80:1–6.
[147] Adcock DM, Kressin DC, Marlar RA. Effect of 3.2 % vs 3.8 % sodium citrate concentration on routine coagulation testing. Am J Clin Path 1997; 107:105–10.
[148] Adcock DM, Kressin DC, Marlar RA. Minimum specimen volume requirements for routine coagulation testing: dependence on citrate concentration. Am J Clin Pathol 1998; 109:595–9.
[149] Chuang J, Sadler MA, Witt DM. Impact of evacuated collection tube fill volume and mixing on routine coagulation testing using 2.5-ml (pediatric) tubes. Chest 2004; 126:1262–6.
[150] Raijmakers MT, Menting CH, Vader HL, van der Graaf F. Collection of blood specimens by venipuncture for plasma-based coagulation assays: necessity of a discard tube. Am J Clin Pathol 2010; 133:331–5.
[151] Smock KJ, Crist RA, Hansen SJ, Rodgers GM, Lehman CM. Discard tubes are not necessary when drawing samples for specialized coagulation testing. Blood Coagul & Fibrinol 2010; 21:279–82.
[152] Lippi G, Salvagno GL, Montagnana M, Lima-Oliveira G, Guidi GC, Favaloro EJ. Quality standards for sample collection in coagulation testing. Semin Thrombosis and Hemostasis 2012; 38:565–75.
[153] Favaloro EJ, Lippi G, Raijmakers MT, Vader HL, van der Graaf F. Discard tubes are sometimes necessary when drawing samples for hemostasis. Am J Clin Pathol 2010; 134:851.
[154] Kunicki TJ, Williams SA, Salomon DR et al. Genetics of platelet reactivity in normal, healthy individuals. J Thrombos Hemostas 2009; 7:2116–22.
[155] Ernst DJ, Ernst C. Phlebotomy tools of the trade: part 4: proper handling and storage of blood specimens. Home Healthcare Nurse 2003; 21:266–70.

[156] Li G, Cabanero M, Wang Z et al. Comparison of glucose determinations on blood samples collected in three types of tubes. Ann Clin Lab Sci 2013; 43:278–84.
[157] Peake MJ, Bruns DE, Sacks DB, Horvath AR. It's time for a better blood collection tube to improve the reliability of glucose results. Diabet Care 2013; 36:e2.
[158] Gambino R. Sodium fluoride: an ineffective inhibitor of glycolysis. Ann Clin Biochem 2013; 50:3–5.
[159] Norman M, Jones I. The shift from fluoride/oxalate to acid citrate/fluoride blood collection tubes for glucose testing - the impact upon patient results. Clin Biochem 2014; 47:683–5.
[160] del Pino IG, Constanso I, Mourin LV, Safont CB, Vazquez PR. Citric/citrate buffer: an effective antiglycolytic agent. Clin Chem Lab Med 2013; 51:1943–9.
[161] Ridefelt P, Akerfeldt T, Helmersson-Karlqvist J. Increased plasma glucose levels after change of recommendation from NaF to citrate blood collection tubes. Clin Biochem 2014;47:625–8.
[162] Toulon P, Abecassis L, Smahi M, Ternisien C. Monitoring treatments with unfractionated heparin: CTAD must be used instead of citrate as the anticoagulant solution when using partial-draw collection tubes. Results of a multicenter evaluation. Thrombos Res 2010; 126:536–42.
[163] Eriksson E, Tengborn L, Risberg B. The effect of various anticoagulant/antiplatelet mixtures on determination of plasminogen activator inhibitor, platelet proteins and hemostasis parameters. Thrombos Haemostas 1989; 61:511–6.
[164] Gigliello JF, Kragle HA, inventors; Method and apparatus for multiphase fluid collection and separation patent 3,920,549. 1975.
[165] Bush V, Cohen R. The Evolution of Blood Colection Tubes. Lab Med 2003; 34:304–10.
[166] Faught RC, Marshall J, Bornhorst J. Solution densities and estimated total protein contents associated with inappropriate flotation of separator gel in different blood collection tubes. Arch Pathol Lab Med 2011; 135:1081–4.
[167] Spiritus T, Zaman Z, Desmet W. Iodinated contrast media interfere with gel barrier formation in plasma and serum separator tubes. Clin Chem 2003; 49:1187–9.
[168] Gerin F, Ramazan DC, Baykan O, Sirikci O, Haklar G. Abnormal gel flotation in a patient with apperant pneumonia diagnosis: a case report. Biochem Med(Zagreb) 2014; 24:180–2.
[169] Fatas M, Franquelo P, Franquelo R. Anomalous flotation of separator gel: density or viscosity? Clin Chem 2008; 54:771–2.
[170] Bergqvist Y, Eckerbom S, Funding L. Effect of use of gel-barrier sampling tubes on determination of some antiepileptic drugs in serum. Clin Chem 1984; 30:465–6.
[171] Streete PJ, Flanagan RJ. Ethylbenzene and xylene from Sarstedt Monovette serum gel blood-collection tubes. Clin Chem 1993; 39:1344–5.
[172] Ritter D, Mayo MM. Erroneous detection of hypercalcemia in specimens stored in Greiner Bio-One Vacuette Plasma Separator Tubes and analyzed by the Arsenazo III methodology. Arch Pathol Lab Med 2009; 133:1363–4.
[173] Wang F, Wang J, Zhang Z, et al. Falsely elevated troponin I attributed to collection tubes using the Vitros ECiQ system. Clin Chem Lab Med 2009; 47:1577–8.
[174] Chance J, Berube J, Vandersmissen M, Blanckaert N. Evaluation of the BD Vacutainer PST II blood collection tube for special chemistry analytes. Clin Chem Lab Med 2009; 47:358–61.
[175] Babic N, Zibrat S, Gordon IO, Lee CC, Yeo KT. Effect of blood collection tubes on the incidence of artifactual hyperkalemia on patient samples from an outreach clinic. Clin Chim Acta 2012; 413:1454–8.
[176] Ji SQ, Evenson MA. Effects of contaminants in blood-collection devices on measurements of therapeutic drugs. Clin Chem 1983; 29:456–61.
[177] Shi RZ, van Rossum HH, Bowen RA. Serum testosterone quantitation by liquid chromatography-tandem mass spectrometry: interference from blood collection tubes. Clin Biochem 2012; 45:1706–9.

[178] Gounden V, Soldin SJ. Clinical use of reference intervals derived from some CALIPER studies questioned. Clin Chem 2014; 60:416–7.
[179] Okamoto R, Togawa K, Isogawa H, inventors; Blood Coagulation Promoter and Blood Collection Tube. US- patent EP 1 619 497 A1. 2006.
[180] Dimeski G, Masci PP, Trabi M, Lavin MF, de Jersey J. Evaluation of the Becton-Dickinson rapid serum tube: does it provide a suitable alternative to lithium heparin plasma tubes? Clin Chem Lab Med 2010; 48:651–7.
[181] Masci P, de Jersey J, Lavin M, Phillips J, inventors; Serum Preparation. US-patent 0,273,583 A1. 2013.
[182] Setoguchi Y, Okamoto R, Togawa K, inventors; Blood Collection Container. US-patent EP 2 320 208 B1. 2013.
[183] Ng WY, Yeo CP. Thrombin-accelerated quick clotting serum tubes: an evaluation with 22 common biochemical analytes. Adv Hematol 2013; 2013:769479.
[184] Sampson M, Ruddel M, Albright S, Elin RJ. Positive interference in lithium determinations from clot activator in collection container. Clin Chem 1997; 43:675–9.
[185] Wang C, Shiraishi S, Leung A, et al. Validation of a testosterone and dihydrotestosterone liquid chromatography tandem mass spectrometry assay: Interference and comparison with established methods. Steroids 2008; 73:1345–52.
[186] Pilny R, Bouchal P, Borilova S et al. Surface-enhanced laser desorption ionization/time-of-flight mass spectrometry reveals significant artifacts in serum obtained from clot activator-containing collection devices. Clin Chem 2006; 52:2115–6.
[187] Mannello F. Serum or plasma samples? The "Cinderella" role of blood collection procedures: preanalytical methodological issues influence the release and activity of circulating matrix metalloproteinases and their tissue inhibitors, hampering diagnostic trueness and leading to misinterpretation. Arterioscler Thromb Vasc Biol 2008; 28:611–4.
[188] Brady AM, Spencer BL, Falsey AR, Nahm MH. Blood collection tubes influence serum ficolin-1 and ficolin-2 levels. Clin Vaccine Immunol 2014; 21:51–5.
[189] Gadow A, Fricke H, Strasburger CJ, Wood WG. Synthesis and evaluation of luminescent tracers and hapten-protein conjugates for use in luminescence immunoassays with immobilised antibodies and antigens. A critical study of macro solid phases for use in immunoassay systems, Part II. J Clin Chem Clin Biochem 1984; 22:337–47.
[190] Kasai M, inventor Blood Collecting Tube. US-patent 5,033,476. 1991.
[191] Wickus GG, Mordan RJ, Mathews EA. Interference in the Allegro immunoassay system when blood is collected in silicone-coated tubes. Clin Chem 1992; 38:2347–8.
[192] Morovat A, James TS, Cox SD et al. Comparison of Bayer Advia Centaur immunoassay results obtained on samples collected in four different Becton Dickinson Vacutainer tubes. Ann Clin Biochem 2006; 43:481–7.
[193] Yin P, Peter A, Franken H, et al. Preanalytical aspects and sample quality assessment in metabolomics studies of human blood. Clin Chem 2013; 59:833–45.
[194] Heppel LA, Makan N. Methods for rapidly altering the permeability of mammalian cells. J Supramolecular Structure 1977; 6:399–409.
[195] Bowen RA, Vu C, Remaley AT, Hortin GL, Csako G. Differential effect of blood collection tubes on total free fatty acids (FFA) and total triiodothyronine (TT3) concentration: a model for studying interference from tube constituents. Clin Chim Acta; 2007; 378:181–93.
[196] Wang X, Zhao G, Teng W, Song H, He H. Blood Compatibility of PET Fabric Modified by Surface Grafting. J Fiber Bioengin Inform 4 2011:329–36.
[197] Ottinger W. Q & A. Lab Med 2006; 37:40–1.
[198] Bush V, Cohen R. The Evolution of Blood Collection Tubes. Lab Med.2003; 34:304–310.
[199] Young DS, Bermes EW, Haverstick DM. Specimen collection and processing. In Tietz: Textbook of Clinical Chemistiry and Molecular Diagnostics. 4th ed. St. Louis: Elsevier Saunders; 2006.

Raffick A. R. Bowen, Dorothy M. Adcock-Funk

4.6 Influences and Interferences from Blood Sampling Device Materials on Clinical Assays: II Special Devices and Procedures; Recommendations

4.6.1 Order of draw

The importance of the "order of draw" in obtaining accurate laboratory tests has been known for many decades. Calam and Cooper [1] demonstrated that the initial drawing of blood into potassium-EDTA tubes falsely decreased calcium values and increased potassium levels in blood collected into consecutive tubes containing no anticoagulants [1]. These alarming findings prompted the establishment of the Clinical and Laboratory Standards Institute (CLSI), which created guidelines to standardize tube sequences and syringe usage for blood collection to minimize the carryover of tube additives [2]. When laboratories switched from glass to plastic tubes, the CLSI order of draw guideline changed, because plastic serum tubes were considered equivalent to gel separator tubes with clot activators [2]. The current CLSI guideline for glass and plastic tubes with respect to order of draw is as follows: blood culture tubes; sodium citrate tubes; serum tubes with and without clot activator and with or without gel separator; heparin tubes with or without gel separator; EDTA tubes; tubes containing acid citrate dextrose; and glycolytic inhibitor (fluoride, iodoacetate) tubes [2]. The adoption of order of draw guidelines has recently been questioned, and studies by Salvagno et al [3] and Fukugawa et al [4] demonstrated negligible effects of the order of draw on sample quality for some routine chemistry and coagulation tests [3, 4]. Sulaiman et al [5] investigated whether incorrect order of draw of blood samples during phlebotomy causes in vitro EDTA contamination of blood samples. The findings from this study showed that the order of draw using the Sarstedt Safety Monovette™ system had no effect on serum biochemistry results. Sulaiman et al [5] also suggested that the order of draw of blood specimens during phlebotomy should be as follows: blood culture/sterile tubes, plain tubes, gel tubes, sodium citrate tubes, lithium heparin tubes, EDTA tubes, and finally fluoride/EDTA or fluoride/oxalate tubes. Overall, most tube manufacturers color-code tube closures for easy identification of tube additives. To improve accuracy in testing, laboratorians will need to incorporate the associated additives, proper order of draw, and carryover effects of additives on clinical assays into their practice.

4.6.2 Protease inhibitors

Protease inhibitors are among the most abundant plasma protein components, far outnumbering active proteases except where activation occurs by surfaces or other stimuli [6]. Chelating agents, such as EDTA and citrate, do not directly inhibit serine proteases, but they do limit the activation of proteases in the coagulation system by interfering with calcium-mediated reactions. Direct inhibitors of thrombin or coagulation Factor Xa serve as alternative anticoagulants, but they have not achieved widespread use because of cost [7]. Such products, however, can increase protein stability and allow chemistry and hematology tests on a single specimen. Small bioactive peptides such as parathyroid hormone and insulin are more stable in EDTA-anticoagulated plasma compared to citrate-anticoagulated plasma or serum [8]. Because aprotinin increases the stability of brain-type natriuretic peptides, some reference laboratories recommend the collection of specimens for bioactive peptide analysis in tubes containing aprotinin or other protease inhibitors [9]. Many peptides, such as glucagon-like peptide 1, undergo rapid cleavage by the exopeptidase dipeptidyl peptidase IV; thus, collection tubes must contain exopeptidase inhibitors to recover the intact peptide [10]. EDTA-containing tubes are generally recommended for proteomic analyses to minimize protein component changes [11]; small peptide components can also undergo rapid degradation by exopeptidases [12]. Yet, the addition of chemically reactive protease inhibitors, such as sulfonyl halides, can covalently modify proteins [12]. An alternative approach is to inhibit protease activity by decreasing pH [12]. In general, small peptides are less stable than proteins, because proteases sequestered in an α_2-macroglobulin inhibitor retain peptidolytic activity even though they are sterically hindered from cleaving full-size proteins [13]. Furthermore, peptides lack a globular structure and are more accessible to exopeptidase action. Although endogenous protease inhibitors are quite abundant in plasma, most are found against serine-dependent endoproteases and exhibit relatively little activity against exopeptidases. Therefore, the addition of exogenous, low-molecular-weight protein inhibitors or small synthetic compounds to blood specimens is a common way to stabilize samples.

Protease activity may be accentuated by the release of intracellular proteases from white or red blood cells (RBCs). For example, insulin is substantially less stable in hemolyzed blood because of the thiol proteases from RBCs [14]. The use of protease inhibitors has a limited effect on the recovery of chemokines and cytokines from plasma, but the rapid processing of blood can limit this problem, because most cytokines and chemokines are degraded by intracellular protease [15].

The addition of exogenous protease inhibitors depends on the intended use of specimens. There is wide variability in protein and peptide stability, and as a result, each laboratory ought to analyze the stability of components of interest; where protein or peptide stability problems are identified, protease inhibitors may be considered. Fortunately, manufacturers have developed blood collection systems (e. g. BD P100™) containing a cocktail of protease inhibitors that enable preservation of plasma proteins for proteomic investigation (Becton Dickinson website: www.bd.com). Additionally, the use of sodium citrate with protease inhibitor(s) such as D-phenylalanine – proline – arginine – chloromethylketone can protect the integrity of plasma samples from protease activity prior to performing nonroutine coagulation assays [16]. The package insert associated with the special evacuated tube should be referenced to determine the assays for which these different anticoagulant tubes are recommended. If these special collection tubes are used for routine hemostasis assays, reference range must be determined based on samples collected in the same type of special collection tube in order to understand potential matrix effect.

4.6.3 Microcollection devices

Analytical instrumentation advancements allow many diagnostic tests to be performed on small quantities of blood (those obtained by spring-loaded puncture of the finger, heel, or earlobe). Microcollection with capillary tubes and microcollection tubes is typically used for infants, geriatric patients, and those with veins not amenable to venipuncture [17, 18]. Various sizes, volumes, and shapes of capillary tubes are commercially available with or without heparin, EDTA, and citrate [18]. To minimize breakage, shattering, and exposure to blood-borne pathogens and to provide flexibility, a Mylar™ film is added to the glass or plastic tubes, although plastic capillary tubes are recommended [19]. Microcollection tubes have virtually replaced Caraway/Natelson tubes, which cannot be individually labeled, must be cut open to separate the serum from the RBCs, and produce lower serum yields [20]. Microcollection tubes can be designed to protect neonatal specimens from the visible light degradation of bilirubin (amber-colored tubes) and may include an integrated collection scoop to improve capillary blood collection [21]. Manufacturers of microcollection tubes include BD (Franklin Lakes, NJ, USA), Kendall Co. (Mansfield, MA, USA), Sarstedt Inc. (Nuembrecht, Germany and Newton, NC, USA), and Greiner Bio-One (Kremsmuenster, Austria and Monroe, NC, USA), to name a few [22, 23]. Compared to larger evacuated blood collection tubes (BCTs), the collection, handling, and processing of blood specimens from microcollection devices is more time-consuming [23, 24].

Plastic microcollection devices are recommended to reduce the risk of injury and blood exposure (Food and Drug Administration website [25]). The tube wall is usually made from clear thermoplastics like polypropylene (preferred), polyethylene, and polyvinylchloride so that the blood is easily visualized by the health-care professional [26]. The statistically significant, but clinically insignificant, differences in analyte levels collected by microcollection versus evacuated tubes may be attributable to the tube wall material [24]. The disadvantages of microcollection include lower serum or plasma yields and increased hemolysis from the flicking of the blood down the tube [27].

As with venous blood specimens, platelets, fibrin, and clots in capillary blood may adhere to the plastic tube walls [26, 28]. This may be enhanced in microcollection because of the smaller tube diameters [26, 28]. Therefore, microcollection tubes may be coated with surfactants (SFs) to enhance the blood flow into the tube and minimize protein and cell adhesion to the tube wall [26, 29]. The same immunoassay interference from SFs in SST™ tubes that occurs with venous blood may also occur with capillary specimens in microcollection tubes [30]. Separator gels in microcollection tubes are the same as those used in venous BCT, and studies have shown that microcollection tubes with or without separator gel are suitable for specimens intended for clinical assays, including therapeutic drug levels [20, 23, 24, 29]. Although plastic screw caps are commonly used to cover microcollection tubes for transport, centrifugation, and storage, there is no indication that these materials interfere with clinical chemistry assays.

The effect of anticoagulants in microcollection devices has not been well described [31]. Two recent neonatal cases showed that BCTs containing lithium heparin resulted in elevated serum lithium concentrations [32]. Underfilling of the microcollection devices can lead to erroneously high lithium levels, in the toxic range [33]. Thus, health-care personnel should be aware of the importance of proper filling, mixing, and additive use with respect to microcollection devices. Manufacturers and laboratory technicians must be aware of all components of microcollection devices and understand their potential effects on clinical assays.

Recently, Microtube for Automated Process (MAP™) tubes has been developed. These devices are specially designed, 13x75 mm plastic tubes with penetrable color-coded closure and, for direct barcode labeling, a larger physical area than conventional microcollection tubes; they are amenable to streamlined processing on automated hematology instruments [34]. These newly designed MAP™ BCTs have resulted in improved turnaround times for hematologic profiles on neonatal specimens [34]. In clinical laboratory tests, these types of BCTs can potentially be used to reduce blood volume and improve the timeliness of test results.

4.6.4 Molecular testing

Advances in scientific research and technologies have resulted in significant improvements both in patient health conditions for which genetic defects can be detected with molecular methods and in the spectra of the molecular testing methods. As the number of molecular genetic tests performed on patients has risen, so too have the number of laboratories that perform such testing. As discussed above, exogenous contaminants can be inadvertently introduced during specimen collection from sources such as the tube components. It is critical that specimens are collected in appropriate containers and that the containers' interactions with blood are noted and adjusted for. This will prevent tube components from interfering with molecular test results in unpredictable ways. It is also important that samples are labeled with the correct patient identification information and that they are free of contamination from other specimens.

Whole blood is the most common source for DNA and RNA isolation in clinical laboratories. For the collection of DNA, a variety of BCTs can be used, but EDTA and acid citrate dextrose are preferred because they are the most compatible for molecular testing [35]. Sodium citrate is also acceptable, as is frozen whole blood [35, 36]. Gel separators cannot be used when the source of the DNA is within the cellular constituents and can be used only when nucleic acid is present in the plasma. However, a recent study by Sun et al [37] demonstrated that the formulation of gel separators in some commercially available BCTs creates an imperfect barrier between plasma and cellular layers. These findings have important implications for molecular testing, especially cell-free DNA and RNA analysis in which it is critical that intracellular nucleic acids do not contaminate the plasma layer, which will lead to potentially erroneous molecular test results that are used for diagnosis, prognosis, and monitoring therapy. It has been reported that BCTs with heparin should not be used for amplification reactions because heparin is a potential inhibitor of polymerase enzymes, leading to false-negative test results [35]. However, this finding has been disputed, because equivalent results have been demonstrated when samples are collected into either heparin or EDTA [38].

Blood plasma is extremely high in ribonuclease (RNase) activity, which must be minimized for successful RNA isolation. This can be achieved by using BCTs designed for this purpose or by adding an appropriate RNA stabilizing reagent. For example, PAXgene™ (Qiagen™, Venlo, Limburg, Netherlands) blood RNA tubes can be used, because they contain additives to stabilize in vivo gene transcription and have already been used in several studies for transcript analysis of blood samples [39]. Introduction of these new BCT additives will require investigation regarding their influence on specific molecular assays.

Cell-free DNA (cf DNA) has been extensively studied over the past few decades, and many studies have investigated the use of cf DNA as a biomarker in various clinical fields such as prenatal diagnosis and oncology [40]. Several techniques have been developed to detect and characterize cf DNA, including direct sequencing, cold PCR, digital PCR, high-resolution melting analysis, and restriction fragment length polymorphism [41]. Numerous cancer studies using cf DNA as a possible biomarker have shown conflicting data [42]. The discrepancies in cf DNA concentrations among these studies may be due to preanalytical factors, because there is no standard operating procedure for cf DNA analysis [43]. BCTs have been evaluated as a source of the variability [44]. The results of such studies have demonstrated that BCT additives can alter both the quantity and the quality of cfDNA in blood specimens [44]. An excellent review paper examined the preanalytical factors affecting cf DNA analysis [45]. During the last several years, there has been a burgeoning interest in circulating microRNAs (miRNAs) as potential novel biomarkers in many clinical fields. About 22 nucleotides in length, miRNA is a type of single-stranded, noncoding, small ribonucleic acid, located within introns of protein-coding genes that regulate gene expression by causing a block in translation or mRNA degradation [46]. A complete understanding of the impact of BCT components as a source of preanalytical variability on miRNA measurements is needed in order to obtain accurate and reliable results.

The laboratory plays a key role in validating the compatibility and acceptability of all products and methods for specified molecular tests. It is the responsibility of each laboratory to determine the equivalency of test results before switching to new collection devices. Validation, as a formal requirement to meet accreditation, is necessary when converting from one BCT type to another or when switching from using serum to plasma or vice versa. CLSI has published a step-by-step guide to help with validation of BCTs [47].

4.6.5 Proteomic studies

Clinical proteomics focuses on the identification and quantitation of protein disease biomarkers in biological fluids or tissue samples. Such biomarkers can function as indicators of abnormalities before clinical symptoms arise, enabling initiation of therapy at early stages. Analyzing the proteome requires specialized quantitative proteomic technologies and expertise. There are different techniques for visualization, identification, and quantitation of proteins, like two-dimensional electrophoresis and two-dimensional difference gel electrophoresis combined with matrix assisted laser desorption/ionization time-of-flight mass spectrometry, surface enhanced laser desorption/ionization time-of-flight mass spectrometry, liquid chromatography tandem mass spectrometry, and the isotope-coded affinity tags and isotope tags for relative and absolute quantification technologies that are used for quantitative analysis [48]. A study showed that the collection tube wall material itself can alter the cerebrospinal fluid measurements of β-amyloid and tau proteins [48]. These CSF protein levels were lowest in polystyrene tubes, and polypropylene tubes should therefore be used to prevent erroneous results [49]. Studies have suggested that the type of blood collection tube (e. g., serum, EDTA plasma, heparin plasma) can alter the observed proteomic profiles in as little as four hours [50, 51, 52]. Furthermore, polymers from different brands of BCTs and heparin have been described as inducing ion suppression effects [53]. Different blood tube components such as gel or non-gel separator in BD tubes have been shown to alter plasma peptide profiles [54, 55]. This alteration in plasma proteome profiles in different BD tubes changed over time [55]. For this reason, BD P100 tubes that contain a patented non-gel mechanical separator that is released from the stopper have been developed to reduce cellular contamination of the plasma layer and thus minimize preanalytical variability [54, 55]. Although proteomics techniques are able to accurately detect and quantify small differences in abundance between samples, they are unable to distinguish whether differences arose through biological means or from so-called "preanalytic variables." To the authors' knowledge, there are no commercially available serum collection tubes that can reproducibly produce serum samples for proteomic studies. It is for this reason that plasma instead of serum has been recommended by the Human Proteome Organization [56]. In view of the potential interference of components shed from BCTs in the mass spectrometric analysis of serum polypeptides, researchers examining patterns of

low-molecular-weight peptides and proteins for diagnostic purposes are advised to take the following precautions:
1. The type of collection tube used for a diagnostic application should be standardized, as should the procedure for specimen processing and handling. Use of a diverse range of collection tubes and procedures may complicate the interpretation of mass spectrometric analysis of serum banks collected from multiple sites or over an extended period of time during which types of collection tubes have changed.
2. Collection tubes should be tested for interference in the analysis of interest. Examination of the components eluted from tubes into a saline solution can be used as a simple initial check for potential interference from components.

4.6.6 Recommendations for clinical laboratories and manufacturers

In this section, we offer some practical suggestions to clinical laboratories and manufacturers in an effort to reduce errors during the preanalytical phase.

In order to evaluate interferences from collection device components in clinical tests, laboratory personnel should:
1. Test the same analyte with an alternative assay;
2. Incubate the sample with the different parts of the collection device to identify the source of potential interferences;
3. Contact both the collection device and assay manufacturers;
4. If necessary, file a medical device alert with the Food and Drug Administration or other national and international institutions; and
5. If possible, switch to a new blood collecting tube manufacturer.

We recommend that tube manufacturers implement Design for Six Sigma, and similar methodologies to reduce variation of blood collection device components. This would ultimately produce higher quality laboratory specimens [57].

For any new or substantially modified BCTs introduced to the clinical laboratory, a well-planned validation protocol should be written and reviewed for scientific validity by qualified individuals. The protocol should describe the validation procedure in detail and should include predefined acceptance criteria and statistical methods, in addition to following the policies and procedures related to testing human subjects established by the institution's review board or ethics committee. Blood specimens from both patients and apparently healthy individuals should be collected and included in the tube validation study. The entire blood collection system (needles, holders, tubing, etc.) rather than a particular tube or component should also be included in the study.

To determine the accuracy of assay results obtained from new or substantially modified BCTs, a tube comparison study similar to that described in the CLSI EP9-A [58] guideline should be conducted. Specimens that cover the reportable range for each analyte should be evaluated with an adequate number of samples. This will provide sufficient power to conduct statistical analyses of the data. Linear regression analysis or a similar type of regression method and Bland-Altman type plots should be used to analyze the tube comparison data.

To assess imprecision of assay results collected in new or substantially modified BCTs, the clinical laboratory can compare the variability of results for the samples collected in new tubes with the variability obtained from samples collected in their current BCT. This can also be achieved by replicate testing of quality control material and/or patient specimens, as described in the CLSI EP5-A [59] guideline.

For analytes that are physiologically undetectable or those that exist in low concentrations in healthy individuals, or to generate samples that cover the reportable range, samples should be spiked with the analyte of interest when feasible. The total number of assays for tube validation studies will depend on the intended use of the blood collection device. Laboratories can select representative assays from different testing methodologies (e. g. ion-specific electrode, immunoassay and spectrophotometry) to be evaluated.

Quality control (QC) evaluates the measurement procedure by periodically assaying QC material for which the correct result is known in advance. If the results for QC materials are within acceptable limits of the known value, the measurement procedure is verified to be performing as expected, and results for patient samples can be assumed to be valid. However, if QC results are not within acceptable limits, patient results are not reported, and corrective action is necessary [60–62]. Good laboratory practice requires verification that a method is performing correctly at the time that the patient results are measured. Blood collection device problems are difficult for laboratory technicians to recognize in a timely manner, because routine QC testing may not use the problematic collection devices [60–62].

The evaluation of method performance by an external entity is referred to as proficiency testing. Proficiency testing allows a laboratory to verify that its results are consistent with other laboratories using the same or similar methods for an analyte and thus to confirm that the methods are being performed correctly. Typically, proficiency testing providers send a set of samples to a group of laboratories. Each laboratory includes the proficiency testing specimens along with the patient samples in the usual assay process. The results for the proficiency specimens are reported to the proficiency testing provider for evaluation. However, unlike patient specimens, proficiency specimens do not require collection with routinely used blood collection devices; hence, proficiency testing will also fail to detect blood collection tube-related problems [60, 61].

The comparison of the results for the control sera exposed and unexposed to BCTs should reveal adverse effects from tube additives, [60, 61] but this testing is uncommon in most clinical laboratories. It is also impractical for most clinical laboratories because of the diversity of tubes used and frequent changes in tube lots. It may therefore be more appropriate for manufacturers to expose quality control sera to BCTs on a lot-by-lot basis. When a clinical laboratory changes the tubes it uses, a well-planned tube verification protocol should be implemented. Routine monitoring of moving averages based on patient data may be potentially useful for identifying future tube-related problems [63].

It is important for clinical trials and research studies to offer a rationale for their selection of BCTs, because the nature of materials and additives in the tubes can possibly interact with the specimens and affect measurements. We advise that the same type of tubes from a particular manufacturer be utilized throughout clinical trials and research studies to prevent or minimize tube-related interferences in assay results. This is especially relevant for emerging technologies in the clinical laboratory where greater sensitivities and lower concentrations of analytes are measured, hence, becoming more susceptible to analytical interference; even small amounts of interferents from BCT components may alter assays and therefore potentially produce erroneous test results. Thus, a thorough evaluation of the effect of blood sampling device materials on these high sensitivity methods is warranted before their release into the marketplace. The ultimate goal of tube validation studies performed by clinical laboratories is to demonstrate that both new tubes and those currently in use are clinically acceptable.

Conclusions

We have examined BCTs and their components, discussed known flaws, and offered recommendations to reduce errors related to blood specimen collection and testing. For the most part, because current BCTs work as designed, the effects they can have on research findings and test results are often overlooked. Because BCTs are "taken for granted" medical devices, it is important that laboratorians become more aware of the potential problems BCTs can cause in the analysis of specimens. BCTs are medical devices and, as such, have inherent limitations. When BCTs are improperly used, when their limitations are not fully understood, or when there are issues that manufacturers overlook or fail to address, laboratory results for blood specimen tests can be adversely affected. Inaccuracies in test results related to BCTs may decrease laboratory efficiency, delay test results, and increase the cost per test due to the need for re-collection and retesting. Most importantly, these preanalytical errors cost patients in need of treatment valuable time they may not have to spare. For these reasons, it is vital to optimize and standardize BCTs. We urge laboratorians, tube manufacturers, diagnostic companies, other researchers and stakeholders to continue investigating the subject and to remain vigilant by doing what is necessary to protect against the adverse effects that sampling device materials can have on clinical assays and laboratory sciences more generally.

References

[1] Calam RR, Cooper MH. Recommended "order of draw" for collecting blood specimens into additive-containing tubes. Clin Chem 1982; 28:1399.
[2] Clinical and Laboratory Standards Institute (CLSI). Procedures for the Collection of Diagnostic Blood Specimens by Venipuncture; Approved Guideline.5th ed. Wayne, PA; CLSI: Document H3-A5; 2007.

[3] Salvagno G, Lima-Oliveira G, Brocco G, Danese E, Guidi GC, Lippi G. The order of draw: myth or science? Clin Chem Lab Med 2013; 51:2281–5.
[4] Fukugawa Y, Ohnishi H, Ishii T, Tanouchi A, Sano J, Miyawakilt, et al. Effect of carryover of clot activators on coagulation tests during phlebotomy. Am J Clin Pathol 2012; 137:900–3.
[5] Sulaiman RA, Cornes MP, Whitehead SJ, Othonos N, Ford C, Gama R. Effect of order of draw of blood samples during phlebotomy on routine biochemistry results. J Clin Pathol 2011; 64:1019–20.
[6] Hortin GL, Sviridov D, Anderson NL. High-abundance polypeptides of the human plasma proteome comprising the top 4 logs of polypeptide abundance. Clin Chem 2008; 54:1608–16.
[7] Menssen HD, Brandt N, Leben R, Muller F, Thiel E, Melber K. Measurement of hematological, clinical chemistry, and infection parameters from hirudinized blood collected in universal blood sampling tubes. Semin Thromb Hemost 2001; 27:349–56.
[8] Joly D, Drueke TB, Alberti C, Houillier P, Lawson-Body E, Martin KJ, et al. Variation in serum and plasma PTH levels in second-generation assays in hemodialysis patients: a cross-sectional study. Am J Kidney Dis 2008; 51:987–95.
[9] Evans MJ, Livesey JH, Ellis MJ, Yandle TG. Effect of anticoagulants and storage temperatures on stability of plasma and serum hormones. Clin Biochem 2001; 34:107–12.
[10] Kim BS, Hrkach JS, Langer R. Biodegradable photo-crosslinked poly(ether-ester) networks for lubricious coatings. Biomaterials 2000; 21:259–65.
[11] Rai AJ, Gelfand CA, Haywood BC, Warunek KDJ, Yi J, Schuchard MD, et al. HUPO Plasma Proteome Project specimen collection and handling: towards the standardization of parameters for plasma proteome samples. Proteomics 2005; 5:3262–77.
[12] Nilsson TK, Boman K, Jansson JH, Thogersen AM, Berggren M, Broberg A, Granlund A. Comparison of soluble thrombomodulin, von Willebrand factor, tPA/PAI-1 complex, and high-sensitivity CRP concentrations in serum, EDTA plasma, citrated plasma, and acidified citrated plasma (Stabilyte) stored at –70 degrees C for 8–11 years. Thromb Res 2005; 116:249–54.
[13] Hortin GL, Warshawsky I, Laude-Sharp M. Macromolecular chromogenic substrates for measuring proteinase activity. Clin Chem 2001; 47:215–22.
[14] Sapin R, Ongagna JC, Gasser F, Grucker D. Insulin measurements in haemolysed serum: influence of insulinase inhibitors. Clin Chim Acta 1998; 274:111–7.
[15] Ayache S, Panelli M, Marincola FM, Stroncek DF. Effects of storage time and exogenous protease inhibitors on plasma protein levels. Am J Clin Path 2006; 126:174–84.
[16] Vogler EA, Siedlecki CA. Contact activation of blood-plasma coagulation. Biomaterials 2009; 30:1857–69.
[17] Meites S, Levitt MJ. Skin-puncture and blood-collecting techniques for infants. Clin Chem 1979; 25:183–9.
[18] Dale JC, Ruby SG. Specimen collection volumes for laboratory tests. Arch Pathol Lab Med 2003; 127:162–8.
[19] Young DS, Bermes EW, Haverstick Dm. Specimen collection and processing in Tietz: Textbook of Clinical Chemistiry and Molecular Diagnostics. 4th ed. St. Louis: Elsevier Saunders; 2006.
[20] Norling LL, Smith CH, Landt M. Heparinized "microtainer" tube evaluated for collection of capillary blood. Clin Chem 1984; 30:810–1.
[21] Desjardins PR, Mactavish G, Schoemperlen J. Storage of specimens in "Microtainer Amber Tubes" protects bilirubin from degradation by light. Clin Chem 1987; 33:712.
[22] Bush V, Leonard L, Szamosi D. Advancements in blood collection devices. Lab Med 1998; 29:616–22.
[23] Kupke IR. Evaluation of the polyester gel barrier Microtainer. Clin Chem 1987; 33:1468–9.
[24] Landt M, Norling LL, Steelman M, Smith CH. Monoject Samplette capillary blood container with serum separator evaluated for collection of specimens for therapeutic drug assays and common clinical-chemical tests. Clin Chem 1986; 32:523–6.

[25] Glass Capillary Tubes: joint safety advisory about potential risks. (Accessed May, 1999, at www.osha.gov/pls/oshaweb/owadisp.show-document.)
[26] Nugent E, Losada R, Conway H, Montgomery D, Williams J, inventors; Platelet stable blood collection assembly. US-patent 4,967,763. 1989.
[27] Meites S. Skin-puncture and blood-collecting technique for infants: update and problems. Clin Chem 1988; 34:1890–4.
[28] Cohen R, Keusch P, inventors; Medical article having blood-containing surface. US-patent 6,551,267. 1993.
[29] Hicks JM, Rowland GL, Buffone GJ. Evaluation of a new blood-collecting device ("microtainer") that is suited for pediatric use. Clin Chem 1976; 22:2034–6.
[30] Wang S, Ho V, Roquemore-Goins A, Smith FA. Effects of blood collection tubes, including pediatric devices, on 16 common immunoassays. Clin Chem 2006; 52:892–3.
[31] Yip PM, Chan MK, Zielinski N, Adeli K. Heparin interference in whole blood sodium measurements in a pediatric setting. Clin Biochem 2006; 39:391–5.
[32] Tanaka T, Moretti ME, Verjee ZH, Shupak M, Ivanyi KE, Ito S. A pitfall of measuring lithium levels in neonates. Therap Drug Monit 2008; 30:752–4.
[33] Coyle JT. Psychiatric drugs in medical practice. Med Clinics North Am 1977; 61:891–905.
[34] Park SH, Chi HS, Choi MO, Park BG, Jang S, Park CJ. Improved turnaround time for neonatal hematology profile tests (complete blood count) using a new microcollection tube. Clin Chem Lab Med 2011; 49:1083–5.
[35] Elliott P, Peakman TC. The UK Biobank sample handling and storage protocol for the collection, processing and archiving of human blood and urine. Int J Epidemiol 2008; 37:234–44.
[36] Adcock DM, Hoefner DM, Kottke-Merchant K, Marlar RA, Szamosi DI, Warunek DJ. Collection, Transport, and Processing of Blood Specimens for Testing Plasma-Based Coagulation Assays and Molecular Hemostasis Assays: Approved Guideline- 5th ed. Wayne, PA: Clinical Laboratory Standards Institute, 2008.
[37] Sun K, Oh H, Emerson JF, Raghavan SR. A new method for centrifugal separation of blood components: Creating a rigid barrier between density-stratified layers using a UV-curable thixotropic gel. J Mater Chem 2012; 22:2378–82.
[38] Vaught JB. Blood collection, shipment, processing, and storage. Cancer Epidemiology, Biomarkers & Prevention: a publication of the American Association for Cancer Research, cosponsored by the Am Soc Preventive Oncology 2006; 15:1582–4.
[39] Gunther K, Malentacchi F, Verderio P et al. Implementation of a proficiency testing for the assessment of the preanalytical phase of blood samples used for RNA based analysis. Clin Chim Acta 2012; 413:779–86.
[40] Swarup V, Rajeswari MR. Circulating (cell-free) nucleic acids--a promising, non-invasive tool for early detection of several human diseases. FEBS Letters 2007; 581:795–9.
[41] Gormally E, Caboux E, Vineis P, Hainaut P. Circulating free DNA in plasma or serum as biomarker of carcinogenesis: practical aspects and biological significance. Mutation Res 2007; 635:105–17.
[42] Jung K, Fleischhacker M, Rabien A. Cell-free DNA in the blood as a solid tumor biomarker--a critical appraisal of the literature. Clin Chim Acta 2010; 411:1611–24.
[43] Srinivasan A, Bianchi DW, Huang H, Sehnert AJ, Rava RP. Noninvasive detection of fetal subchromosome abnormalities via deep sequencing of maternal plasma. Am J Human Gen 2013; 92:167–76.
[44] Fernando MR, Chen K, Norton S et al. A new methodology to preserve the original proportion and integrity of cell-free fetal DNA in maternal plasma during sample processing and storage. Prenat Diagn 2010; 30:418–24.
[45] El Messaoudi S, Rolet F, Mouliere F, Thierry AR. Circulating cell free DNA: Preanalytical considerations. Clin Chim Acta 2013; 424:222–30.

[46] Bartel DP. MicroRNAs: genomics, biogenesis, mechanism, and function. Cell 2004; 116:281–97.
[47] Clinical and Laboratory Standards Institute (CLSI). Validation and Verification of Tubes for Venous and Capillary Blood Specimen Collection; Approved Guideline. Wayne, PA; CLSI: 2010.
[48] Wodzig WKea. Proteomics as a tool for biomarker detection using SELDI-TOF-MS and 2-D DIGE W. K. W. H. Klin Chem Labgeneesk 2012; 37:82–7.
[49] Lewczuk P, Beck G, Esselmann H, Bruckmoser R, Zimmermann R, Fiszer M et al. Effect of sample collection tubes on cerebrospinal fluid concentrations of tau proteins and amyloid beta peptides. Clin Chem 2006; 52:332–4.
[50] Banks RE, Stanley AJ, Cairns DA, Barret JH, Clarke P, Thompson D, Selby PJ. Influences of blood sample processing on low-molecular-weight proteome identified by surface-enhanced laser desorption/ionization mass spectrometry. Clin Chem 2005; 51:1637–49.
[51] Rai AJ, Zhang Z, Rosenzweig J, Shi IeM, Pham T, Fung ET et al. Proteomic approaches to tumor marker discovery. Arch Pathol Lab Med 2002; 126:1518–26.
[52] Bowen RA, Chan Y, Cohen J, Rehak NN, Hortin GL, Csako G, Remaley AT. Effect of blood collection tubes on total triiodothyronine and other laboratory assays. Clin Chem 2005; 51:424–33.
[53] Mei H, Hsieh Y, Nardo C, Xu X, Wang S, Ng K, Korfmacher WA. Investigation of matrix effects in bioanalytical high-performance liquid chromatography/tandem mass spectrometric assays: application to drug discovery. Rap Comm Mass Spectrom 2003; 17:97–103.
[54] Hsieh SY, Chen RK, Pan YH, Lee HL. Systematical evaluation of the effects of sample collection procedures on low-molecular-weight serum/plasma proteome profiling. Proteomics 2006; 6:3189–98.
[55] Effect of Plasma Separators on Peptide MALDI-TOF MS Spectra. at http://www.bd.com/proteomics/pdfs/Effects-of-Plasma-Separators-on-Peptide-MALDI-TOF-MS-Spectra.pdf)
[56] Omenn GS, States DJ, Adamski M et al. Overview of the HUPO Plasma Proteome Project: results from the pilot phase with 35 collaborating laboratories and multiple analytical groups, generating a core dataset of 3020 proteins and a publicly-available database. Proteomics 2005; 5:3226–45.
[57] Stankovic AK, Romeo P. The role of in vitro diagnostic companies in reducing laboratory error. Clin Chem Lab Med 2007; 45:781–8.
[58] Clinical and Laboratory Standards Institute (CLSI). Method Comparison and Bias Estimation Using Patient Samples; Approved Guideline. Wayne, PA: CLSI; Document EP 9-A, 1992.
[59] Clinical and Laboratory Standards Institute (CLSI). Evaluation of Precision performance of Clinical Chemistry Devices. Approved guideline. Wayne, PA: Clinical and Laboratory Standards Institute; Document EP 5-A, 1992.
[60] Bowen RA, Chan Y, Ruddel ME, Hortin GL, Csasko G, Demoshy SJ jr, Remaley AT. Immunoassay interference by a commonly used blood collection tube additive, the organosilicone surfactant silwet L-720. Clin Chem 2005; 51:1874–82.
[61] Kricka LJ, Park JY, Senior MB, Fontanilla R. Processing controls in blood collection tubes reveals interference. Clin Chem 2005; 51:2422–3.
[62] Bowen RA, Sattayapiwat A, Gounden V, Remaley AT. Blood collection tube-related alterations in analyte concentrations in quality control material and serum specimens. Clin Biochem 2014; 47:150–7.
[63] Cembrowski GS, Engebretson MJ, Hackney JR, Carey RN. A systems approach to assure optimal proficiency testing in the hematology laboratory. Clin in Lab Med 1993; 13:973–85.

5. Sampling Materials and Techniques*

* The opinion expressed and claims made by the manufacturers are solely their own and the editors and the Publisher neither endorses or claim responsibility for the content expressed in their chapters.

Helene Ivanov, Jaqueline Präuer, Melanie Schimpl, Gabriele Castelo-Rose, Claire Wiesner, Thomas Ehrenfellner

5.1 Materials and Techniques of Sampling Blood and other Body Fluids. Contributions of Greiner – Bio One

5.1.1 Disposable tourniquet

By restricting venous blood flow, veins become more prominent, are easier to palpate and are clearer to differentiate from pulsating arteries. The use of a disposable tourniquet is a hygienic way to achieve this. Ideally, the tourniquet is placed approximately 7.5–10 cm above the intended puncture site and should not be left tightened on the patient for more than one minute. It is also important to apply the correct pressure. As it states in the literature, if using a blood pressure cuff to restrict venous blood flow, it is recommended that 40 mmHg is not exceeded to ensure arterial flow to the extremity [1]. Obstruction of the arterial flow can lead to changes in the laboratory parameters [1, 2] or insufficient filling of the blood collection tubes.

A single-use tourniquet is the preferred practice, as re-use could lead to the spreading of nosocomial and blood-borne infections due to contamination with pathogenic bacteria/viruses and/or blood [3, 4]. For ease of use, there are products available in the market that enable one-handed release. Latex-free products are preferable to help prevent allergic reactions of both users and patients [1] and products made out of lightly powdered silicon material ensure comfortable application for the patient.

Fig. 5.1.1: Disposable Tourniquet.

5.1.2 Sharps & holders for blood collection

Ideally, blood collection is carried out using a closed system incorporating an integrated safety mechanism. OSHA [5] recommends tube holders should always be single-use devices, since this enables the disposal as one unit, without the need to separate contaminated sharps. OSHA goes on to further recommend the application of sharps with an engineered sharps injury protection mechanism. Safety devices are available on the market in the form of holders with attached safety shields for manual activation which need to be activated intentionally. There are other safety sharps devices which are considered semi-automated in terms of activation, such as, for example a tube to be inserted into the holder to activate the safety mechanism.

Blood collection needles are daily routine items intended exclusively for single-use. Manufactured from stainless steel for smooth insertion into the patient's skin, with a rubber valve at the end to enable blood collection from multiple tubes, a label seal around the cap ensures the integrity of the product prior to use. Needles, holder and tubes that are compatible as a system should always be selected.

Needles are available in varying sizes, which is indicated by the "G", meaning the gauge size. The higher the number, the thinner the needle (i. e. the smaller the diameter). Needles within the range of 19–25G are recommended for venipuncture. A color-coded needle cap is generally used to indicate the gauge size:

Table 5.1.1: Blood collection needle sizes (gauges) and color codes.

Color code [6]	Gauge [7]
Orange	25
Blue	23
Black	22
Green	21
Yellow	20
Brown	19

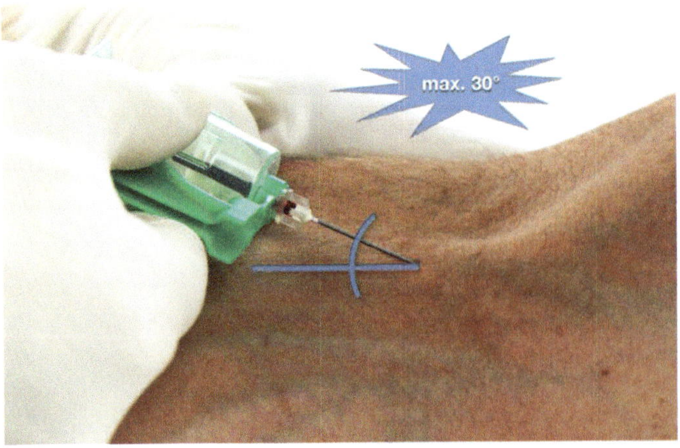

Fig. 5.1.2: Safety holder and recommended penetration angle.

The needle length should be 1 to 1 ½". A winged needle is usually ¾" long. Needle preference depends on the user and the depth of the vein being punctured. Today, the application of safety products is standard technology.

Needles with flash chamber are available which allows for an immediate optical control of successful vein entry.

A standard tube holder should be ergonomically designed and made of a stable non-breakable material that is sufficiently transparent to enable visual control of the attached evacuated blood collection tube being filled. An optimal barrel length is long enough to protect the user from exposure to the back-end needle whilst short enough to minimize excessive waste and ensuring stable positioning of the tube during collection. Tube holders are disposed of immediately after blood collection, as one unit with the needle or winged blood collection set.

According to literature it is recommended that the optimal angle for blood collection be 30° or less [8]. There are specially designed holders on the market featuring an eccentric Luer connector on the top and a stainless steel needle inside the holder body to facilitate this best practice procedure. Another aspect that can be attributed to this design is a reduction in haemolysis when collecting from a catheter. Studies are available on this product type [9], confirming that the design can prevent damage to erythrocytes as the pressure between vein and tube is equalized.

The exclusive obligatory use of safety products in blood collection is more and more widespread. In addition to OSHA guidelines in the USA, since 11th May 2013, the sharps safety EU Directive 2010/32 [10], has been compulsory in national law for all EU countries, stating that a safety engineered device shall be provided if any risks of sharps injuries are determined. The Biosafety Network has published guidelines [11] on the features of such devices. Holders with integrated safety mechanisms are on the markets that are compatible with existing blood collection needles. Ideally, if a needle with flash chamber is used, the user additionally has visual vein entry control.

When a needle is assembled with a holder, it becomes one device which according to regulations and recommendations must be disposed of as a single unit after use. If the needle were first unthreaded, this would override the safety mechanism feature, which has to remain an integral part of the safety device, according to the sharps safety EU Directive 2010/32. If a higher level of safety is desired, safety holders with pre-assembled needles are available ready to use. Sterility of the product is ensured either in an individual blister pack or in the form of a sterile fluid pathway.

5.1.3 Evacuated specimen collection tubes

For collection of blood, urine and saliva, it is important to have a closed system to ensure hygienic and clean sample collection. Evacuated, standardized collection systems are the preferable sampling method, because the tubes automatically fill to the pre-defined nominal volume, ensuring the correct sample to additive ratio [12]. Ideally, tubes are made of virtually unbreakable plastic material such as PET (polyethylene terephthalate), which is as clear as glass and ensures an unobstructed view of the specimen.

One important aspect of collection tubes is their validated compatibility with clinical analysers and centrifuges. Hundreds of instruments are available on the worldwide market and it should be verified that the systems function with the chosen primary tubes [13].

Table 5.1.2: Most common tube sizes.

Diameter	Length	Max. volume
13	75	Up to 4.5 mL
13	100	Up to 6.0 mL
16	100	Up to 10.0 mL

Depending on requirements, the tubes can contain various additives which have the function of either stabilizing the specimen or enhancing/inhibiting coagulation. These additives can be sprayed directly on the inner tube walls or added in form of a powder or liquid. The additive concentration is adjusted to the nominal fill volume of the tube, which has to be accurate to +/-10 % [14].

If the tubes need to be transported and are used as the primary receptacle, they have to be leak-proof and able to withstand a pressure of 95 kPa (0.95 bar), since blood samples are category B biological substances and classified as UN3373 [14].

The use of specially designed safety tube caps (available in various standard colors) contributes to clean and hygienic sample collection. The cap should allow easy piercing and ensure that the vacuum is maintained during the entire shelf-life of the product. They are ideally designed with a thread to enable easy manual opening which may occasionally be necessary. The cap colors also give immediate visual identification of the additives contained, conforming to EN 14820 [15] standards.

The tube labels, available optionally in paper or transparent plastic material, should contain all the information according to international standards [16]; this includes additive description, nominal volume and space for patient identification. A technological advance is the availability of prelabeled barcode tubes, which, when used in combination with an appropriate software system, give the added

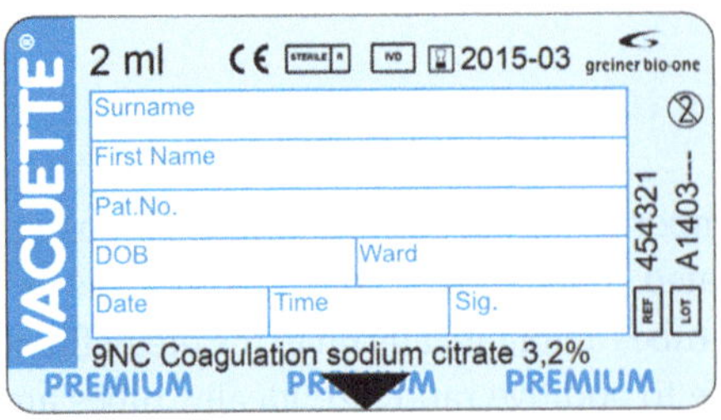

Fig. 5.1.3: Specimen container labeling according to ISO 6710 and EN14820 [15].

value of increased safety in the preanalytical phase, as manual identification errors can be avoided.

Sample collection into vacuum tubes is carried out using a needle and tube holder system. The tube is inserted, cap first, into the holder until the back end needle fully pierces the rubber stopper. The sample is then automatically drawn into the tube via the vacuum. As soon as the fill volume is achieved, blood flow stops and the tube can be removed from the holder and should be inverted several times to mix the sample with the additive, when applicable.

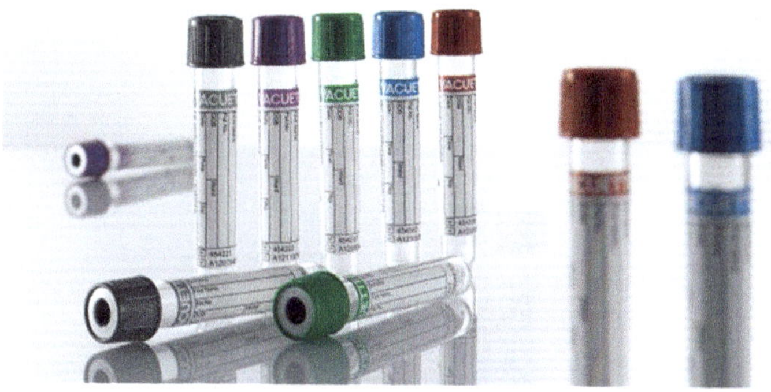

Fig. 5.1.4: Evacuated specimen collecting tubes of Greiner Bio One.

5.1.3.1 eHealth and Pre-barcoded tubes

A relatively new but expanding area is eHealth. When pre-barcoded sample collection tubes are used in combination with the appropriate electronic IT system, the preanalytical process is supported and streamlined. By using an eHealth system, additional value is added and potential sources for error can be reduced, for example in the following areas [16]:

- Making the appropriate test requisition
- Ensuring safe patient identification via scan of patient wristband
- Proposing correct order of draw
- Supporting of whole collection workflow by IT
- Making collection and transportation recommendations (e. g. cooled sample)
- Checking that correct tubes are being taken during collection procedure (scan)
- Efficient processing of samples in the laboratory

The pre-barcoded tubes from the manufacturer are labeled with a unique barcode 128, as recommended by CLSI [17] and readable by most common analyzers. Additional labelling is no longer required, hence reducing turnaround time and additional costs. As there is no need for manual labelling, in addition to savings on time and increasingly significant costs, an important factor is the increased transparency of the collection procedure, for example the time of collection is automatically recorded. This provides the ideal basis for processing on laboratory instrumentation, thereby raising productivity whether the samples come from GPs, hospitals or laboratories.

This comprehensive systematic solution supports the accreditation of laboratories by providing information about collection time, sample transportation (e. g. on ice, water bath, ...), identification of the employee collecting the sample, name of requesting person, requested tests, description of sample container, etc. In addition to facilitating and easing documentation steps, the use of prebarcoded tubes (barcode quality and position is based on ISO [18] and CLSI requirements) and the scan of the patient ID can help prevent misidentification and increase patient safety.

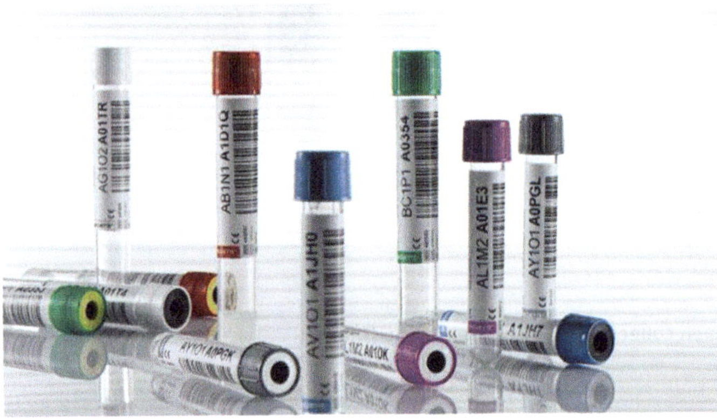

Fig. 5.1.5: Pre-barcoded specimen collection tubes from Greiner Bio One.

5.1.4 Safety winged blood collection sets

When blood collection is necessary from patients with smaller, difficult veins, for example with the elderly or children, or if it is required to use the back of the hand for a sample, the use of a safety blood collection set is recommended. When the safety mechanism is activated whilst the needle is still in the vein of the patient, the risk of medical staff coming into contact with a contaminated needle is completely avoidable.

This is a winged needle (3/4" in length) connected to the collection adapter via flexible tubing [1]. The device is sterile, single-use and individually packed in a blister. Depending on user requirements and the situation, sets are available, for example with a pre-assembled tube holder or Luer adapter on the end of the tubing. In addition to blood collection, these sets can be used for short-term infusion purposes by removing any male adapter prior to use for infusion. Special pre-assembled sets are even available for the collection of blood into blood culture bottles.

When using the blood collection sets, the wings of the device should be held between the thumb and the index finger to ensure an angle flat enough to enter small, delicate veins. Successful venipuncture is indicated when the transparent flash chamber fills with blood. If the first tube to be filled contains additives, a discard tube should be used prior to collection to flush out air in the dead space of the tubing. The next additive tube that is used will then be filled to the correct volume, which is important to ensure the correct mixing ratio of blood to additive [1].

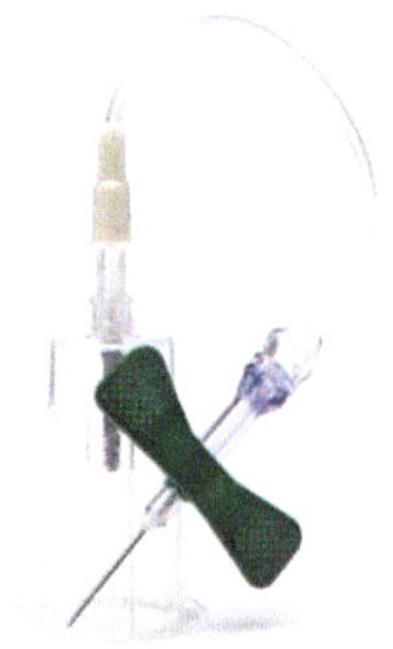

Fig. 5.1.6: Safety winged blood collection set of Greiner Bio One.

5.1.5 Safety lancets

Since the introduction of the EU sharps safety directive 2010/32 [19], the use of safety engineered devices is compulsory wherever possible and applicable. This means that a safety mechanism is activated after use, irreversibly preventing further re-use. There are currently many types of safety lancets available on the market offering various features, that is, with different activation mechanisms, with a choice of needle or blade, different lengths/diameters/widths.

The purpose of a lancet is to make an incision in the skin, or to make a puncture if using a needle lancet, in order to obtain a capillary blood sample. Sterile single-use lancets should be used to prevent injury or infection [20].

When selecting the specification of the device, it is important to consider the patient. Age of patient, collection site and volume of specimen required are the essential criteria, and the device should be selected so as to ensure collection without causing injury.

Safety lancets can also be selected according to the different activation mechanism: either the user has to manually activate the lancet, for example by pushing the release button, or the activation is automatic, for example by pressure activation when the lancet is pushed against the skin.

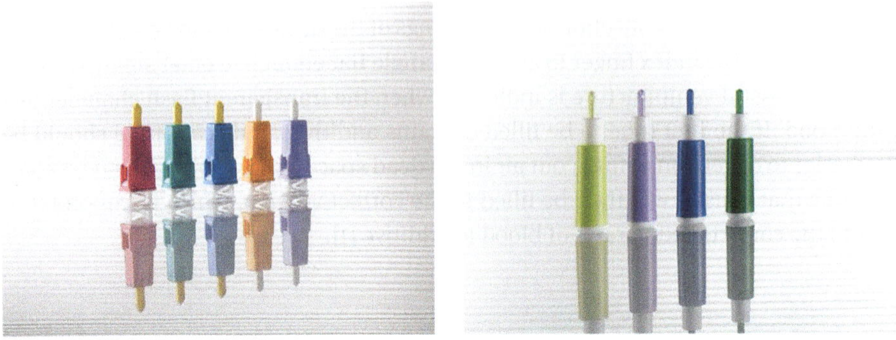

Fig. 5.1.7: Safety Lancets of Greiner Bio-One.

5.1.6 Specimen collection tubes for capillary blood samples

Special smaller tubes are available for the collection of capillary blood samples. This kind of sample is suitable mainly if the patients are babies, infants, and elderly or just have difficult, fragile veins [20]. Further cases where a capillary sample could be preferable to a venous sample could be:

- Patients with extreme burns
- Obese patients
- Patients at risk of thrombosis
- Point of care tests
- Patients with a fear of needles
- Anaemic patients
- Patients subject to frequent blood collection (e. g. oncology patients) [20, 21].

Capillary blood collection tubes are available with different additives and in volumes up to 1 ml. Please note that capillary blood is not suitable for testing in sodium citrate tubes.

Capillary blood collection tubes are not evacuated, they are filled with the aid of a scoop/funnel or via capillary action. The fill mark on the tube or tube label must be used to ensure the correct mixing ratio with the additives. For easy identification of the additives, the cap colors are ideally coded similar to venous blood collection tubes which corresponds with ISO 6710 [15]. To be sure that the sample is mixed sufficiently with the additive, tubes should be tapped lightly after collection and gently inverted several times [20].

In routine sample collection, a lancet is usually applied on the finger pad or in the case of babies on the heel. After the stick has been made, the drops of blood can be collected. It is recommended to avoid milking the puncture site, as this could dilute the sample with tissue fluids [20]. If the blood flow is insufficient, it can help to warm up the extremity and try again. As a rule, the puncture site should be below heart level to make blood flow easier.

With capillary blood, it should be noted, that this is a mixture of blood from capillaries, venules, arterioles and interstitial as well as intercellular liquid. Due to this mixture, normal values are different to those of venous blood. This means glucose values could be raised, whilst potassium, total protein and calcium values could be lower. In order to prevent false interpretation, the sample must be clearly marked as capillary blood [20, 22]. The smaller tubes for capillary blood collection can be inserted into a 13 mm carrier tube or tube adapter so that they can be centrifuged and analyzed in standard sized instrumentation.

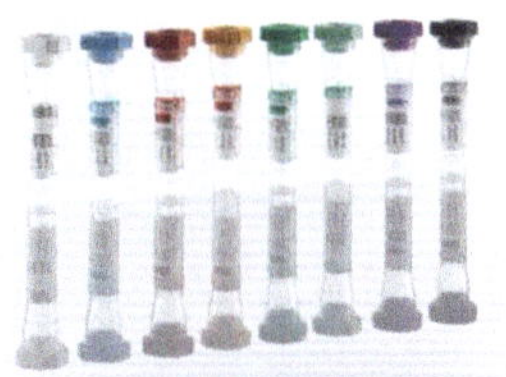

Fig. 5.1.8: Capillary collection system from Greiner Bio-One.

5.1.7 Urine system

Only a clean and correctly collected urine sample can provide accurate results [23].

For everyday routine urine sampling and analysis, a comprehensive urine system covering all needs, from collection in beaker to transport in tube to 24 h urine collection is essential. Mid-stream urine samples are used for most routine urine collection procedures. These are currently collected in a wide variety of collection containers, however, it is recommended that to be sure of a clean and hygienic collection, a closed beaker system, for example with an integrated transfer device be used. This is a urine beaker for volumes up to 100 ml, with an attached lid integrating a transfer adapter for immediate sample collection into a vacuum tube. All that is required is to hold the beaker on a level surface and insert the vacuum tube(s) into the opening on the lid [24]. The tube will fill automatically to the pre-defined volume. The rubber needle sleeve of the transfer adapter prevents any leakage of the urine specimen during transfer. The beaker remains leak-proof even when mixing to homogenize the sample.

A further option could be, for example a standard beaker. A transfer straw can be used to transfer the urine sample into the evacuated urine tube. The transfer straw has a needle inside which pierces the vacuum tube stopper. Transfer straws are available in different lengths, depending on height of container used.

Depending on user requirements, sterile urine beakers either with integrity seal or packed in a sterile single blister can be used as well as sterile transfer straws in single blister packs. This sterile system helps prevent bacterial contamination [25].

For timed collection needs, any container with sufficient volume capacity – usually around 3 litres - can be used; however, containers specifically intended for this application are available. These containers are ideally acid resistant in case stabilizers are required, amber colored to protect light sensitive analytes, graduated to easily check volume, and have a wide diameter for hygienic filling. Also available are urine collection containers with integrated collection straws, so that when collection is complete, the tube(s) only need be inserted without any need for opening the container.

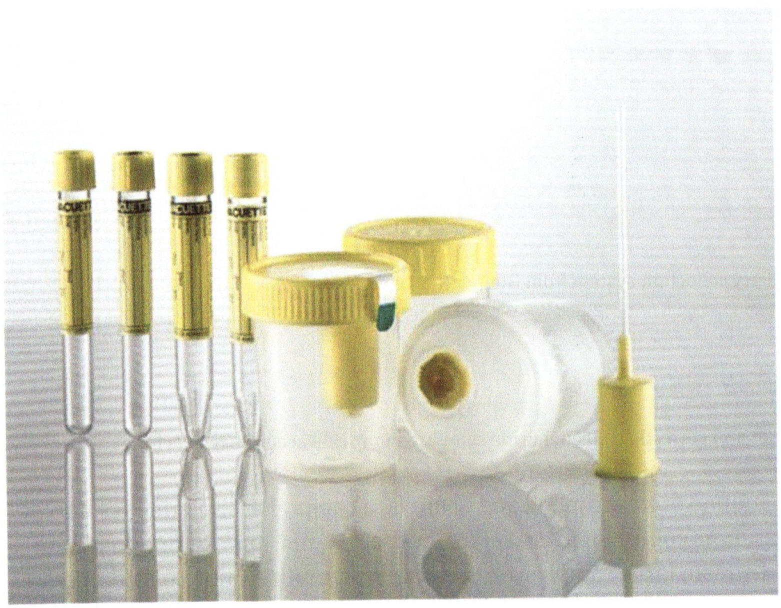

Fig. 5.1.9: Urine collection system from Greiner Bio-One.

5.1.8 Saliva collection

For decades, urine and blood have been the matrices of choice for drugs of abuse testing. Oral fluid (OF) as a sample matrix offers significant advantages: collection can be performed in almost any location, is noninvasive and it can be taken under direct observation, thus reducing the risk of adulteration and substitution. See Chapter 2.8.2 for further information.

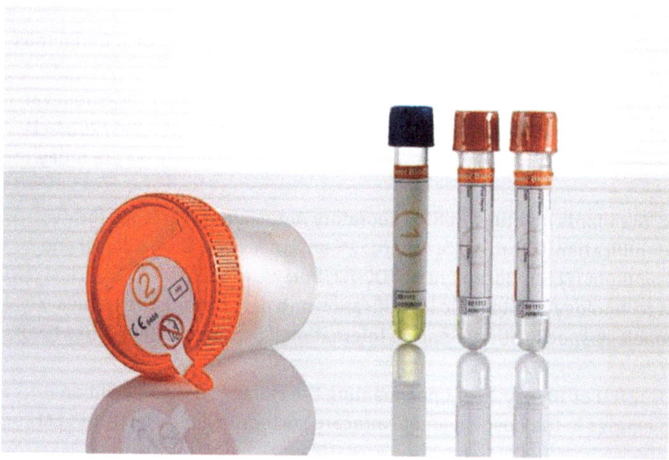

Fig. 5.1.10: Oral fluid collection system from Greiner Bio One.

References

[1] Clinical Laboratory Standardization Institute (CLSI), Procedures for the Collection of Diagnostic Blood Specimens by Venipuncture; Approved Standard, 6th ed. Wayne, PA: Document GP41-A6, 2007, p 4, p 10.

[2] Hallbach J. Klinische Chemie und Hämatologie. 3rd ed. Stuttgart; Georg Thieme: 2011, p 6.

[3] Golder M, Chan C L H, O´Shea S, Corbett K, Chrystie I L, French G. Potential risk of cross-infection during peripheral-venous access by contamination of tourniquets. The Lancet 2000, 355: 44.

[4] Rourke C, Bates C, Read R C. Poor hospital infection control practice in venepuncture and use of tourniquets. J Hosp Infect 2001, 59–61.

[5] OSHA Disposal of contaminated needles and blood tube holders used for phlebotomy, https://www.osha.gov/dts/shib/shib101503.pdf, 1910.1030(d)(2)(vii)(A)

[6] ISO 6009. Hypodermic needles for single-use – Colour coding for identification. Geneva: International Organization for Standardization (ISO), 1992.

[7] ISO 9626. Stainless steel needle tubing for the manufacture of medical devices. Geneva: International Organization for Standardization (ISO), 2002–03.

[8] Clinical and Laboratory Standards Institute (CLSI). Procedures for the Collection of Diagnostic Blood Specimens by Venipuncture; Approved Standard, 6th ed. Wayne, PA, USA; CLSI: Document GP41-A6, 2007, p 159.

[9] Lippi G, Avanzini P, Aloe R, Cervellin G. Reduction of gross hemolysis in catheter-drawn blood using Greiner Holdex® tube holder. Biochemia Medica (Zagreb) 2013; 23: pages?

[10] Council Directive 2010/32/EU of 10 May 2010 implementing the Framework Agreement on prevention from sharp injuries in the hospital and healthcare sector concluded by HOSPEEM and EPSU.

[11] European Biosafety Network: Implementation Guidance for the EU Framework Agreement, p 7, www.europeanbiosafetynetwork.eu

[12] World Health Organization (WHO) Guidelines on drawing blood: best practices in phlebotomy, Geneva; WHO: 2010.

[13] Clinical Laboratory Standardization Institute (CLSI) Laboratory Automation: Specimen Container/ Specimen Carrier; Approved Standard, Wayne, PA,USA: Document AUTO1-A, 2000, p 13.

[14] United Nations (UN) (IATA) Recommendations on the transport of dangerous goods. Dangerous goods regulations 55th Edition, Document UN 3373, 2014.
[15] ISO/EN/DIN Single use containers for human venous blood specimen collection. 1995Geneva: International Organization for Standardization (ISO) Document 6710/EN 14820.1995/2004–11
[16] Hammerling JA. A Review of medical errors in laboratory diagnostics and where we are today, Lab Med. 2012; 36:42 f
[17] Clinical and Laboratory Standards Institute (CLSI). Laboratory Automation: Barcodes for Specimen Container Identification; Approved Standard, 2nd ed. Wayne, PA, USA: Clinical and Laboratory Standards Institute (CLSI), Document AUTO2-A2, Vol. 25, No. 29, 2005.
[18] ISO 15417. Information technology – Automatic identification and data capture techniques – code 128 bar code symbology specification. Geneva: International Standardization Office (ISO), Document 15417. 2007.
[19] Council Directive 2010/32/EU of 10 May 2010 implementing the Framework Agreement on prevention from sharp injuries in the hospital and healthcare sector concluded by HOSPEEM and EPSU.
[20] Clinical and Laboratory Standards Institute (CLSI). Procedures and Devices for the Collection of Diagnostic Capillary Blood Specimen. Approved Standard, 6th ed. Wayne, PA; CLSI. Document GP42-A6, 2008, p 4.
[21] World Health Organization (WHO), WHO guidelines on drawing blood: best practice in phlebotomy, 2010, p 36.
[22] Bruhn H D, Schäfer H, Junker R, Schreiber S. Labor Medizin, Indikationen, Methodik und Laborwerte, Pathophysiologie und Klinik, 3. Auflage, Ort: Verlag ? 2011, p 7.
[23] Preanalytics Manual. VACUETTE® Kremsmünster: Greiner-Bio-one, P 51, http://www.gbo.com/documents/980183_Preanalytikfibel_108x190_rev04_08_2012_e_lowres.pdf(25).
[24] See Instructions for use 980205 Rev. 05 11-2011 http://www.gbo.com/documents/980205_Urine_rev05_GB.pdf
[25] Clinical and Laboratory Standards Institution (CLSI) Urinalysis; Approved Guideline – 3rd ed. Wayne PA USA: CLSI, Document GP16-A3, 2009 5.6.4.

Provided by Christa Seipelt, Sarstedt

5.2 Materials and Techniques of Sampling Blood by Sarstedt

5.2.1 Sarstedt venous blood collection system: S-Monovette®

The invention of the S-Monovette® was a pioneering innovation in venous blood collection which led to quality improvements in the field of preanalytics. The S-Monovette® was the first system to combine two blood collection techniques, the aspiration and vacuum technique.

The S-Monovette® allows for traditional vacuum-style collection for robust veins, while also allowing gentle, syringe-style collection directly into the primary tube for the fragile veins of pediatric, geriatric and oncology patients. This special collection technique eliminates the dangerous and expensive "syringe and transfer" technique, and has been shown to improve the quality of samples collected from intravenous catheters [1, 2].

The vacuum in the S-Monovette® is created at the point of collection which ensures accurate and precise filling in every tube. As a result of this innovation, fill accuracy in the S-Monovette® is unaffected by altitude [3].

The S-Monovette® System requires only two components for use: the S-Monovette® and the Safety-Needle or Safety-Multifly®-Needle. The system offers tubes for both routine and specialty testing with standard color coding. S-Monovette® needles are pre-assembled for ease of use and individually wrapped, providing absolute confidence in product quality. Sarstedt takes needle safety seriously; this is why every Sarstedt Safety-Needle comes with a simple, one-handed safety closure device.

(a)

(b)

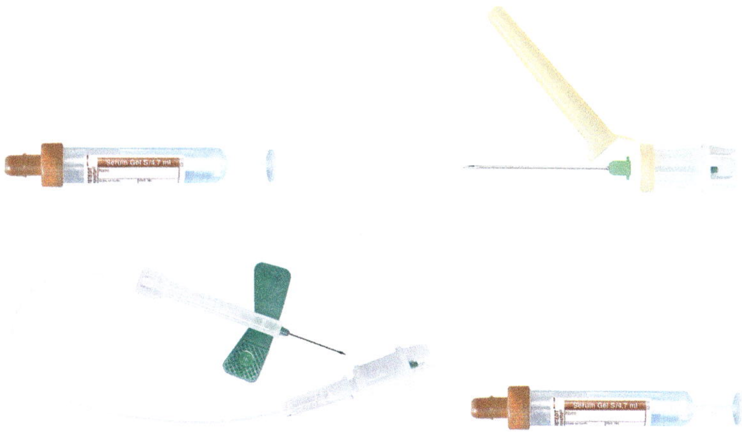

Fig. 5.2.1: Sarstedt S-Monovette® with Safety-needle (a) and Safety-Multifly®-Needle (b).

Reduction of diagnostic blood loss becomes more and more important [4]. The S-Monovette® with reduced dimensions and low nominal volume about 1 ml, also called S-Monovette® Pediatric, fulfills the requirements for patients and modern analyzers as follows:

- These low volume tubes can be used to reduce diagnostic blood loss due to phlebotomy by more than 70 % [5], which is especially beneficial in pediatric and critically ill patient populations.
- The sensitivity of modern analyzer systems increases more and more so that the minimum sample volume for routine tests can be reduced.

Sarstedt especially developed an 80 mm tubing on the Safety-Multifly®-Needle in order to further minimize phlebotomy-related blood loss (Fig. 5.2.2).

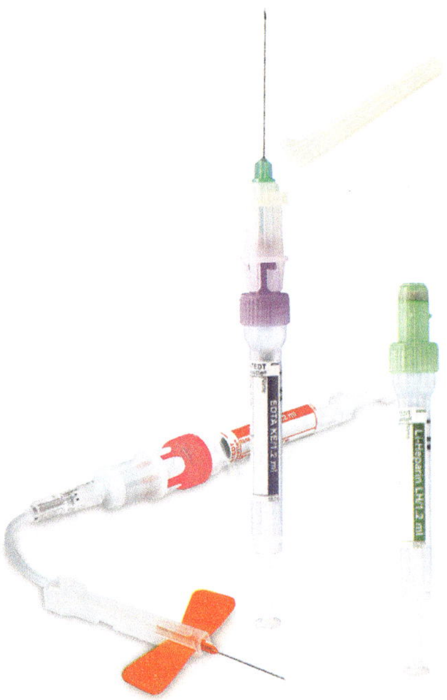

Fig. 5.2.2: S-Monovette® with Safety-Multifly®-Needle.

The S-Monovette® is available in a variety of diameters and heights to accommodate various automation and analyzer systems and with either paper or clear labels to accommodate global standards, and every label features a clear fill volume line. S-Monovette® caps can be safely and easily removed for open tube sampling or may be left in place to accommodate cap-piercing sampling.

The Sarstedt S-Monovette® tubes are in accordance with DIN ISO 6710 [6], EN 14820 [7] and CLSI H1-A6 [8] and H21-A5 [9]. In different publications and CLSI H1-A6 the equivalent use for both K_2EDTA and K_3 EDTA is documented [10, 11, 12, 13].

5.2.2 Sarstedt capillary blood collection: Microvette®, Multivette®, Minivette® POCT

The unique requirements of capillary blood collection shaped the development of our capillary blood collection systems. Capillary collection for different patient populations (neonates, infants, adults and elderly) demands special designs and performance requirements to ensure excellent patient care and superior sample quality. Sarstedt's capillary systems, including the Microvette®, Multivette®, Minivette® POCT as well as Incision -Lancet and Safety-Lancet meet these challenges.

Microvette® 100/200

The Microvette® 100/200 is available in a standard cylindrical tube or in a conical tube for easy processing (Fig. 5.2.3). Both 100 μl and 200 μl versions offer the option of blood collection using tube rim or assembled End-to-End capillary, are color coded to indicate the additive, and feature a fill line for volume accuracy.

Fig. 5.2.3: Microvette® 100/200.

Microvette® 300/500

The Microvette® 300/500 is ideal for open-tube, "gravity flow" collection (with rim). Common additives are available in both the 300 µl and 500 µl options, and serum gel is also available in the 500 µl tube (Fig. 5.2.4).

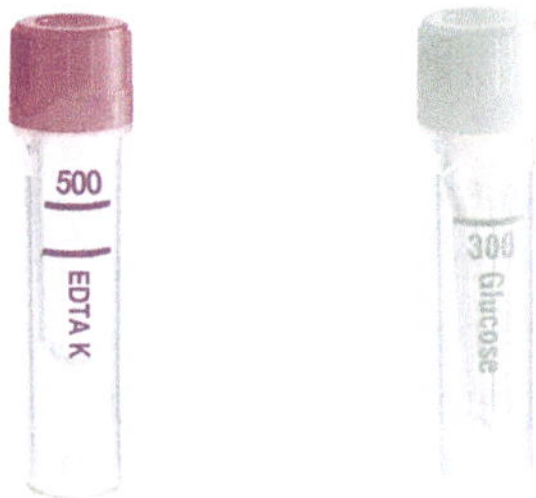

Fig. 5.2.4: Microvette® 300/500.

All Microvette® tubes come with a twist cap for easy opening and secure transport. The Microvette® CB 300 is a unique, specially-designed, automation-friendly, 300 µl capillary system that provides the maximum possible serum/plasma recovery (Fig. 5.2.5).

Fig. 5.2.5: Microvette® CB 300.

Multivette® 600

The Multivette® 600 accommodates venous or capillary blood collection with the same device. This unique all-in-one, 600 µl blood tube allows ultra low volume, venous collection or convenient and hygienic capillary collection. The Multivette® 600 comes with standard additives and color codes, including a serum gel option (Fig. 5.2.6).

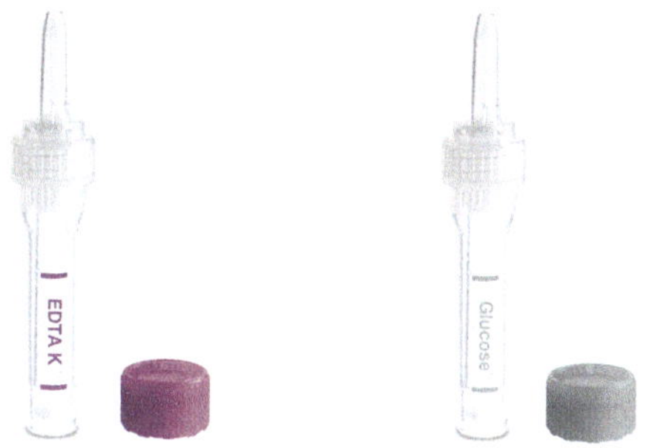

Fig. 5.2.6: Multivette® 600.

Safety Lancets

Sarstedt Safety Lancets are sterile, primed and ready-to-use, single-use devices that feature an automatically retracting blade or needle, and are offered in a range of penetration depths and widths for personalized blood collection. Safety Lancets are easy to use and are color coded to indicate type (Fig. 5.2.7).

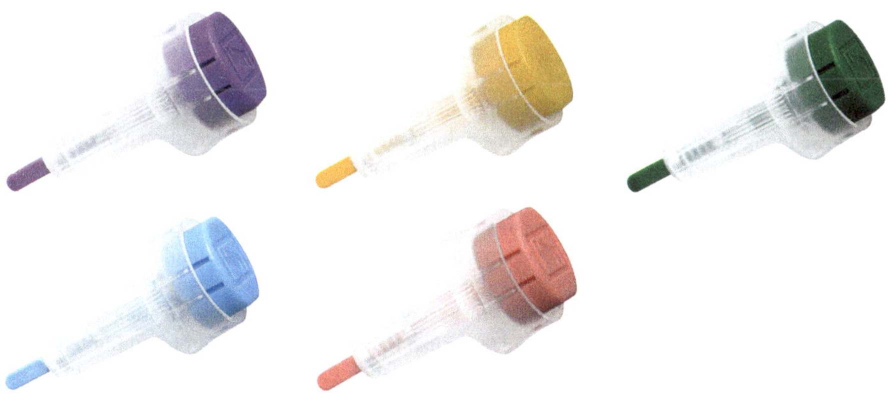

Fig. 5.2.7: Safety lancets.

Minivette® POCT

The Sarstedt Minivette® POCT is an advance in capillary collections for POCT applications. The special design of the Minivette® POCT provides for precise volume whole blood collection and easy dispensing from the same device. The Minivette® POCT comes with color-coded Heparin or EDTA treatments or without additive and is available in volumes from 10 µl to 200 µl (Fig. 5.2.8).

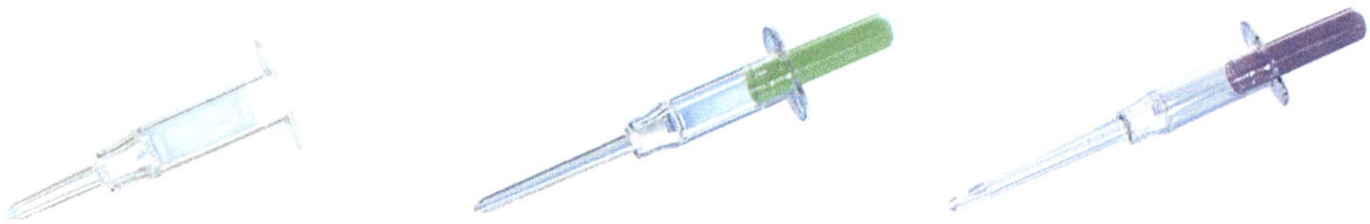

Fig. 5.2.8: Minivette® POCT.

Sarstedt offers special carrier tubes for the Microvette® and Multivette® tubes to allow for automated sample processing and analysis in the laboratory.

References

[1] Lippi G, Avancini P, Cervellin G. Prevention of hemolysis in blood samples collected from intravenous catheters. Clin Biochem 2013; 46:561–4.

[2] Lippi G, Cervellin G, Matiuzzi C. Critical review and meta-analysis of spurious hemolysis in blood samples collected from intravenous catheters. Biochemia Medica 2013; 23:193–200.

[3] Gros N. Evacuated blood-collection tubes for haematological tests – a quality evaluation prior to their intended use for specimen collection. Clin Chem Lab Med 2013; 51:1043–51.

[4] Wisser H, van Ackeren K, Knoll E, Wisser H, Bertsch T. Blood loss from laboratory tests. Clin Chem 2003; 49:1651–5.

[5] Sanchez-Giron F, Alvarez – Mora F. Reduction of blood loss from laboratory testing in hospitalized adult patients using small-volume (pediatric) tubes. Arch Pathol Lab Med 2008; 132:1916–9.

[6] ISO, DIN 6710 Single use containers for human venous blood collection (Gefäße zur einmaligen Verwendung für die venöse Blutentnahme beim Menschen). Geneve: International Organization for Standardization (ISO) and Berlin: Beuth-Verlag; 2007.

[7] European Standard (EN) 14820.Single-use containers for human venous blood collection specimen collection. Bruxelles: European Committee for Standardization (CEN) 2004.

[8] CLSI H1-A6. Tubes and Additives for Venous Blood Collection; Approved Standard – 6th ed. Wayne Pa, USA: Clinical and Laboratory Standards Institute (CLSI), Document H1-A6, 2010.

[9] CLSI H21-A5 Collection, Transport, and Processing of Blood Specimens for Testing Plasma-Based Coagulation Assays and Molecular Hemostasis Assays; Approved Guideline – 5th ed. Wayne PA, USA: Clinical and Laboratory Standards Institute (CLSI), Document H21-A5,2008.

[10] Philips J, Coiner J, Smith E, Becker D, Leong J. Performance of K_2EDTA- vs K_3EDTA-collected blood specimen on various hematology analyzers. Lab Hematol 1998; 4:17–20.

[11] Leathem S, Zantek ND, Kemper M, Korte A, Langeberg A, Sandler SG. Equivalence of spray-dried K_2EDTA, spray dried K_3EDTA, and liquid K_3EDTA anticoagulated blood samples for routine blood center or transfusion service testing. Immunohematology 2003; 19:117–21.
[12] Brunson D, Smith D, Bak A, Przyk E, Sheridan B, Muncer DL. Comparing hematology anticoagulants: K_2EDTA vs. K_3EDTA. Lab Hematol 1995; 1:112–9.
[13] Goosens W, van Duppen V, Verwilghen RL. K_2- or K_3EDTA: The anticoagulant of choice in routine haematology. Clin Lab Haematol 1991; 13:291–5.

Kathrin Schlueter, Stephen Church

5.3 BD Preanalytical Systems – Diagnostic Sample Collection

5.3.1 BD Vacutainer® Blood Collection System for venous blood sampling

The blood collection system is the connecting element in the complex process of laboratory diagnostics, from the patient's bed to archiving samples for add-on testing. The quality of the sample has huge consequences for health care systems and patients, as it can lead to inefficiencies in the workflow, delay in diagnosis and even inappropriate therapy.

While compromised sample quality is mainly caused by non-compliances during blood collection and subsequent handling during transport and preanalytical sample preparation, manufacturers can incorporate design features in their products that have the potential to minimize the impact of these errors.

The BD Vacutainer® Blood Collection System consists mainly of three elements: the evacuated tubes (in different sizes and with different additives), the cannula (either straight or a wingset with tubing), and the holder (a plastic cylinder ensuring correct positioning and safe blood collection) (Fig. 5.3.1).

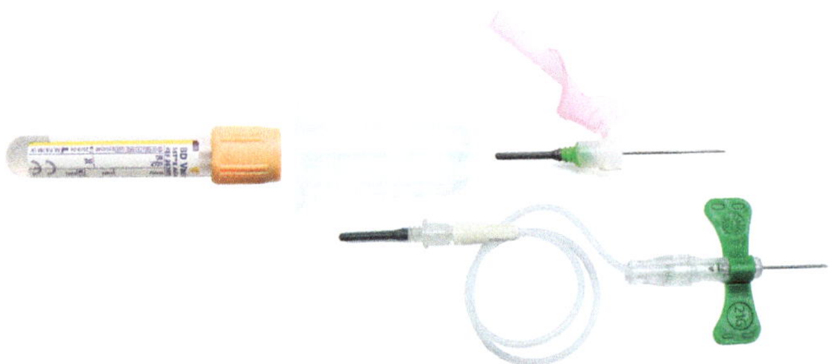

Fig. 5.3.1: BD Vacutainer® Blood Collection System: tube–holder – cannula, straight resp. wingset.

The standardized, closed vacuum blood collection method enables hygienic handling and ease of use. The blood flows into the tube steadily with decreasing flow rate until the vacuum is exhausted and the blood to additive ratio is correct for optimal sample quality. All BD Vacutainer® tubes are sterilized.

To ensure instrument compatibility, tube dimensions are restricted to two lengths and two diameters: 100 mm × 16 mm, 100 × 13 mm, 75 mm × 13 mm. The exceptions are specialized tube sizes for specific applications such as erythrocyte sedimentation rate (ESR) analysis with citrate anticoagulant (dilution 1:5) or Peripheral Blood

Mononuclear Cell (PBMC) isolation. The level of vacuum is set during manufacturing and defines the nominal fill volume of the tubes, ranging from 2 to 10 mL. Tubes are defined as those that completely fill 'full draw tubes' or those that partially fill 'partial draw tubes'. BD Vacutainer® tubes have an indicator for the nominal fill line, with the exception of the plastic coagulation tubes. These have a 360° minimum fill line imprinted into the plastic for improved visibility. The vacuum is maintained to ensure that the correct blood to additive ratio is achieved throughout the shelf life of the tubes. The BD Hemogard™ closure with its inner rubber stopper and outer plastic shield ensures not only that the vacuum is maintained within shelf life, but also minimizes the potential exposure to the blood sample. The tube closure is suitable for automatic cap piercing by laboratory instrumentation. Tube closures are color coded according to the international standard ISO 6710 [1] and the WHO guidelines [2] on drawing blood: best practices in phlebotomy, 2010. Different label formats are available, paper labels with and without predefined identification fields, pre-barcoded labels and transparent labelling of the tubes.

The standard material used for BD Vacutainer® tubes is a medical grade PET (polyethylene terephthalate), which is virtually unbreakable and clear, allowing visual assessment of the sample. For BD Vacutainer® coagulation tubes, two different plastic materials are combined to ensure the best surface for coagulation assays: an outside PET layer and an inner polypropylene layer. In addition, with this specific tube design headspace is minimized even with small nominal blood volumes. In coagulation assays, it has been shown that too much headspace can lead to falsely shortened clotting times, specifically for patients undergoing unfractionated heparin therapy [3, 4]. For coagulation analysis tests with special requirements and for serum analysis, there are still glass BD Vacutainer® tubes available. The PET BD Vacutainer® tubes are compliant with international transport regulations such as IATA and P650.

In BD Vacutainer® tubes, the additives and their ultimate concentrations in the blood sample are compliant with ISO 6710 [1], EN 14820 [5] and CLSI H1-A6 [6] and H21-A5 [7]. Furthermore, they are compliant with the recommendations of the GEHT (Groupe d'Etude sur l'Hémostase et la Thrombose: buffered citrate solution for coagulation analysis is preferred) and ICSH (International Council Society of Haematology: K_2EDTA for hematology is preferred) [8, 9].

For the additives to function efficiently, it is important that they mix easily with the blood. Therefore, a special spray dry technology is used for most of the additives in BD Vacutainer® tubes. In BD Vacutainer® PET serum tubes, spray dried silica particles are used to efficiently initiate and speed up the clotting. Thrombin-based clot activators tubes are also available, and which provide further accelerated clotting with or without barrier. In coagulation tubes, the additive is a buffered citrate solution with a dilution factor of 1:9. The citrate concentration is 0.109 M, in some regions 0.129 M is also available. The ultimate concentration of K_2EDTA in a blood sample is 1.8 mg/mL. For special needs, K_3EDTA is available. For heparin anticoagulated samples, the lithium or the sodium salt of heparin can be chosen for clinical chemistry and cell analysis. Other special additive formulations are available, for example to prevent release of platelet factor 4, inhibit glycolytic enzymes or stabilize erythrocytes.

There are BD Vacutainer® tubes that contain a gel like thixotropic material at the bottom of the tube. This material gel moves during centrifugation and builds a stable barrier between the supernatant and the cells/clot. As cells are separated, analyte stability is improved in the sample after centrifugation, for example potassium is stable for up to one week at 4–8 °C in BD SST™ II tubes. Depending on the tube type, different gel materials are used such as polyester based or polyacrylic based. It should be noted that depending on the gel material, the storage time and temperature, the volume of the blood and the type of analyte (mainly its hydrophobicity), adsorption of the analyte to the gel can occur, altering the concentration of the analyte in the blood sample away from its actual in vivo concentration. It has been shown that an acrylic based gel, as used in BD SST™ II and BD PST™ II, is much less prone to adsorption than a polyester based gel [10]. BD SST™ II produces a robust gel barrier formation under a variety of centrifugation conditions, which can also be used to shorten turnaround time with shorter centrifugation times at a higher g-force [11].

Fig. 5.3.2: BD Vacutainer® tubes.

Special BD Vacutainer® blood collection tubes for more demanding downstream applications are available. BD CPT™ tubes can be used for isolation of PBMC in a single-step protocol within the primary tube. These tubes contain a gel with a specific density differing from the routine tubes´ gel, plus anticoagulant and FICOLL® solution[1]. After centrifugation, the PBMC are separated from erythrocytes and granulocytes by the gel barrier and can be remixed with the plasma. The PBMC are then stable for up to 24h, depending on the application. RNA is known to undergo changes upon blood collection, caused by gene induction and simultaneously, RNA degradation [12]. The additive in PAXgene® RNA tubes lyses the cells immediately after blood collection and stabilizes the RNA so that it can be stored in the primary tube before RNA isolation, allowing the analysis of the in vivo RNA profile. Further specific tubes are available

1 FICOLL is a registered trade mark of GE Healthcare companies.

for research use only for the stabilization of the plasma proteome or for stabilization of specific peptides important in the context of diabetes.

Depending on the vein conditions and the puncture site, different needle formats and sizes may be preferred for venipuncture, preferably with a safety mechanism to minimize the risk of needle stick injuries. The BD Vacutainer® Eclipse™ Blood Collection Needle (Fig. 5.3.3) minimizes the risk by featuring immediate one-handed activation at the puncture site. The safety shield is an integral part of the needle and no hard surface is needed for activation. An audible 'click' signal indicates that the safety shield has been mooved to its safe position. A variant of this needle provides a 'flash' indication that venipuncture has been successful. The BD Vacutainer® Safety-Lok™ Blood Collection Set is a safety engineered blood collection set that is simple and easy to use. The safety mechanism can be activated immediately after the blood draw and helps protect against needle stick injury (Fig. 5.3.3). The BD Vacutainer® Push Button Collection Set has push-button activation technology (Fig. 5.3.3). The push-button safety mechanism instantly helps protect against needlestick injury. Its in-vein activation reduces risk of healthcare worker exposure to a contaminated needle, provides easy activation without patient discomfort, and is ideal for use in high-risk environments. The one-handed safety activation of the push button allows for activation of the safety mechanism while still attending to the patient/venipuncture site. Both safety mechanisms have an audible signal that indicates that the safety mechanism has been activated. Studies have proven that these devices are efficient in reducing needle stick injuries (Fig. 5.3.3). Special straight needles for passive activation of the safety feature are available in some regions.

BD Vacutainer® Eclipse™ blood collection needle	BD Vacutainer® Safety-Lok™ blood collection set	BD Vacutainer® Push Button blood collection set

Study on efficiency in reducing needle stick injuries: 80% reduction compared to conventional needle during 1st year of use, no needle stick injuries during 2nd year anymore	Study on efficiency in reducing needle stick injuries: 81% reduction compared to conventional winged blood collection set	Study on efficiency in reducing needle stick injuries: 88% reduction compared to 1st generation safety winged blood collection set
[Visser 2006 (13)]	[Rogues et al, 2004 (14)]	[Hotaling 2009(15)]

Fig. 5.3.3: Safety needles for venous blood collection and their efficiency in reducing needle stick injuries.

The needles (preassembled with the BD Vacutainer® Luer-Adapter) have to be attached to the holder which is compatible with all BD Vacutainer® Blood Collection Tubes. In addition, the containers of the BD BACTEC™ Blood Culture System are compatible with the holder, omitting the need to transfer blood from a syringe into a blood culture bottle with its associated risk of needle stick injuries and compromising hygiene. For all needles, single packed, sterile pre-attached versions with the BD Vacutainer® Holder are available. All winged blood collection sets are also suitable for short term infusion (without a BD Vacutainer® Luer-adapter). There are also options to connect the holder with a BD Vacutainer® Luer-Adapter for access to Luer systems.

For easier location of the vein, a tourniquet is usually used for venous stasis. During recent years, there has been increasing evidence that re-using tourniquets is an infection risk as these tourniquets are frequently contaminated with blood and may carry bacteria like MRSA [16–19]. Adopting a single-use tourniquet like the BD Vacutainer® Stretch Latex-Free Tourniquet can reduce the cross-contamination between patients. Also, this safeguards both health care workers and patients from latex allergies.

5.3.2 BD Microtainer® Blood Collection System for capillary blood sampling

A special set of blood collection tubes for very small blood volumes, intended for capillary blood collection, are available. Primarily, they are used whenever the blood volume for diagnostic tests has to be kept at a minimum, for example, for babies, infants, elderly patients or anaemic patients. Another reason to choose capillary blood sampling is when the vein conditions are very difficult, for example, for patients with very thin or fragile veins like oncology patients, obese patients or patients with severe burns. There are tubes available with the same additives as the venous blood collection tubes, except citrate. Capillary blood is a mixture containing undetermined proportions of blood from arterioles, venules, capillaries, plus interstitial and intracellular fluids. Due to this mixture, normal analyte values may differ from those of venous blood. Also, puncturing the skin releases thromboplastin, which activates the coagulation process; platelets aggregate at the puncture site forming a platelet plug. For these reasons, capillary blood is not suitable for coagulation testing with citrate plasma. Color coding of the BD Microtainer® Blood Collection Tubes is the same as for the previously described venous blood collection tubes for easier identification of the additives. The additives are spray-dried on the wall of the tube using a specific method, enabling the additives to mix with the blood as soon as it flows into the tube. Thus, microclots in EDTA and heparinized samples are minimized. The collector end of the BD Microtainer® Blood Collection Tubes ensures an easy flow of the blood into the tube where it immediately comes into contact with additives. Fill line indicators simplify identification of optimal blood to additive ratio for the user. The tubes have fill volumes ranging from 200 µL to 600 µL. After the BD Microgard® Closure has been used to close the tube, mixing is important for optimal dispersal of the additive in the blood. Adapters are available for the BD Microtainer® Blood Collection Tubes to enable processing in standard sized instrumentation. For the BD Microtainer® MAP Microtube for Automated Process, no additional adapters are needed: this tube has the outer dimensions of a venous blood collection tube of 13x75 mm and a pierceable closure for direct sampling by the instrument. No manual processing is required. The tube is compatible with direct processing on most hematology analysers. Also, the size of the tube allows for adequate space for labelling with a patient's information, avoiding sample misidentification. It has been shown that Turnaround Time (TAT) was improved significantly by using BD Microtainer® MAP Microtubes because six process steps were not necessary with these tubes, including relabelling, remixing prior to analysis, change of the instrument program to manual mode, uncapping of tube, rechecking for clot and recapping after analysis [20].

Fig. 5.3.4: BD Microtainer® Blood Collection Tubes including BD Microtainer® MAP Microtube for Automated Process (top left).

There are two types of lancing devices that are used for collection of capillary blood: puncture devices and incision devices. Puncture devices such as BD Microtainer® Contact-Activated Lancets puncture the skin by inserting either a needle or blade vertically into the tissue. Puncture devices are preferable for sites that are repeatedly punctured. Incision devices like BD Microtainer® Quikheel™ Lancets slice through the capillary beds and are less painful than puncture devices. They require fewer repeat incisions and shorter collection times and are recommended especially for infant heelsticks [21]. Both types of devices are available in a variety of styles, sizes and depths. Lancets for puncture are available from 30 G to 21 G and a 1.5 mm blade with depth from 1.5 mm to 2.0 mm, leading to a single drop of blood up to 500 µL. Incision devices are available with blades of 1.75 mm to 2.5 mm and depths of 0.85 mm to 1.0 mm, optimal for low birth-weight, premature babies or full-term infants. Features common to all are sterility, single-use and permanently retracted blade/needle to reduce possibility of accidental needle stick injuries or reuse.

Fig. 5.3.5: BD Microtainer® Contact-Activated Lancet and BD Microtainer® Quikheel™ Lancet.

5.3.3 BD Blood Gas Syringes for arterial blood sampling

Blood gas syringes are offered in two different designs: BD A-Line™ for arterial blood collection from indwelling lines with aspiration and BD Preset™ with a venting membrane that fills under arterial blood pressure to a user-predefined volume, minimizing the presence of air in the sample. All BD blood gas syringes contain spray dried, calcium-balanced lithium-heparin that is mixed with the blood to prevent microclots. The sample can be used for blood gas analysis as well as for other emergency parameters. The use of Calcium-balanced lithium heparin ensures that the BD blood gas syringes can be used for the assessment of electrolytes. The syringes are available in two different sizes and with a Luer connection as well as Luer-Lok™ and different closures. Pre-assembled versions of the BD Preset™ with collection needle are available. The the one-hand activation of the BD Eclipse™ safety mechanism is identical to the venous blood collection needles and therefore simplifies handling.

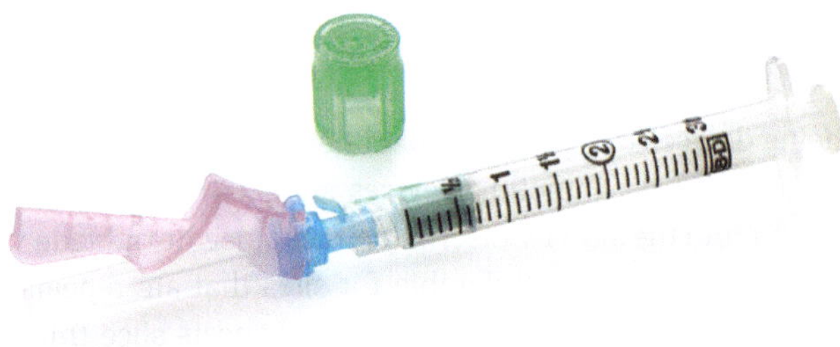

Fig. 5.3.6: BD Preset™ Eclipse™ Blood Gas Syringe.

5.3.4 BD Vacutainer® Urine Collection System

The BD Vacutainer® Urine Collection System range of products offers a closed system that benefits healthcare workers by reducing their need to come into contact with potentially hazardous specimens. The product portfolio includes a range of evacuated urine collection tubes utilising the same design and principles as the BD Vacutainer® Blood Collection tubes, a 24h urine container (depending on the region), a urine cup with sterile interior and the BD Vacutainer® Urine Collection Straw. Tubes are available without additive, and with preservatives. The BD Vacutainer® UAP Tubes contain a mercury-free preservative formulation with Ethyl Paraben, Sodium Propionate and Chlorhexidine. The BD Vacutainer® UAP & Non Additive tubes are designed for automated and manual chemistry dipstick urinalysis and to obtain sediment for examination. The BD Urine Preservative Tube (UAP) is designed to inhibit the metabolism of/or render nonviable the bacteria present in urine while maintaining their cellular integrity. The preservative allows for transport, testing and storage of the specimen up to 72 h at room temperature. Without the presence of a preservative, the bacteria continue to metabolize and reproduce, causing changes in the urine chemistry components measured in a routine urinalysis. Erythrocytes (RBC), leucocytes (WBC) and casts will breakdown over time, resulting in a decrease in numbers of these elements in the sample. The rate of this degradation is dependent on various factors such as patient pathology, urine pH, urine specific gravity, storage conditions, etc. These factors will limit the stability achieved for these formed elements when conducting urine microscopy. Samples collected in these tubes are not suitable for microbiological analysis.

In urine samples, bacteria may multiply at the same rate as in nutrient broth [22]. Therefore, a urine sample without preservative that is delayed in transit or left at room temperature for an extended period of time may give an erroneous result [23]. All BD Vacutainer® C&S Preservative Urine Tubes are intended for the collection and transport of urine samples for culture and sensitivity (C&S) testing. The tubes contain a lyophilized preservative of Boric Acid, Sodium Borate and Sodium Formate which maintains the bacterial population in the urine specimen for a period of up to 48 h at room temperature.

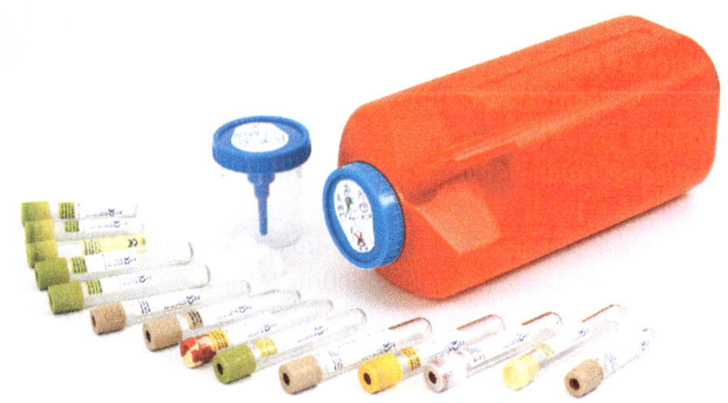

Fig. 5.3.7: BD Vacutainer® Urine Collection System.

A potential cause of needlestick injury is urine sampling through a Foley catheter. “Ideally, the most effective way of removing the hazard of a contaminated needle is to eliminate the needle completely by converting to a needleless system [22].” By using the BD Vacutainer® Luer-Lok™ Access Device, the sample can be transferred directly from the Foley catheter to the tube. This means fewer steps, less sample manipulation, and reduced risk of contamination.

The BD Vacutainer® Urine Collection Straw is a transfer device that allows for the transfer of urine from any open specimen collection cup or paediatric bag to one or more BD Vacutainer® urine tubes without exposure to the specimen. The BD Vacutainer Urine Collection cup with its sterile interior, can be used for subsequent C&S testing. The transfer device is an integral part of the lid of the cup. By using this closed system, pouring off, and potential leakage are avoided. Depending on the region, 3 L 24 h urine collection containers are available, similar to the urine cup with an integral transfer device. Once sampled from the various patient collection sites, the BD leak proof evacuated urine tubes can be safely transported to the laboratory for analysis.

References

[1] ISO/EN/DIN 6710 Single-use containers for human venous blood specimen collection. Geneva, Bruxelles, Berlin 2007.

[2] World Health Oranization (WHO) Guidelines on drawing blood: best practices in phlebotomy,Geneva; WHO: 2010.

[3] PA Toulon, V Eschvege, M Dreyfus, T Boutekedjiret, V Proulle: Are citrated partial-draw polymer collection tubes adequate for basic coagulation tests? Journal of Thrombosis and Haemostasis 2009; Volume 7, Supplement 2: Abstract PP-TH-470.

[4] JB Lawrence: Preanalytical Variables in the Coagulation Laboratory. Lab Med, Jan 2003; 34:49–57.

[5] EN 1482: Single use container for human veneous blood specimen collection. Bruxelles 2004.

[6] Clinical and Laboratory Standards Institute (CLSI) Tubes and Additives for veneous and Capillary Blood Collection. Approved Standard H1-A6.Wayne:CLSI 2010.

[7] CLSI Collection, Transport and Processing of Blood Specimen for Testing Plasma-based Coagulation Assays and Molecular Hemostasis; approved guidelines H21-A5, Wayne:CLSI 2008.

[8] Polack B, Schved JF, Boneu B, Groupe d´Etudes sur l´Hemostase et la Thrombose: Preanalytical recommendation of the Groupe d´etude de l´Hemostase et la Thrombose (GEHT) for veneous blood testing in hemostasis laboratories. Haemostasis 2001; 31:61–8.

[9] International Council for Standardization in Haematology. Recommendations of the ICSH for ethylenediamine tetraacetic acid anticoagulation of blood for blood cell counting and sizing. Expert panel on cytometry. Am J Clin Pathol. 1993; 100:371–2.

[10] Bush V, Blennerhasset J, Wells A, Dasgupta A. Stability of therapeutic drugs in serum collected in vacutainer serum separator tubes containing a new gel (SST II). Ther Drug Monit, 2001; 23:259–62.

[11] Mensel B, Wenzel U, Roser M, Lüdemann J, Nauck M. Considerably reduced centrifugation time without increased hemolysis: Evaluation of the new BD Vacutainer® SST™II Advance. Clin Chem 2007; 53:794–5.

[12] Rainen L, Oelmueller U, Jurgensen S, Wyrich R, Ballas C, Schram J, et al.. Stabilization of mRNA expression in whole blood samples. Clin Chem 2002; 48:1883–90.

[13] Visser L. Toronto hospital reduces sharps injuries by 80 %, eliminates blood collection injuries. A case study: Toronto East General Hospital pioneers healthcare worker safety. Healthc Q, 2006; 9:68–70.

[14] Rogues AM, Verdun-Esquer C, Buisson-Valles I, Laville MF, Lasheras A, Sarrat A, et al. Impact of safety devices for preventing percutaneous injuries related to phlebotomy procedures in health care workers. Am J Infect Control, 2004; 32:441–4.
[15] Hotaling M. A retractable winged steel (butterfly) needle performance improvement project. Jt Comm J Qual Patient Saf, 2009; 35:100–5, 61. Index
[16] Golder M, Chan CL, O'Shea S, Corbett K, Chrystie IL, French G. Potential risk of cross-infection during peripheral-venous access by contamination of tourniquets. Lancet 2000; 355: 44.
[17] Rourke C, Bates C, Read RC. Poor hospital infection control practice in venepuncture and use of tourniquets. J Hosp Infect, 2001; 49:59–61.
[18] Hensley DM, Krauland KJ, McGlasson DL. Acinetobacter baumannii and MRSA contamination on reusable phlebotomy tourniquets. Clin Lab Sci 2010; 23:151–6.
[19] Elhassan HA, Dixon T. MRSA contaminated venepuncture tourniquets in clinical practice. Postgrad. Med. J. 2012; 88:194–7.
[20] Park SH, Chi HS, Choi MO, Park BG, Jang S, Park CJ. Improved turnaround time for neonatal hematology profile tests (complete blood count) using a new microcollection tube. Clin Chem Lab Med, 2011; 49:1083–5.
[21] Shah V, Taddio A, Kulasekaran K, O'Brien L, Perkins E, Kelly E. Evaluation of a new lancet device (BD QuikHeel) on pain response and success of procedure in term neonates. Arch Pediatr Adolesc Med, 2003; 157:1075–8.
[22] Perry J, Parker G, Jagger J. EPINet Report: 2001 percutaneous injury rates. Adv Exposure Prev 2003; 6:32–6.
[23] O'Grady F, Catell WR. Kinetics of urinary tract infections. Br J Urol. 1966; 38:149–51.
[24] Hindman R, Tronic B, Bartlett R: Effect of delay on culture of urine. J. Clin. Microbiol. 1976; 4:102–3.

6. Specimen Processing in the Preanalytical Phase

Walter G. Guder, Sheshadri Narayanan

6.1 Sample Transport, Treatment after Arrival, Storage and Disposal

The different aspects of sample handling during transport and storage have been published in a compendium preanalytics, which is translated for the first time here [1] actualizing earlier publications [2].

6.1.1 Preparation of sample for transport

Before samples are to be transported, they should be tested regarding identification, safety conditions and stability (see Annex). Errors in either sample identification, sample preparation before or after transport have an adverse effect on the patient if they are not detected in time. Thus the temperature during storage and transport is of special relevance. The aim is to maintain the biological sample unchanged during these procedures.

6.1.2 Storage conditions during transport

According to the ISO-Guide 30, "Terms and definitions used in connection with reference materials" [3] stability is defined as the ability of sample material to maintain the original conditions of an analyte unchanged in well defined limits over a defined time sequence.

Since sample stability is a complex question, only some aspects of which may be named here: imminent are the internal conditions that are inherent with the sample itself, which depend on the analyte (i. e. aPTT, osmolality, etc.) as well as external influences (like temperature, reagents used, sample containers, stabilizers, light, etc.) may be of major impact. Conditions published in the literature are often contradictory and more often than not, the conditions not clearly described. The working group on extraanalytical quality therefore has published a yearly updated list whose present version is provided in the Annex [4].

6.1.3 Transport of samples

6.1.3.1 Transport systems [5, 6]

Choosing the optimal transport system depends on several factors depending on the organizational structure of health systems (central laboratory or point of care test site?).

- The medical needs regarding turn around time
- Traditional aspects to be considered (like personnel, transport distance, technical equipment)

In general, we differentiate the so-called classical transport systems (like personal transport messengers, courier by car, posting, in seldom cases by train, boat or air-transport) and mechanized transport systems present in the same institution (pneumatic or cast transport systems).

6.1.3.2 Legal and medical rules during transport

Transport of infectious material is regulated by legal conditions described in international agreements on transport of dangerous goods in road traffic, train, air and ships [7–12]. Whenever an infectious substance is suspected or the presence of an infectious disease is confirmed, the sample is classified as dangerous goods and danger group 6.2. The package material class 6.2 is needed using the packing material of prescription P 620 [8, 11].

In Europe there are special details to be considered when dangerous goods are transported. Here the European Standard EN 829 [13] for in vitro diagnostic systems is to be applied. In the standard detailed definitions, requests and test conditions are described for package material, sample container's absorbing materials and protecting containers.

In the United States the Guideline "Procedures for the Handling and Processing of Blood Specimens" of CLSI, H18-A2 is valid [5].

- The responsible sender (either the laboratory chief, the sending physician or the nursing personnel) has to follow the TRBA 250 [14], and carries the responsibility for the respective classification of tthe diagnostic goods including respective categories A or B [15, 16] as well as for the correct package.
- When transport services of the intended laboratory is involved, the correct package is under the responsibility of the driving person, usualy the car driver of the organization.
- The sender or the driver is responsible for correctly following the rules of the dangerous goods transport.
- All persons involved in sampling, storage or transport of infectious samples are to be protected in order that they are not infected.
- All persons involved have to ensure, that the sampling, transport and storage of cultures and other infectious material do not hurt any person involved or the public.

6.1.3.3 Package

The triple packaging system for transport of biological substances and cultures consists of the following parts [17].

6.1.3.3.1 Package directive P650

According to the directive P 650 concerning package [14, 17] packages have to be of high quality and resistance to withstand all changes that may occur during transport, including transfers between different kinds of transport systems while also preventing any loss of sample material. The packages have to be constructed in a way to prevent any loss of contents by either vibration, temperature, pressure or humidity changes.

The package consists of three parts:

a) the primary sample container
b) a secondary container and
c) the outer package.

For transport each piece has to be labeled with a rhomb shaped label according to "UN 3373" (Fig. 6.1.1). The surrounding line has to be 2 mm broad at the minimum and, letters and numbers are to be printed to at least 6 mm height.

When posting samples, the outer package has to be marked with the label "DIAGNOSTIC SPECIMENS" [9].

Package material and composition are described in the P 650 standard for liquid as well as for solid materials.

Fig. 6.1.1: Package label according to package directive P 650.

Transport of potentially infectious material according to danger category 6.2: [8, 14, 15]

Potentially infectious goods according to the ADR definition are materials, which are known or to be assumed to contain pathogenic substances.

Potentially infectious materials are to be classified as belonging to category 6.2 of dangerous goods and may belong to the cases potentially infectious goods with the UN-number 2814, 2900 or 3373.

They are classified into the following categories:

Since January 1 2005, the former risk categories of WHO [19] were replaced by the following risk categories A and B [15].

Category A:
This is a pathogen in either humans or animals, that is capable of causing permanent disability or life theatening or fatal disease.

Remark: An exposition has to be assumed, if potentially infectious material has been released from the package and had physical contact with humans or animals. This category covers all germs of the hitherto risk group 4 of WHO [19], that is ebola-, lassa- and smallpox-viruses.

Potentially infectious materials, which fulfill these criteria and are able to cause diseases in either humans or animals, fall under the UN-Number 2814. Substances of the same kind, which cause diseases only in children are summarized under the UN-number 2900. They are to be marked as follows:

UN 2814 POTENTIALLY INFECTIOUS MATERIALS; DANGEROUS FOR HUMANS

UN 2900 POTENTIALLY INFECTIOUS MATERIALS; DANGEROUS ONLY FOR ANIMALS

These substances are to be packed with proven package materials according to P 620 and are to be transported according to the directive for dangerous goods [8, 14].

Category B:
All potentially infectious materials which are not classified as belonging to category A. All materials falling under category B fall under UN-number 3373, with the exception of newly defined cultures, which, depending on the individual case, fall UN-numbers 2814 or 2900.

Category B includes all pathogenic germs which before belonged to risk group 2 (i. e. influenza virus, salmonella types) and risk group 3 (i. e. mycobacterium tuberculosis, HIV or hepatitis B and C).

The official name to be labeled is for UN 3373: DIAGNOSTIC SAMPLES (BIOLOGICAL SUBSTANCE) or UN 3373 CLINICAL SAMPLES

They are to be packed according to the packaging regulations of P 650, which are followed up by directives for dangerous goods, like those of ADR [8, 12] or IATA-DGR [11].

Examples from the European Standard EN 829 [17]

- When posting diagnostic samples, injection needles have to be separated.
- Glass slides (like blood smears) have to be packed in such a way as not to be damaged if knocked, by high pressure or shaking.
- When stool samples are posted, sample tubes are to be covered by a second screw capped container.
- When mailing dried blood specimen on filter paper, place in a stronger paper envelope and then seal in plastic lined, padded post bag.This provides protection against potentially infectious dried blood specimen and ensures the integrity of specimen during transport.

- Samples are to be protected from direct light, to keep light sensitive analytes like bilirubin unchanged.
- For posting or shipping frozen or refrigerated specimens, an insulating material such as polystyrene container is adequate. Dry ice is to be used for freezing. Caution should be taken to insure the container packed with dry ice is able to release carbon dioxide gas so as to avoid a build up of pressure that could cause the package to explode.

Cultures

Cultures are the result of a process, where pathogens (infectious germs causing disease) are multiplied and enlarged in amounts, which normally do not appear in natural surroundings. For this reason they are especially infectious and present a high risk of potential infection when contacted.

This includes especially subcultures, which usually consist of diagnostic samples of isolated microorganisms. These are usually transported in stick- or flat agar plates if not in special transport media and can serve as confirming culture for further diagnostic procedures. Subcultures for standardization- and quality-assurance purposes likewise fall under these definitions [8, 10].

Cultures for diagnostic or clinical purposes are not included in this definition. Cultures for diagnostic purposes of the former risk group 2 and 3 are now the same as general diagnostic specimen. For these cultures category B UN-Nr. 3373, to be marked with the label "DIAGNOSTIC SAMPLE" and the packaging directive P 650 is to be applied. This classification, valid since January 1 2005, offers considerable reliefs for the posting of culture standards by special and reference labratories which are needed for further diagnostic steps like subtyping, resistance determination and epidemiological purposes [16].

Liquid materials

For liquid materials the package material has to contain three parts:

a. Inner package (the primary sample container): This is to be fluid tight.
b. The second part of package (secondary vessel) is a water tight protecting vessel consisting of plastic with a screw drive closure. This secondary vessel for liquid samples has to contain sufficient absorbing material to absorb the total liquid in case of a leak. It is possible to put several primary tubes into one secondary vessel. Each primary tube is to be surrounded by absorbing material or neeed to be separated in a way to prevent contact with each other in order to prevent contamination of one sample with the liquid from the others. The outer packaging material should remain untouched and uncontaminated. (The primary tube or the secondary container should withstand a pressure difference of at least 95 kPa (0.95 bar) without any loss of filling liquid).
c. The outer package material (Posting cover), secured by its own cushion material.

Solid Material

For solid substances the package likewise consists of three parts:

a. Inner package (the primary sample container): This should be dust tight.
b. The second part of package (secondary vessel): This should be dust tight. In case several primary tubes are covered by only one secondary tube, they have to be separately covered with adequate material or separated from each other, thus preventing any contact between them.
c. Outer package (posting cover).

Infectious material as pathogens

Pathogens are microorganisms (including bacteria, viruses, rickettsiae, parasites and fungi) and other infectious substances like priones, which can cause diseases in animals and men.

Package material according to P-620 and P-650 differs in the secondary and outer packaging. The secondary container or the outer package should be rigid.

For liquids, absorbent material in sufficient capacity to absorb the total liquid in case of a leak should be placed between the primary and secondary container. When multiple fragile primary containers are placed in a secondary packaging, they should be either individually wrapped or separated to avoid contact between them.

For mailing except human samples, an envelope of tear-proof paper or plastic sheet is sufficient. In Germany mailing of all infectious materials and cultures of risk Category A and cultures of WHO risk group 3 is not allowed.

The triple packaging system for packing infectious substances Category B or except human specimen should consist of:

1. primary leak proof, water tight receptacle: absorbent material;
2. secondary leak proof, rigid, watertight container;
3. outer package consisting of cardboard (wood, suitable plastic or metal) for biological substance Category B; 3a = envelope for except human specimen.

Always remove injection needles when mailing blood sampling systems [18]. Package glass slides adequately to ensure they do not get damaged if knocked, dropped or if pressure is applied.

6.1.4 Disposal

Although the use of safe needle containers has increased since the 1980s and safety products are increasingly in use, one third of the needle stick injuries occur during neddle disposal [18].

6.1.4.1 Needles and other sharp objects

The safe disposal of all sharp objects like injection needles, cannula, etc. is accomplished by placing them into tight and leak-proof, puncture-resistant containers with appropriate labels shown in Fig. 6.1.2 [19].

Fig. 6.1.2: Disposal of needle into sharps container (kindly provided by BD-Heidelberg, Germany).

6.1.4.2 Tube and sample disposal

Specimen-collection tubes containing blood or other body fluids should be disposed using special safety devices. This is possible by using breakage and puncture resistant containers, so called biohazard bags, which can withstand autoclaving and can be transported into burning or sterilizing devices without infectious hazard for the persons involved [18].

6.1.4.3 Chemicals

Toxic, corrosive and inflammable or reactive reagents should not be used as stabilizers in the preanalytical phase [5].

6.1.5 Sample handling after arrival

6.1.5.1 Centrifugation

Centrifugation of serum tubes with coagulated blood should be done after the coagulation process is complete. This usually takes 30 min after sampling. In patients under anticoagulants therapy or having disturbances in coagulation process, longer coagulation time is to be expected. Centrifugation usually is to be performed at 20–22 °C. Only samples with thermolabile analytes to be measured should be centrifuged at 4–6 °C.

Blood samples should not be recentrifuged after serum or plasma separation. This may change the plasma water cell volume relationship leading to changes in analyte concentration in plasma/serum. Samples with separator gels likewise should be cenrifuged only once.

Centrifugation of blood samples to separate plasma or serum usually requires a little over 10 min at 1500–2000 *g*. To obtain platelet poor plasma (<10.000 platelets/µL) centrifugation for 10–15 min at 2000–3000 *g* is recommended. Anticoagulated blood for either coagulation tests or measurement of plasma constituents should be centrifuged over 10–15 min at 2000–2500 *g* [2, 23]. After centrifugation the quality of centrifuged samples should be controlled by a visual inspection.

The relative centrifugal force (RCF) can be calculated using an empirical formula or be read from a respective normogram [2]. The formula for calculating RCF is:

$$\text{RCF} = 1{,}118 \times 10^{-5} \times r \times rpm^{2}$$

In this formula r is the distance of the centriguged tube's bottom to the centrifugal axis in cm (= centrifugal radius) and rpm the centrifugal speed in rotations per min. $1{,}118 \times 10^{-5}$ is a constant which is derived from the centrifugal force as multiple of the *g* constant, deriving centrifugal force as multiple of gravity (*g*).

6.1.5.2 Sample handling before and after centrifugation

When samples are to be distributed into secondary vessels, they should be handled with the same rules used for primary tubes regarding identification, sample storage and safety rules.

To reduce direct contact with blood or other potentially infectious materials sample aliquoting should be reduced to a minimum. Separator gels are especially helpful in reducing infectious risks. Likewise sample distributor systems have been shown to be of similar help in reducing infectious risks.

Special rules and further recommendations

- Avoid storage of whole blood. Refer to information on special storage-sensitive analytes
- Whole blood samples should arrive in the laboratory within 45 min after sampling to ensure centrifugation one hour after sampling.

- Prevent glycolysis to maintain stability of glucose, lactate and pH. This is usually achieved by a combination of respective anticoagulants and glycolysis inhibitors. After fluoride was seen to be not stopping glycolysis rapid enough, tubes containing acidified citrate was recently used to keep glucose stable before and after centrifugation [21]. Alternatively, centrifugation of blood collected in plasma gel separator tubes immediately after blood collection and storing plasma on the gel barrier ensures preservation of glucose over 2 days [4].
- Avoid exposure of samples to direct light. Otherwise bilirubin, vitamin C , porphyrins, creatine kinase or folic acid may decrease.
- Avoid direct air contact of sample as much as possible. Otherwise concentration of all nonvolatile constituents may increase because of water evaporation or sublimation.
- Do not store whole blood in refrigerators. When urine samples are cooled, salts may precipitate irreversibly (calcium, magnesium phosphate, uric acid).
- Some analytes should not be deep frozen, because this changes their concentration.

6.1.5.2.1 Correct freezing and thawing of samples

Recommendations regarding transport and storage of samples:

The working group on preanalytical quality of the German society for clinical chemistry and laboratory medicine published “Quality of Diagnostic Samples” which is a collection of validated data and recommendations regarding stability during transport and storage of samples and the use of stabilizers. These recommendations have been published by WHO and appeared woldwide in several languages [4, 20].

To ensure optimal sample stability, the following general rules should be considered:

1. The stability of an analyte in the sample matrix determines type and speed of action taken. Several important mechanisms cause changes in analyte’s concentration:
   - Metabolism of blood cells
   - Evaporation and sublimation
   - Chemical reactions in the matrix
   - Microbiological decomposition
   - Osmotic processes
   - Light effects
   - Gas diffusion
2. Rapid transport and short storage times improve the reliability of laboratory results.
3. Specimen and samples are stable for longer time, the cooler they are stored (remark exceptions) (see Table 6.1.1 and Annex).

4. Specimen and samples should be stored in closed containers to prevent evaporation.
5. The danger of evaporation also exists in refrigerators (condensation of liquid along the wall of the refrigerator).
6. Storage problems are minimized, when one-way sample containers are used for sampling.
7. Seperators (mostly plastic gels) improve yield of serum/plasma and allow to maintain serum/plasma above gels during storage if not frozen.
8. Prevent shaking of sample tubes and rapid deceleration during transport (pneumatic transport systems!) to prevent hemolysis.
9. Blood containing samples should always be kept upright to allow coagulation process to proceed effectively and build a small clot.
10. Infectious material is to be marked especially and to be handled with special care.

Table 6.1.1: Examples of analytes in blood and urine, which should not be deeply frozen.

Sample	Analyte
Serum/plasma	Lipoprotein-electrophoresis
	Lipoprotein X
	Apolipoprotein A I and B
	LDL-Cholesterol
	(can be prevented by adding glycerol)
	Fibrin monomer-positive plasma*
EDTA-blood	All hematological analytes
Urine	IgG, sediment, uric acid
	(precipitations!)

6.1.5.2.2 Correct thawing procedure of frozen samples

A deep-frozen sample should be rethawn at room temperature. A very common source of error is inadequate mixing after rethawing a deep-frozen sample. Concentration gradients are produced during thawing, as the concentrated solution melts first and then runs down inside the vessel along the inner wall. After thawing, the sample should be inverted several times, avoiding the formation of foam. After the thawing process is complete, the sample should be examined for detection of undissolved material. This may be redissolved by carefully warming the sample [2].

After analysis, samples should be stored in such a way as to permit the confirmation of results, checking the identity of samples or performing additional tests for medical or legal reasons.

6.1.5.3 Safe disposal

Using sharp objects is a daily routine in the medical area. Careful handling of these products is a prerequisite to avoid infectious risks. Smaller pricks and incisions into the skin or scratching with articles which were in contact or contaminated with blood or other body fluids, can have fatal medical and social consequences.

Therefore not only safe use but also safe disposal of sharps [18] is of great importance in the preanalytical phase.

Recently, several producers of tubes and needles offer so called safety needles and respective sharp containers, which are shown elsewhere in this book (see Chapter 5.1–5.3).

6.1.5.3.1 What characterizes safe sharp containers?

Experts demand:

- The container has to have a wide opening which allows one to directly dispose the needles and other sharps into the container without skin contact.
- A closure should permit the capping of the container between several uses to prevent loss of materials and any contact of infectious material or sharps with other persons.
- The container should be unpiercable for needles to a high degree, and should not allow loss of either liquid or sharps and is non breakable.

A new generation of sharps container should fulfill the requests of OSHA (Occupational Safety & Health Administration, USA), BSI (Brithish Standards Institute), NIOSH (National Institute for Occupational Safety & Health, USA), AFNOR (Agence Francaise de Normalisation) and the German institutions BauA (Bundesanstalt für Arbeitsschutz und Arbeitsmedizin) and HVBG (Hauptverband der gewerblichen Berufsgenossenschaften).

In detail the European Standard BS 7320, French Standard NF X30–500, the terms UN 3291 and in Germany the TRBA 250 (technical rules for biological working materials in healthcare and wellfare services (Technische Regeln für Biologische Arbeitsstoffe im Gesundheitsdienst und in der Wohlfahrtspflege)) / BGR 250 (Employers protective rules for safety and health during work [Berufsgenossenschaftliche Regeln für Sicherheit und Gesundheit bei der in Arbeit]) have to be met.

Table 6.1.2: Sample stability and external influences.

External Influence Factors	Examples of Analytes Concerned
Time and temperature	all analytes
Material of sample containers	tetrahydrocannabinol (THC) effect on coagulation markers, i. e. aPTT

Table 6.1.2: (continued)

External Influence Factors	Examples of Analytes Concerned
Closure of sample container	ethanol, pCO_2, pO_2
Gel separators, stabilizers	Various analytes, assay dependent
Surface/volume ratio of sample	all analytes, especially in pediatric samples
Centrifugal force (rcf)	all analytes, especially if sample containers are not closed, not cooled and centrifuged at high speed
Reagents	Coagulation tests like prothrombin time or aPTT
Antibodies	C_3-complement
Light	bilirubin, porphyrins, vitamin C
Sample preparation	
Freezing and rethawing	calcium, phosphate in urine
Mixing after rethawing	all analytes

Table 6.1.3: Sample stability and examples of influencing factors which are part of samples.

Sample immanent influences	Examples of analytes concerned
pH	Cells in urine
Osmolality	Cells in urine
Number and kind of microorganisms	Glucose in CSF and/or urine
Number, kind and degree of lysis of cells in samples (erythrocytes, platelets)	Potassium in serum
"Matrix effects" (i. e.proteins, lipids, etc.)	Difference between "normal" and "pathological" samples

References

[1] Guder WG, Hagemann P, Wisser H, Zawta B. Fokus Patientenprobe; Kompendium Präanalytik. CD Rom Heidelberg: BD 2007.

[2] Guder WG, Narayanan S, Wisser H, Zawta B. Diagnostic Samples: From the Patient to the Laboratory. 3rd ed. Weinheim: Whiley-Blackwell 2009. pp 40–1.

[3] ISO Guide 30. Terms and definitions used in connection with reference materials. Geneva: ISO 2007; 30, 2nd ed.

[4] Guder WG, Fiedler M, da Fonseca - Wollheim F, Schmitt Y, Töpfer G, Wisser H, Zawta B, Quality of Diagnostic Samples. 4th ed., BD-Diagnostics Oxford 2015. World Health Organization (WHO). Use of Anticoagulants in Diagnostic Laboratory Investigations. Geneva: WHO 1999 and WHO/Dil/Lab/99.1, 2002.

[5] Clinical and Laboratory Standards Institute (CLSI). Procedures for the Handling and Processing of Blood Specimes; Approved Guideline Wayne, PA: CLSI H18-A2, 1999.

[6] Clinical and Laboratory Standards Institue (CLSI). Collection, Transport, and Processing of Blood Specimens for Coagulation Testing and General Performance of Coagulation Assays, Wayne PA CLSI H21-A3, 1999.

[7] World Health Organization. Guidance on Regulations for Transport of Infectious Substances. Geneva: WHO 2008.
[8] ADR-Handbook. The European Agreement concerning the International Carriage of Dangerous Goods by Road (ADR). ADR-Handbuch 2003. Europäisches Übereinkommen über die internationale Beförderung gefährlicher Güter auf der Straße (ADR) und Gefahrgut-Beföderungsgesetz (GGBG) in the version of the GGBG-suplementary 2001.
[9] United Nations (UN); Convention concerning International Carriage of Dangerous Goods by Rail (RID) 14th revised ed. New York/Geneva: United Nations 2005.
[10] Ordnung über die internationale Eisenbahnbeförderung gefährlicher Güter (RID Rahmenrichtlinie 96/49 EG), ABL der EG Nr. L 235, S 25 vom 17.09. 1996.
[11] International Air Transport Association (IATA). Dangerous Goods, Regulation Montreal/ Geneva: IATA 2007.
[12] ADM; European Agreement concerning International Carriage of Dangereous Goods by Inland Waterways (ADM).
[13] European Committee for Standardization (CEN). EN 829. In Vitro Diagnostic Systems. Transport Packages for Medical and Biological Specimens. Requirements, Tests. Bruxelles: European Committee for Standardization (CEN), 1996.
[14] Technische Regeln für Biologische Arbeitsstoffe (TRBA): www.baua.de
[15] ADR. Anlage zur 17. ADR-Änderungsverordnung vom 27. August 2004, Anlageband zum Bundesgesetzblatt Nr. 28 vom 14.09.2004, Teil II.
[16] United Nations (UN). Recommendations on the Transport of Dangereous Goods. 14th revised ed. New York: UN, 2005.
[17] European Comittee for Standardization (CEN). EN 829. In Vitro Diagnostic Systems. Transport Packages for Medical and Biological Specimens. Requirements, Tests. Brussels: European Committee for Standardization (CEN), 1996.
[18] National Institute for Occupational Safety and Health (NIOSH). Selecting, evaluating and using sharps disposal containers. National institute for occupational safety and health, USA 111–1997.
[19] World Health Organisation (WHO). Laboratory Biosafety Manual, 2nd ed. Geneva: WHO, 1993.
[20] World Health Organization (WHO). Use of Anticoagulants in Diagnostic Laboratory Investigations. Geneva:WHO 1999 and WHO/Dil/Lab/99.1, 2002.
[21] Yagmur E, van Helden J, Koch A, Jadem J, Tacke F, Trautwein, C. Effektive Glykolyse-Inhibierung im Citrat-gepufferten venösen Vollblut und Plasma. (Effective inhibition of glycolyis in venous whole blood and plasma samples.) J Lab Med 2012;36:169–77.

Thomas Streichert, Alexander von Meyer

6.2 Preanalytical Workflow Techniques and Procedures

Health Care Providers are in the midst of a shift from conventional paper records to electronic health records (eHR) combined with electronic order systems (computerized, physicians order entry [CPOE]). In parallel, laboratories are achieving high levels of automation. These changes lead to new preanalytical workflows and hence new influences possibly causing errors.

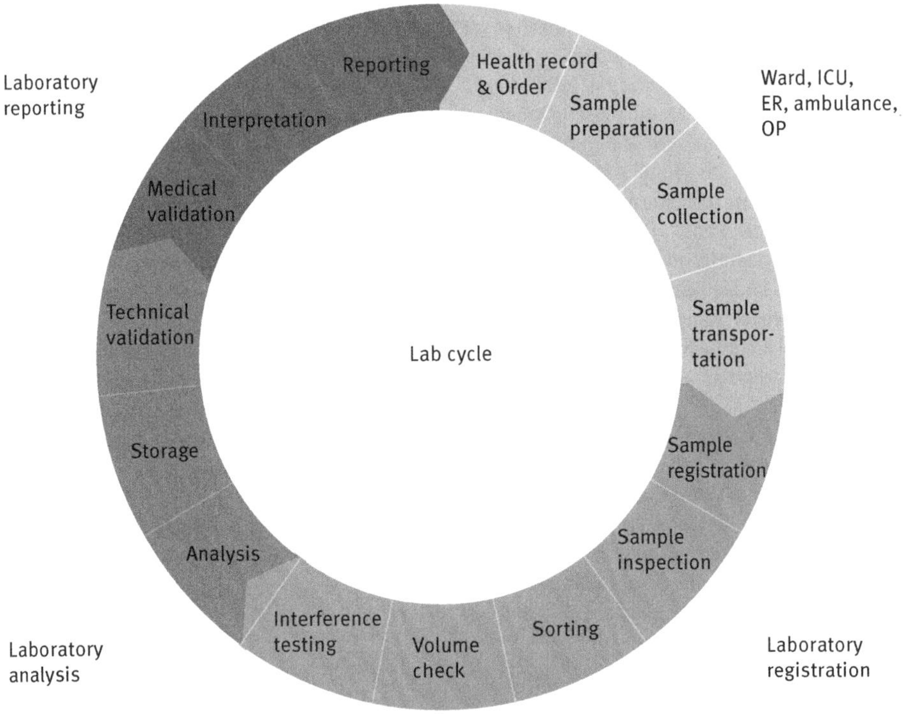

Fig. 6.2.1: The preanalytical phase in the laboratory cycle of total turnaround time.

6.2.1 Order entry

Traditionally, orders were sent to the laboratory with an order requisition slip. First, they were filled out and marked by hand with all the difficulties handwriting can cause. In the next evolution step, the paper requisition slips were created with peel-off barcodes for the specimen and the patient information was on a printed label affixed on the slip. These order forms are well suited for scanning devices to automate the recording of orders. Some laboratories used requisition slips filled out on a computer, which got printed out after completion. This meant that the patient information and the order information was already digital before being transferred to paper, which was then scanned in the laboratory and digitized again. This media disruption was chaotic, costly and labour-intensive.

Hence, CPOE became an essential tool in laboratories. In the US, the healthcare reform signed into law in 2010 promotes the use of digital health records and of CPOE [1, 2]. In Germany, there are no regulations and thus the health care provider and the laboratory are free to decide whether they use CPOE and digital health records or traditional order requisition slips. Of course, CPOEs could improve quality and patient safety [3] but they also have the potential to contribute to new types of errors. Compared to the paper order CPOE significantly reduces specimen-labelling errors (relative reduction of 74 %, absolute 0.31 %) [4], reduces documentation errors in the electronic medical record as well as in the laboratory documentation [4]. CPOEs could positively influence the turnaround times leading to a decrease in laboratory result reporting times [5]. Nevertheless, the CPOE itself might introduce new error types like incorrect logging of test dates (e. g. test requests are logged as taken on the date of request) [6]. In addition CPOEs might increase other types of ordering errors not detected or reported to the laboratory (e. g. selecting the wrong patient is fairly easy) [7]. The bundling of tests could help to standardize patient care and the use of algorithms has been shown to decrease the number of test order errors [8]. Unnecessary and thus erroneous duplicate blood drawing can be identified and blocked by CPOEs reducing the iatrogenic anemia and health care costs [9]. On the other hand bundled test panels might lead to an overuse and could prevent the targeted laboratory follow up [8].

6.2.2 Sample registration and sorting

Highly automated laboratories often use a bulk sorter or an input/output module for sorting while accomplishing the sample receipt and the registration process. This could reduce the number of sorting and routing errors [10], but bulk sorting processes itself could cause preanalytical influences and specimens are particularly sensitive to the forces in these systems (e. g. paternoster lift with sorting boxes). Whereas the clinical relevance is negligible for most parameters, the deviations of e. g. neuron specific enolase (NSE) after sorting may affect clinical decisions [11].

A problem inherent to bulk sorting is the negligence of the “first in – first out principle”. If specimens are loaded continuously, the timespan before they get sorted could be quite variable.

After the sorting of samples they get, if needed, centrifuged (see Chapter 6.1), decapped and aliquoted. This can be combined with cap recognition, volume check, clot detection, and the recognition of haemolytic, icteric or lipemic samples.

For the sample registration a complete order and a correct labeled specimen is needed. The minimal information for the order is defined in the ISO 15189 (see Table 6.2.1 and Chapter 1.2).

Table 6.2.1: Minimal order information for sample registration.

a)	**Patient**	– unique identification – name – sex – date of birth – residence – contact information
b)	**Requester**	– name or other unique identifier of physician or other person legally authorized to request examinations – medical information – destination for the report/requesting clinician's address
c)	**Sample**	– type of sample – anatomic site of origin where appropriate
d)	**Requests**	– examinations requested – urgency
e)	**Clinical Information**	– clinical information relevant to the patient and the requests
f)	**Collection**	– date and time of primary sample collection
g)	**Time stamp**	– date and time of receipt of samples by the laboratory

The samples must be labeled correctly to match the order. A specimen label should at least contain the following information: Patient name, unique patient identifier, date of birth, specimen collection time and date, collector's identification and a specimen identifier, usually bar-coded, like the Laboratory Information System (LIS) accession number [12].

Besides increasing the throughput, the total lab automation could help to reduce turnaround times and the number of lost samples [13].

There are different techniques for the automated sample registration. The two most common techniques entail the use of a barcode reader or a camera to identify the barcode on the sample. Camera systems offer the opportunity to readout additional information.

Undeniably, even the automated sample receipt and registration is prone to errors. There are different types of errors which might occur, starting from tests that are on the order but not in the Laboratory Information System (LIS) or vice versa, discrepancies in the physicians name, incorrect test priority and lastly the abovementioned misidentification errors [14]. Mislabeled specimens and patient identification errors are a well-recognized and serious problem in clinical laboratories with error rates ranging from 0.39/1000 to 1.12%. Automated camera systems using optical character recognition (OCR) technology to identify potential mislabeled samples could help to reduce the number of misidentifications [15].

Even though 1D barcodes are broadly used by CPOEs for sample labelling, it is difficult to provide a reliable estimation of misidentification errors [16], especially of technical barcode reading errors inside the laboratory or even in the analyser due to bad printing quality, contamination or missing parts of the barcode. 2D barcodes which are only sporadically in use have a high level of redundancy, with the data dispersed in several locations throughout the barcode [17] and might help to overcome these problems. Whilst the Clinical and Laboratory Standards Institute (CLSI) defined a standard for specimen labels [12] there is still a lack of industry acceptance [18] and as a consequence a deficiency of implementation in the laboratories.

6.2.3 Sample inspection

Phlebotomy errors like improper labelling at the time of collection, inadequate vial filling, or selection of an inappropriate vial (e. g. citrate vs EDTA to prevent clotting) [19] should be detected in the specimen inspection step. The discrepancy between the labeled material and the material in the tube could cause analytical problems (e. g. measuring calcium in EDTA-plasma instead of in heparin-plasma) as well as interpretation problems because of different reference ranges (potassium out of serum or heparin-plasma). Automated sample inspection modules use cameras to distinguish between different tube types and vendors. The tube is captured from different sides and an image analysis system extracts the information, like shape, (cap) color and barcode from the picture. These image analysis systems could be trained to the needs of the local laboratory (special tubes and labels). Regarding the detection rate of the barcodes, these camera systems perform as well as conventional barcode readers (personnel communication). These detection systems can help to identify discrepancies of the labeled material and the actual material and trigger, for example a sort out process.

6.2.4 Volume check

Many preanalytical sorting machines have at least the option of a so-called volume check. At least three different principles for acquiring the sample volume are available. These three principles differ relevantly from each other, so the results of these methods are eventually not comparable.

The first method is the volume check by measuring the surface of the sample. It does not require an additional volume unit and can be easily implemented in a general aliquoting module. A conductive tip determines the top liquid level. Together with the additional information of the sample tube size the complete blood volume can be calculated. The disadvantage of this principle is the late time point of acquiring the sample volume, because only for aliquoted samples a sample check is performed. Low volume samples can therefore not be sorted out before aliquoting starts. A low throughput is the consequence. Additionally only the top level can be determined. Even with a known tube size, due to the unknown hematocrit, the real serum or plasma volume cannot be quantified. The second technique is a volume check by optical means with a separate camera unit. For the optical assessment of the liquid level, a picture of the sample is taken and in parallel, the sample tube type is recognized. This procedure needs a gap between the different labels on the tube. The system recognizes the level of the top surface and after centrifugation the level of the pelleted blood cells. Again, with the formula for a cylinder the volume of the serum or plasma is determined. The disadvantage of this system is the mandatory 3–5 mm gap between labels. If there is no gap left between the labels, the system cannot calculate the volume. Depending on the settings of the automation, these samples are sorted out to an error position. The third way to determine the volume is measurement by an infrared camera system. This principle offers the lowest error rates while gathering the most information. This way of quantification of plasma volume allows a precise determination of underfilling of coagulation sample tubes, based on real plasma volume and not on top liquid level [20].

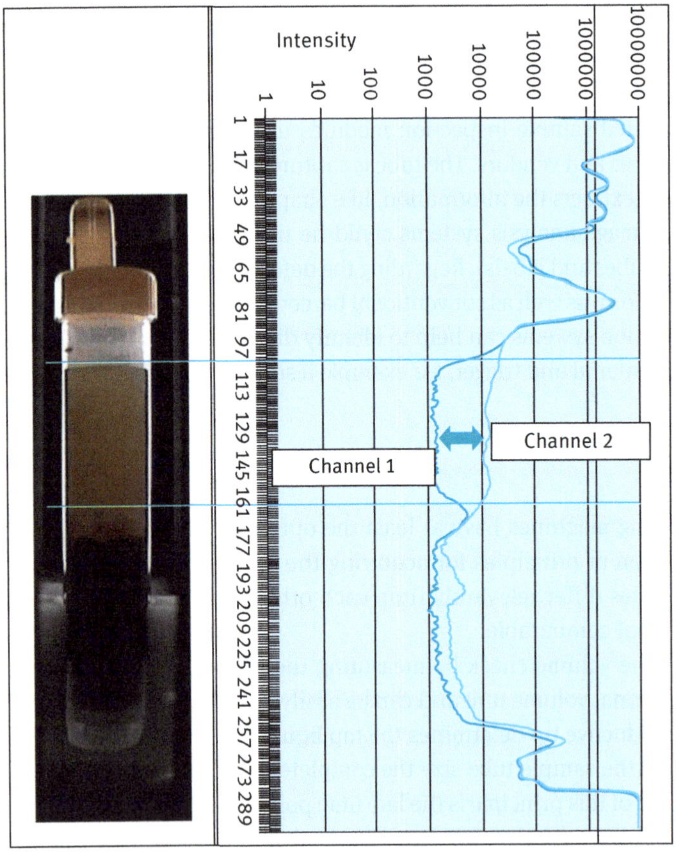

Fig. 6.2.2: Wavelength-dependent transmission of a serum-sample [20].

6.2.5 Hemolysis, Icterus, Lipemia (HIL) interference testing

Various preanalytical influences can alter specimen quality. The main cause of unsuitable samples is the presence of interfering substances [21]. The main interferences are hemolysis, icterus and lipemia [22]. In former times, checking sample quality by visual inspection was the most common way to check for interferences. Today, a number of clinical chemistry analysers allow a systematic assessment of interferences by optical measurement using different wavelengths. This more precise, reliable and reproducible way of measuring interferences has been suggested by the Clinical and Laboratory Standards Institute (CLSI) here [23]. Routine measurement of HIL-indices does not affect turnaround time significantly, as it is mentioned sometimes as a reason for not testing every sample [24]. Most of the tests for HIL-indices simply use saline as "reagent" and are therefore affordable. These automated tests for interferences are mainly semi quantitative, though some of them are customizable to obtain quantitative results. E. g. tests for hemolysis from the implemented HIL assessment can fulfil precision criteria for several clinical questions (e. g. transfusion adverse events) and can be handled like normal photometric assays, including internal and external QC. These tests can be easily requested even weeks later, because the initial hemolysis-index was performed and documented within the first testing [25]. A general testing for interferences is therefore highly recommended for every sample. If samples are not routed to a clinical chemistry analyser, like specimen for coagulation or immunology testing, the assessment of interferences might be possible on automated sorters. Due to regulation, the precision and quality of this technique is limited. According to our experience, every automated assessment of interference is better than manual visual inspection, which is the remaining alternative.

References

[1] Goldman D. Obama's big idea: Digital health records. 2009 [cited 2014 3.7.]; Available from: http://money.cnn.com/2009/01/12/technology/stimulus_health_care/

[2] Litton SJ. Computerized Physician Order Entry: Coming to a Hospital Near You. 2012 [cited 2014 3.7.]; Available from:http://www.physicianspractice.com/blog/computerized-physician-order-entry-coming-hospital-near-you

[3] Rothschild J. Computerized physician order entry in the critical care and general inpatient setting: a narrative review. J Crit Care 2004; 19:271–8.

[4] Kim JY, Kamis IK, Singh B, Batra S, Dixon RH, . Dighe AS. Implementation of computerized add-on testing for hospitalized patients in a large academic medical center. Clin Chem Lab Med. 2011; 49:845–50.

[5] Mekhjian, H. S., Immediate benefits realized following implementation of physician order entry at an Academic Medical Center. J Am Med Informat Ass 2002; 9:529–39.

[6] Appleton A,. Sadek K, Dawson IG, de Lusignan S, Clinicians were oblivious to incorrect logging of test dates and the associated risks in an online pathology application: a case study. Inform Prim Care, 2012; 20: 241–7.

[7] Hill PM, Mareiniss D, Murphy P, Gardner H, Hsieh YH, Levy F, et al., Significant reduction of laboratory specimen labeling errors by implementation of an electronic ordering system paired with a bar-code specimen labeling process. Ann Emerg Med 2010; 56:630–6.
[8] Yeh, DD, A clinician's perspective on laboratory utilization management. Clin Chim Acta 2014; 427:145–50.
[9] Procop GW, Yerian LM, Wyllie R, Harrison AM, Kottke-Marchant K. Duplicate laboratory test reduction using a clinical decision support tool. Am J Clin Pathol 2014; 141:718–23.
[10] Holman JW, Mifflin TE, Felder RA, Demers LM. Evaluation of an automated preanalytical robotic workstation at two academic health centers. Clin Chem 2002; 48:540–8.
[11] Beil B, Nordholt G, Beil FT, Otto B, Streichert T, Jung R et al. Sorting of closed primary blood sample tubes by an automatic sorter may cause haemolysis. Clin Chem Lab Med 2011.
[12] Clinical and Laboratory Standards Institute (CLSI), Specimen Labels: Content and Location, Fonts, and Label Orientation; Approved Standard. Wayne PA; CLSI: document AUTO12-A, 2011.
[13] Hawker CD, Garr SB, Hamilton LT, Penrose JR, Ashwood ER, , Weiss RL. Automated transport and sorting system in a large reference laboratory: part 1. Evaluation of needs and alternatives and development of a plan. Clin Chem 2002; 48:1751–60.
[14] Bonini P, Plebani M, Ceriotti F, Rubboli F. Errors in laboratory medicine. Clin Chem 2002; 48:691–8.
[15] Hawker CD, McCarthy W, Cleveland D, Messinger BL. Invention and validation of an automated camera system that uses optical character recognition to identify patient name mislabeled samples. Clin Chem, 2014; 60:463–70.
[16] Lippi, G. Plebani M, Identification errors in the blood transfusion laboratory: a still relevant issue for patient safety. Transfus Apher Sci 2011; 44:231–3.
[17] Hattie, M, Bar coding: labs go 2D. MLO Med Lab Obs 2009; 41:32–3.
[18] Coffman SK. What constitutes a correctly labeled specimen? LabMed. 2011; 42:630–1.
[19] Felder, RA: Automated specimen inspection, quality analysis, and its impact on patient safety: beyond the bar code. Clin Chem, 2014; 60:433–4.
[20] V. Meyer, A, Frenz C, Schmolke M, Preanalytical automated plasma volume determination in coagulation testing by automate 2500. Clin Chem Lab Med 2011a, 49: S514(abstract).
[21] Plebani, M, Lippi G, To err is human. To misdiagnose might be deadly. Clin Biochem, 2010; 43:1–3.
[22] Lippi G, Plebani M, Favaloro EJ. Interference in coagulation testing: focus on spurious hemolysis, icterus, and lipemia. Semin Thromb Hemost 2013; 39:258–66.
[23] Clinical and Laboratory Standard Institute (CLSI). Hemolysis, Icterus, and Lipemia/Turbidity Indices as Indicators of Interference in Clinical Laboratory Analysis; Approved Guideline. 2012, Wayne, PA: CLSI document C56-A.
[24] Lippi G, Avanzini P, Campioli D, Da Rin G, Dipalo M, Aloe R, et al. Systematical assessment of serum indices does not impair efficiency of clinical chemistry testing: a multicenter study. Clin Biochem 2013; 46:1281–4.
[25] V. Meyer A, Ahting U, Schmidt S, Rolinski B, Pick KH, Hallbach J. Evaluation of free hemoglobin determination in plasma samples on the Architect CI8200 (in technology, instrumentation and methods), in IFCC – EFCC – EUROMEDLAB 2009. Innsbruck: Clin Chem Lab Med 2009; 47:(abstract).

7. Preanalytical Variables and Rules in Specific Fields of Medical Laboratory Diagnostics

Carl-Erik Dempfle, Gottfried Töpfer

7.1 Hemostaseology

A meticulous sample preparation and handling are of utmost importance for reliable coagulation diagnostics. Hemolysis and clotting, as well as inappropriate volume of blood are among the most frequent preanalytical problems [1, 2].

7.1.1 Blood sampling for analyses of blood coagulation

The standard material for analysis of blood coagulation is venous blood, typically drawn from the antecubital veins. In some cases, it may be necessary to draw blood from other veins (including the veins of forearm and hands), or even arterial vessels [3].

Blood should be drawn from resting subjects. The relaxed arm should be in a stable position, with neither "pumping", nor clenching. A tourniquet may be applied to the arm, and should be tight enough to stop venous, but not arterial flow [4]. Venipuncture should be successful within 1 minute after application of the tourniquet [5, 6]. If possible, the tourniquet should be released after placement of the needle. Longer venous stasis results in hemoconcentration [7], activation of fibrinolysis, and other changes, such as an increase in fibrinogen, factors VII, VIII und XII [6, 8]. Generally, veins can be palpated easily in the relaxed arm as bulging structure in between the soft muscles and tissue after placement of the tourniquet. In some cases, the use of a near infrared transillumination device may be helpful [5].

To prevent excessive shear force, venipuncture should be performed with needles of at least 21G, the needle should be placed securely in the lumen of the vein, and should not be manipulated after placement. It may be advantageous to use a butterfly system, if several sample tubes are to be drawn [9, 10]. Blood for coagulation analyses should not be drawn from intravenous catheters or port systems. It is possible though to draw blood from short intravenous catheter systems immediately after placement [11]. If intravenous catheter systems are used for blood sampling, the initial 10 ml of blood drawn should be discarded (or used for other analyses). If the catheter might be contaminated with heparin due to heparin infusion or flushing with heparin-containing fluid, only laboratory tests insensitive to heparin may be used [8].

Table 7.1.1: Summary of recommendations.

Blood sampling
– Resting, relaxed subject, relaxed arm, no "pumping" or clenching.
– Venipuncture within 1 minute after placement of tourniquet, release tourniquet after placement of needle, if possible.
– Use needles or butterfly systems of ≥21G for venipuncture.

Table 7.1.1: (continued)

Blood sampling
– Discard tube may be considered to ensure correct filling of sample tubes for coagulation tests.
– No excessive shear force during blood drawing.
– Sample tubes contain anticoagulant (generally trisodium citrate solution). Tubes containing lepirudin, or other specific factor Xa- and thrombin inhibitors may be used for some tests.
– Ensure correct filling of tubes (>90 % filling).
– In case of high hematocrit, it may be necessary to remove part of the citrate solution from the tube prior to drawing of blood.
– Blood should instantly be mixed with the anticoagulant during blood drawing. Tubes may be gently inverted 1–2 times to improve mixing.
– No foam or bubbles in blood samples, no shaking or agitation of blood samples, discard dropped samples.
– Immediately after drawing, transfer capped sample tubes to racks and transport, sort and store tubes vertically and at room temperature (20–25 °C).
Processing of sample tubes
– Time from sampling to centrifugation: < 6 h (optimal: <1 hour).
– Centrifugation for platelet poor plasma (PPP): 1500–2000 *g*, 10–15 min.
– Platelet count in PPP <10/nl (<10 Gpt/L). Some assays require double centrifugation. An alternative might be centrifugation at 5000 *g* for 6 min with centrifuge brake set off.
– Centrifugation for platelet rich plasma (PRP): 200–250 *g*, 10 min.
– PRP: no adjustment of platelet count, if <600/nl (<600 Gpt/L).
– Store samples at room temperature (20–25 °C) until analysis.
– Time from sampling to analysis depends on analyte: aPTT, fibrinogen and factor VIII < 3 hours (in patients treated with unfractionated heparin < 1 hour), PT and factors II, V, VII, IX, XI and XII < 6 h, antithrombin and von Willebrand-factor <48 h.
– Check tubes for presence of clots, precipitates or hemolysis.
Freezing and thawing
– Do not perform PT, aPTT and factor VIII assays from frozen samples.
– Use rapid freezing technique (liquid nitrogen, ...), store samples at −70 °C, if possible.
– Do not re-freeze samples (prepare several aliquots as "backup").
– Thaw samples in water bath at 37 °C.
– After thawing, gently invert samples 5–6 times. Do not vortex or shake.

7.1.2 Sample tubes, anticoagulant

Sample tubes for hemostasis analyses generally contain an anticoagulant. The standard anticoagulant is 1/10 volume of 0.105 to 0.129 mol/L trisodium citrate. For some analyses, especially platelet function assays, buffered citrate solution is used, or other anticoagulants, such as lepirudin [12], or synthetic inhibitors of thrombin and factor Xa [13] are used.

A standardized sequence of blood samples may be helpful [14], but is not strictly enforced, as the carryover , for example of clot activators from serum tubes drawn before the coagulation assay tubes appears to be minimal [15].

As the ratio of blood and anticoagulant is important and underfilling of the tubes severely affects the laboratory results, it may be necessary to use an initial discard tube (or a sample tube for analyses that are not affected by underfilling) before drawing the blood for coagulation analyses, especially, if the blood is drawn using butterfly systems, or other intravenous catheter devices [16]. The use of discard tubes is not necessary for other reasons, if proper filling is ensured [17]. Ver Elst et al. determined that the minimal citrate tube fill volumes are 73 % for PT, 90 % for aPTT and 63 % for fibrinogen [18]. As aPTT is an integral component of coagulation diagnostics, the 90 % fill requirement should be adapted as standard.

An increased hematocrit may also result in inappropriate ratio of plasma to anticoagulant. It might be necessary to remove some citrate solution from the sampling tubes if blood is to be drawn from patients with very high hematocrit levels.

The blood should be drawn in a way that mixing with the anticoagulant occurs immediately during the process of blood sampling [19], and dispersion of the anticoagulant may be enhanced by gently inverting the tube once or twice after detaching it from the needle. Blood sampling should be performed without generating foam or bubbles in the sample and with as little as possible shear stress. For some analyses, such as thrombin generation assays, or some platelet assays, even vacuum tubes may be problematic, and a syringe system, which allows slow manual drawing of the blood, may be superior. Hemolyzed samples may lead to early flow obstruction in the platelet function analyzer (PFA) [20], presumably due to platelet activation and fragmentation. A small degree of hemolysis, though, has little impact on PT and aPTT [21].

Blood sampling for coagulation analyses may be done using siliconized glass tubes, as well as polypropylene tubes, or using syringe-type aspiration tubes, as well as vacuum tubes [22]. The syringe type tubes are generally polypropylene with low protein binding, while the vacuum tubes may be glass, or polypropylene. It was observed that the use of a certain type of vacuumized glass tubes results in severe changes of clotting assay results [23], and this effect was attributed to contamination of the blood samples with a high concentration of magnesium, which stems from the rubber stoppers of the tubes [24, 25, 26].

7.1.3 Transport and processing of samples

Sample tubes should be transferred to racks and all further storage, transport, sorting and processing should be done with the tubes securely standing (vertically) in racks. Tubes should not be subjected to vibration, shaking, vortexing, continuous mixing or agitation. Transport and sorting as bulk goods is not compatible with reliable coagulation diagnostics.

Generally, blood samples for coagulation analyses should not be shaken and dropped samples should be discarded.

Use of pneumatic tube systems for transport of samples may be problematic. Rapid acceleration and deceleration may result in hemolysis, platelet activation, and other effects [27]. Some pneumatic tube transport systems, though, do not seem to damage the blood samples significantly [28].

Blood samples for coagulation diagnostics should be stored at room temperature [20–25 °C] until analysis. Storage at lower temperature, or on ice, may strongly influence some of the coagulation assays [29]. A storage of uncentrifuged samples at room temperature of up to 6 hours may yield acceptable results, although a shorter delay is desirable [30, 31].

Whole blood assays should be performed within 4 hours after blood sampling. For platelet function assays, samples should rest (at room temperature) for 30 minutes before analysis.

For analysis in platelet poor plasma (PPP), which is the standard material for PT, aPTT, fibrinogen, single coagulation factor assays, inhibitors, and many other tests, sample tubes should be centrifuged in capped tubes within four hours after drawing of the blood. The temperature of the plasma should be 20–25 °C at the end of centrifugation. This may require cooling of the centrifuge chamber.

Centrifugation is normally performed at 1500–2000 *g* for 10–15 minutes [32], but higher speed with shorter centrifugation time may be used. The platelet count of the PPP obtained after 15 min 2000–3000 *g* centrifugation should not exceed 10/nL(10 Gpt/L) [33], Some assays, such as functional protein S assays, or lupus anticoagulant assays, may require a second centrifugation (after careful transfer of the PPP to a secondary tube) [34]. Double centrifugation significantly reduces the residual amount of platelets in the sample [31]. Use of the rotor brake may lead to higher residual platelet counts [31], and also has an effect on PT and fibrinogen concentration [35]. Therefore it seems feasible to centrifuge samples for coagulation analyses with the rotor brake set to off [35]. Use of high speed centrifuges may accelerate plasma preparation. The platelet count obtained after 6 minutes, 5000 *g* and rotor brake set off has been shown to be <3 Gpt/L [36], resulting in improved stability of plasma samples upon storage, compared to PPP with higher platelet counts.

Very high centrifugation speed may result in platelet activation, hemolysis or other unwanted effects.

Before analysis, PPP is transferred carefully to a secondary tube without disrupting the red blood cell and buffy coat sediment. Analyses may be performed by pipetting from the primary tube, if this tube is stored vertically in a rack and any disruption of the sediment is prevented. A second round of centrifugation of the primary tube is not permissible.

7.1.4 Platelet-rich plasma for platelet function analysis

The preparation of platelet-rich plasma (PRP) for platelet function assays requires special care and centrifugation speed and duration need to be carefully optimized to ensure optimal results. Generally, centrifugation is performed at 200–250 *g* for 10 minutes without application of a rotor brake [37]. Centrifugation with less than 150 *g* does not yield a sufficient volume of PRP [38]. The PRP should then be transferred to a capped polypropylene tube and a platelet count performed. For light transmission aggregometry, the platelet count is not adjusted (by addition of PPP), unless the platelet count is >600/nl (600 Gpt/L).

7.1.5 Storage and processing of centrifuged plasma samples

PPP is stored at room temperature [20–25 °C] until analysis. The time frame for analysis depends upon the stability of the analyte.

Prothrombin time (PT), and aPTT should generally be performed using fresh plasma within 4 hours after blood sampling, and stored at room temperature [39], especially if the analyses are performed using plasma from the (spun down) primary tubes. If the plasma is left on the spun down blood cells, this may result in shortening of the PT and prolongation of the aPTT [4]. The shortest stability is observed for factors V and VIII, as well as protein S, followed by PT, aPTT, and factor VII, whereas fibrinogen and antithrombin appeared to be stable for a longer period [39].

Van Geest-Daalderop et al suggest a time frame of 6 hours after blood collection for coagulation tests [40], Zhao and Lv show results that relevant changes especially for aPTT occur after storage longer than 4 h [41]. Kemkes-Matthes et al indicate that PT, thrombin time and measurement of D-dimer may be performed up to 24 h after blood sampling, of the samples at room temperature [42]. According to the results of Adcock et al, PT is stable for 24 h in centrifuged blood samples, aPTT for 8 h, with no fundamental difference between storage at room temperature and at 4 °C [43]. According to the analyses of Töpfer et al, aPTT, thrombin time, batroxobin time, fibrinogen and factor VIII activity should be measured within 3 h, PT, and factors II, V, VII, IX, XI and XII within 6 h after blood sampling [8]. VonWillebrand factor, and antithrombin in contrast appear to be stable for 48 h [8].

Samples from patients treated with unfractionated heparins are unstable, resulting in changes in PT and aPTT. Samples from patients treated with unfractionated heparin should be centrifuged and aPTT measured within 1 hour at room temperature, or within 4 h at 4 °C [43].

Before analysis, plasma samples are checked for presence of clots, precipitates and signs of hemolysis.

All assays using whole blood or PRP need to be performed within a short period (normally < 4 h) after blood sampling.

7.1.6 Freezing and thawing

If possible, all coagulation analyses should be performed using fresh material, freezing of samples should be an exception.

PT and aPTT should not be performed in samples that were stored frozen, as freezing results in changes especially of aPTT [39, 44], but also PT [45]. Freezing leads to a marked decrease especially of factor VIII activity [46], and to an apparent increase in fibrinogen levels [45]. In contrast to factor VIII, von Willebrand factor appears to be quite stable, with little loss of antigen, and ristocetin cofactor activity [47].

Some assays though may also be performed from frozen platelet-poor plasma. For freezing of plasma samples it is especially important to ensure that the residual platelet count after centrifugation is <10/nl (<10 Gpt/L). Double centrifugation (after transfer of the PPP from the primary tube to a secondary polypropylene tube) may improve the stability of the samples and assay performance. Alternatively, plasma for storage may be prepared by centrifugation at 5000 *g* for 6 min with the rotor brake set off [36].

For freezing, samples are transferred to capped thin-walled polypropylene tubes with low protein binding.

The effect of freezing on aPTT (and other parameters) increases with time, and may be reduced by snap freezing in liquid nitrogen and storage at –70 °C instead of –30 °C [45]. Rapid freezing, for example in liquid nitrogen or specific devices for snap freezing of plasma and tissue samples, is feasible. The internal temperature should reach –30 °C within 5 min. Samples should be stored at a temperature lower than –30 °C, optimally below –70 °C until analysis. In the case of storage at –30 °C, the analyses should be performed within 4 weeks. At –70 °C, samples are stable for a longer period of time (excluding measurement of PT, aPTT, Factor V and factor VIII).

Before analysis, frozen samples are transferred to a water bath with a temperature of 37 °C. After complete thawing (which should occur within no more than 10 min), samples are carefully inverted 5–6 times to mix. Samples should not be shaken or vortexed. Samples are then checked for the presence of cryoprecipitates. Removal of precipitates by centrifugation may modify assay results, especially for fibrinogen, fibrin and von Willebrand-factor.

As refreezing is not permissible, several (small) aliquots should be frozen rather than one large amount of PPP.

References

[1] Goswami B, Tayal D, Chawla R, Mallika V. Pre-analytical factors that influence the interpretation of prothrombin time in the clinical laboratory: one year experience in a super speciality hospital in India. Clin Chim Acta 2009; 410:93–94. doi: 10.1016/j.cca.2009.1009.1014. Epub 2009 Sep 1013.

[2] Salvagno GL, Lippi G, Bassi A, Poli G, Guidi GC. Prevalence and type of pre-analytical problems for inpatients samples in coagulation laboratory. J Eval Clin Pract 2008; 14:351–3. doi: 310.1111/j.1365–2753.2007.00875.x.

[3] Kafian S, Mobarrez F, Kalani M, Wallen H, Samad BA. Comparison of venous and arterial blood sampling for the assessment of platelet aggregation with whole blood impedance aggregometry. Scand J Clin Lab Invest 2011; 71:637–40.
[4] Lippi G, Franchini M, Montagnana M, Salvagno GL, Poli G, Guidi GC. Quality and reliability of routine coagulation testing: can we trust that sample? Blood Coagul Fibrinolysis 2006; 17:513–19.
[5] Lima-Oliveira G, Salvagno GL, Lippi G, Montagnana M, Scartezini M, Picheth G, Guidi GC. Elimination of the venous stasis error for routine coagulation testing by transillumination. Clin Chim Acta. 2011;412:1482–4. doi:1410.1016/j.cca.2011.1404.1008. Epub 2011 Apr 1413.
[6] Lippi G, Salvagno GL, Montagnana M, Guidi GC. Short-term venous stasis influences routine coagulation testing. Blood Coagul Fibrinolysis 2005; 16:453–8.
[7] Lippi G, Salvagno GL, Montagnana M, Franchini M, Guidi GC. Venous stasis and routine hematologic testing. Clin Lab Haematol 2006; 28:332–7.
[8] Töpfer G.,Schulze M,Lutze G, Ziemer S,Siegert S,Frick U. Determination of coagulation parameters in citrated venous blood, catheter blood, and capillary blood: preanalytical problems. J Lab Med 2000; 24:514–20.
[9] Lippi G, Salvagno GL, Brocco G, Guidi GC. Preanalytical variability in laboratory testing: influence of the blood drawing technique. Clin Chem Lab Med 2005; 43:319–25.
[10] Lippi G, Salvagno GL, Montagnana M, Poli G, Guidi GC. Influence of the needle bore size on platelet count and routine coagulation testing. Blood Coagul Fibrinolysis 2006; 17:557–61.
[11] Zengin N, Enc N. Comparison of two blood sampling methods in anticoagulation therapy: venipuncture and peripheral venous catheter. J Clin Nurs 2008; 17:386–93. Epub 2007 Mar 2030.
[12] Loreth RM, Klose G. Comparison of two different blood sample tubes for platelet function analysis with the multiplate(R) system. Transfus Med Hemother 2010; 37:289–92.
[13] Hellstern P, Sturzebecher U, Wuchold B, Haubelt H, Seyfert UT, Bauer M, Vogt A, et al. Preservation of in vitro function of platelets stored in the presence of a synthetic dual inhibitor of factor Xa and thrombin. J Thromb Haemost 2007; 5:2119–26. Epub 2007 Jul 2131.
[14] Sulaiman RA, Cornes MP, Whitehead SJ, Othonos N, Ford C, Gama R. Effect of order of draw of blood samples during phlebotomy on routine biochemistry results. J Clin Pathol 2011; 64:1019–20. doi: 1010.1136/jclinpath-2011–200206. Epub 202011 Jul 200221.
[15] Fukugawa Y, Ohnishi H, Ishii T, Tanouchi A, Sano J, Miyawaki H, Kishino T, et al. Effect of carryover of clot activators on coagulation tests during phlebotomy. Am J Clin Pathol 2012; 137:900–3. doi: 910.1309/AJCPL1303LZZN1309WJGZT.
[16] Favaloro EJ, Lippi G, Raijmakers MT, Vader HL, van der Graaf F. Discard tubes are sometimes necessary when drawing samples for hemostasis. Am J Clin Pathol 2010; 134:851. doi: 810.1309/AJCP1309HIFFC1306FWKYF.
[17] Smock KJ, Crist RA, Hansen SJ, Rodgers GM, Lehman CM. Discard tubes are not necessary when drawing samples for specialized coagulation testing. Blood Coagul Fibrinolysis 2010; 21:279–82.
[18] Ver Elst K, Vermeiren S, Schouwers S, Callebaut V, Thomson W, Weekx S. Validation of the minimal citrate tube fill volume for routine coagulation tests on ACL TOP 500 CTS(R). Int J Lab Hematol 2013; 35:614–19. doi: 610.1111/ijlh.12099. Epub 12013 May 12013.
[19] Lippi G, Salvagno GL, Montagnana M, Guidi GC. Influence of primary sample mixing on routine coagulation testing. Blood Coagul Fibrinolysis 2007; 18:709–11.
[20] Lippi G, Fontana R, Avanzini P, Aloe R, Ippolito L, Sandei F, Favaloro EJ. Influence of mechanical trauma of blood and hemolysis on PFA-100 testing. Blood Coagul Fibrinolysis 2012; 23:82–6. doi: 10.1097/MBC.1090b1013e32834c32836cb32835.
[21] Laga AC, Cheves TA, Sweeney JD. The effect of specimen hemolysis on coagulation test results. Am J Clin Pathol 2006; 126:748–55.

[22] D'Angelo G, Villa C. Comparison between siliconized evacuated glass and plastic blood collection tubes for prothrombin time and activated partial thromboplastin time assay in normal patients, patients on oral anticoagulant therapy and patients with unfractioned heparin therapy. Int J Lab Hematol 2011; 33:219–25. doi: 210.1111/j.1751–1553X.2010.01271.x. Epub 02010 Oct 01227.
[23] Lippi G, Ippolito L, Zobbi V, Sandei F, Favaloro EJ. Sample collection and platelet function testing: influence of vacuum or aspiration principle on PFA-100 test results. Blood Coagul Fibrinolysis 2013; 24:666–9. doi: 610.1097/MBC.1090b1013e32835fada32837.
[24] van den Besselaar AM, Hoekstra MM, Witteveen E, Didden JH, van der Meer FJ. Influence of blood collection systems on the prothrombin time and international sensitivity index determined with human and rabbit thromboplastin reagents. Am J Clin Pathol 2007; 127:724–9.
[25] van den Besselaar AM, van Zanten AP, Brantjes HM, Elisen MG, van der Meer FJ, Poland DC, Sturk A, et al. Comparative study of blood collection tubes and thromboplastin reagents for correction of INR discrepancies: a proposal for maximum allowable magnesium contamination in sodium citrate anticoagulant solutions. Am J Clin Pathol 2012; 138:248–54. doi: 210.1309/AJCPGSB1305YPJRREEV.
[26] van den Besselaar AM, van Vlodrop IJ, Berendes PB, Cobbaert CM. A comparative study of conventional versus new, magnesium-poor Vacutainer(R) Sodium Citrate blood collection tubes for determination of prothrombin time and INR. Thromb Res 2014; 134:187–91. doi: 110.1016/j.thromres.2014.1004.1016. Epub 2014 Apr 1026.
[27] Evliyaoglu O, Toprak G, Tekin A, Basarali MK, Kilinc C, Colpan L. Effect of pneumatic tube delivery system rate and distance on hemolysis of blood specimens. J Clin Lab Anal 2012; 26:66–9. doi: 10.1002/jcla.21484.
[28] Wallin O, Soderberg J, Grankvist K, Jonsson PA, Hultdin J. Preanalytical effects of pneumatic tube transport on routine haematology, coagulation parameters, platelet function and global coagulation. Clin Chem Lab Med 2008; 46:1443–9. doi: 1410.1515/CCLM.2008.1288.
[29] Engbers MJ, Cushman M, Rosendaal FR, Van Hylckama Vlieg A. The effect of time between venipuncture, processing and freezing on the measurement of coagulation factor levels. J Thromb Haemost 2012; 10:1691–3. doi: 1610.1111/j.1538–7836.2012.04792.x.
[30] Salvagno GL, Lippi G, Montagnana M, Franchini M, Poli G, Guidi GC. Influence of temperature and time before centrifugation of specimens for routine coagulation testing. Int J Lab Hematol 2009; 31:462–7. doi: 410.1111/j.1751–1553X.2008.01058.x. Epub 02008 Mar 01021.
[31] Rodgers SE, Wong A, Gopal RD, Dale BJ, Duncan EM, McRae SJ. Evaluation of pre-analytical variables in a commercial thrombin generation assay. Thromb Res 2014; 134:160–4. doi: 110.1016/j.thromres.2014.1004.1010. Epub 2014 Apr 1022.
[32] Lippi G, Salvagno GL, Montagnana M, Manzato F, Guidi GC. Influence of the centrifuge time of primary plasma tubes on routine coagulation testing. Blood Coagul Fibrinolysis 2007; 18:525–8.
[33] Adcock DM. Collection , transport and processing of blood specimens for testing plasma-based coagulation assays: approved guideline. 5th ed. Wayne, PA: Clinical and Laboratory Standards Institute, 2008.
[34] Suchsland J, Friedrich N, Grotevendt A, Kallner A, Ludemann J, Nauck M, Petersmann A. Optimizing centrifugation of coagulation samples in laboratory automation. Clin Chem Lab Med 2014; 52:1187–91. doi: 1110.1515/cclm-2014–0038.
[35] Daves M, Giacomuzzi K, Tagnin E, Jani E, Adcock Funk DM, Favaloro EJ, Lippi G. Influence of centrifuge brake on residual platelet count and routine coagulation tests in citrated plasma. Blood Coagul Fibrinolysis 2014; 25:292–5. doi: 210.1097/MBC.0000000000000026.
[36] Töpfer, G, Zawta, B, Hornig F, Sauer K, Friedel, G. Verwendung von Zitrat/Theophyllin/Adenosin/Dipyridamol (CTAD)-Blut in Hämostaseologie und Hämatologie (The use of Citrate/Theophylline/ Adenosine/ Dipyrodamol (CTAD)-blood in coagulation and haematology) J Lab Med 2002; 26:296–306.

[37] Femia EA, Pugliano M, Podda G, Cattaneo M. Comparison of different procedures to prepare platelet-rich plasma for studies of platelet aggregation by light transmission aggregometry. Platelets 2012; 23:7–10. doi: 10.3109/09537104.09532011.09596592. Epub 09532011 Jul 09537120.
[38] Merolla M, Nardi MA, Berger JS. Centrifugation speed affects light transmission aggregometry. Int J Lab Hematol 2012; 34:81–5. doi: 10.1111/j.1751–1553X.2011.01360.x. Epub 02011 Jul 01327.
[39] Zürcher M, Sulzer I, Barizzi G, Lammle B, Alberio L. Stability of coagulation assays performed in plasma from citrated whole blood transported at ambient temperature. Thromb Haemost 2008; 99:416–26.
[40] van Geest-Daalderop JH, Mulder AB, Boonman-de Winter LJ, Hoekstra MM, van den Besselaar AM. Preanalytical variables and off-site blood collection: influences on the results of the prothrombin time/international normalized ratio test and implications for monitoring of oral anticoagulant therapy. Clin Chem 2005; 51:561–8. Epub 2005 Jan 2013.
[41] Zhao Y, Lv G. Influence of temperature and storage duration on measurement of activated partial thromboplastin time, D-dimers, fibrinogen, prothrombin time and thrombin time, in citrate-anticoagulated whole blood specimens. Int J Lab Hematol 2013; 35:566–70. doi: 510.1111/ijlh.12113. Epub 12013 May 12130.
[42] Kemkes-Matthes B, Fischer R, Peetz D. Influence of 8 and 24-h storage of whole blood at ambient temperature on prothrombin time, activated partial thromboplastin time, fibrinogen, thrombin time, antithrombin and D-dimer. Blood Coagul Fibrinolysis. 2011; 22:215–20. doi: 210.1097/MBC.1090b1013e328343f328348bf.
[43] Adcock D, Kressin D, Marlar RA. The effect of time and temperature variables on routine coagulation tests. Blood Coagul Fibrinolysis 1998; 9:463–70.
[44] Rao LV, Okorodudu AO, Petersen JR, Elghetany MT. Stability of prothrombin time and activated partial thromboplastin time tests under different storage conditions. Clin Chim Acta 2000; 300:13–21.
[45] Alesci S, Borggrefe M, Dempfle CE. Effect of freezing method and storage at −20 degrees C and −70 degrees C on prothrombin time, aPTT and plasma fibrinogen levels. Thromb Res 2009; 124:121–6. Epub 2009 Jan 2006.
[46] Bach J, Haubelt H, Hellstern P. Sources of variation in factor VIII, von Willebrand factor and fibrinogen measurements: implications for detecting deficiencies and increased plasma levels. Thromb Res 2010; 126:e188–95. Epub 2010 Jul 2013.
[47] Töpfer G, Lutze G, Schmidt LH, Friedel G, Seifert A, Novy EM, Lehmann L. Haltbarkeit des Ristocetin-Cofaktors und des Faktor VIII associierten Antigens in Abhängigkeit von den Lagerungsbedingungen.[Stability of ristocetin cofactor and factor VIII associated antigen in relation to storage conditions]. Z Med Lab Diagn 1983; 24:463–8.

Giuseppe Banfi, Walter G. Guder, Sheshadri Narayanan

7.2 Hematology including Flow Cytometry of Blood Cells

For the counting of blood cells, ethylene diamine tetraacetic acid (EDTA) was recommended as an anticoagulant and stabilizer [1]. The International Council for Standardization in Haematology (ICSH) recommended the dipotassium salt K_2 EDTA in preference to Na_2 EDTA as anticoagulant of choice for the collection of blood specimen intended for blood cell counting and sizing [2]. K_2 EDTA was selected in preference to Na_2 EDTA because of the greater solubility of the potassium salt compared to the sodium salt. Because of its lower osmotic effect this was also preferred to K_3- EDTA, still widely used in US [3].

7.2.1 Interferences caused by EDTA-salts [1, 4, 5]

- EDTA- salts cause osmotically induced shrinkage of blood cells. Owing to the lower pH of Na_2 - and K_2-EDTA compared to K_3- EDTA the cells swell, thus compensating for osmotically induced cell shrinkage. Calibration of electronic blood cell counters for **mean corpuscular volume (MCV)** using the minihematocrit value obtained from blood specimens collected in either Na_2- or K_2- EDTA have been reported to provide acceptable results, in contrast to unacceptable results obtained when microhematocrit values obtained from blood specimen collected in K_3-EDTA [6, 7].
- EDTA cannot completely stabilize platelets. Upon collection in EDTA, they change their discoidal to a spherical shape, thus introducing an **error in the mean platelet volume** (MPV) determination. Therefore the MPV should be measured at a fixed time after blood drawing to allow reliable intra-(longitudinal) and inter-individual comparisons [8]. Anticoagulants other than EDTA have been proposed for obtaining correct and valid measurements of MPV, including adenosine, citrate, dextrose (ACD) and sodium EDTA [9], sodium citrate and prostaglandin E_1 (PGE_1) [10], citrate, theophylline, adenosine, dipyramidole (CTAD) and pyridoxal phosphate [11], which have been validated on impedance-based systems [8]. Unfortunately, none of these mixtures has found broad application.
- Stability of the form of cells in EDTA blood is higher in lymphocytes than that of monocytes and neutrophiles. Several additional problems associated with the use of EDTA have been reported, including **leukocyte clumping** due to a reaction between IgM autoantibody and neutrophils at room temperature [12, 13].
- **Pseudothrombocytopenia** is caused by platelet clumping and platelet adhering to neutrophils (platelet satellism) and has been observed in EDTA anticoagulated blood. This phenomenon elevates the white blood cell count while depressing the platelet count. Since most counters do not identify this effect, platelets may be counted as white blood cells, producing spurious pseudothrombocytopenia and

pseudoleukocytosis [14]. Apparently this form of spurious pseudothrombocytopenia is caused by IgM autoantibodies, directed against glycoproteins IIa and IIIb on the platelet surface. In confirming this concept, platelets from patients with Glanzmann disease, which is characterized by a lack of expression of the IIa/IIIb complex, do not react with autoantibodies from pseudothrombocytopenic patients [15]. This is also confirmed by the observation, that aminoglycosides can avoid pseudothrombocytopenia [16].

7.2.2 Ratio of anticoagulant to blood

The volume of blood drawn should ensure maintenance of the recommended concentration range for EDTA salt of 1.5 g/L (1.2–2.0 g/L = 4.1–6.8 mmol/L based on anhydrous EDTA) [2, 3]. The pH of EDTA varies depending on the salt type: acidity decreases when the number of Na^+ or K^+ ions increases. Free acid 1 % solution has a pH of 2.5 +/− 1.0, whilst the K_3-salt is characterized by a pH of 7.5 +/− 1.0.

At concentrations of 1.5 g/L of blood minor changes appear in neutrophil morphology after one hour storage of EDTA blood. Increasing EDTA concentrations however lead to more serious changes such as loss of bridges between lobules, loss of cytoplasic boundary and early crossover occurs within 1 hour.

EDTA concentrations above the recommended range decreases the centrifuged microhematocrit value, the decrease being more pronounced with K_3-EDTA than with K_2-EDTA.

However, with automated instruments, the MCV was not influenced at K_3-EDTA concentrations up to 10 times the nominal value and results obtained with K_2 EDTA were shown to be dependent on the instrument used [6].

Using the recommended concentration of EDTA salts (K_2- or K_3 EDTA) and performing the analysis in the first 4 h after blood collection, no significant difference was seen in results obtained with blood collected in either of the two anticoagulants [6].

7.2.3 Specimen collection and handling

Thorough mixing of the blood specimen with the anticoagulant by inverting the tube several times after sampling is a prerequisite for exact hematological results. The type of mixer used (rocking versus rotary) may affect the extent of mixing, especially if the tube is overfilled, thus leading to inaccurate results [17, 18, 19].

Blood collection tubes must have an air space representing at least 20 % of the volume of the tube to facilitate proper mixing [2]. On the other hand, under-filling can be a source of errors. If a blood collection tube is drawn to one half of its nominal value, the effective concentration of EDTA would be unsuitable for preparation of a peripheral blood smear intended for white blood cell differential count [20].

Transport, Storage and stability of analytes [21]

Owing to the variations between newer automated instruments, including reagents, it has been recommended that the EDTA anticoagulated blood being analyzed within 6 h of collection [2, 3]. In some cases, however this time is already too excessive to achieve constant results. Hemoglobin, erythrocytes, total leucocytes and platelet numbers are stable for a much longer period (see table in Annex). Stability is excellent over 48 h for hemoglobin and erythrocyte count [22]. Leucocytes numbers are stable over 7 days at room temperature and thrombocytes can be preserved for up to 4 days at room temperature. After the differential count and othe specific morphological parameters are less stable, we recommend to keep transport time below 2 h at room temperature. A blood smear should be prepared within 3 h if transport time exeeds this time. Likewise, stability at refrigerator temperature is analyzer-dependent and therefore not higher [21].

Reticulocytes tend to mature and transform to red blood cells in whole blood, but their stability is 72 h in EDTA, if stored at refrigerator temperature [21, 24].

Recommendation: Count blood cells within six h after collection. Prepare blood smear within 1–3 hours after collection.

7.2.4 Influence and interference factors in hematological investigations

Most often, antibodies and cold agglutinins disturb hematological investigations [23]. Besides, a large number of other interfering factors like cryoglobulins can occur.

Antibodies

Antibodies as interference factors can be overlooked, since neither inspection of samples and other signs only seldomly allow one to be informed about these causes of erroneous results. Analytical procedures in hematology and immuno-hematology, and also in clinical chemistry can be affected by antibodies [23].

Antibodies may affect the cell count of erythrocytes, leukocytes and platelets [25]. The following antibodies are known to interfere with hematological analytes:

Cold agglutinins (erythrocyte specific)

Cryoglobulins

EDTA- dependent antibodies with thrombocyte- and leukocyte specificity

Cold agglutinins

High titers of cold agglutinins directed against erythrocytes lead to agglutination. Such agglutination alters the electronic cell count in the following way: erythrocyte

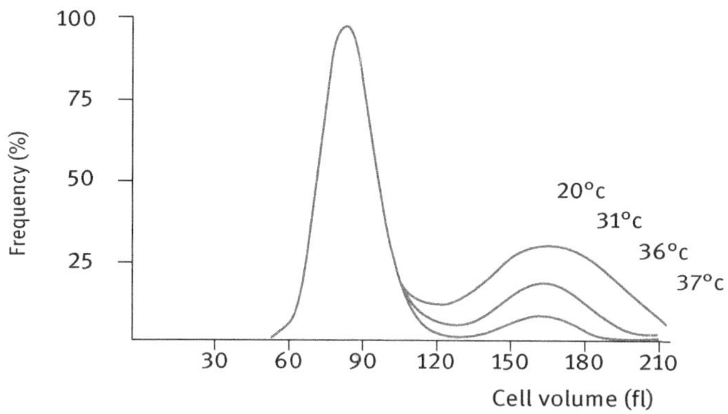

Fig. 7.2.1: Influence of temperature on results of MCV determination of blood in cold agglutinin disease [5].

count is low at normal hemoglobin concentrations; MCV is grossly enhanced (Fig 7.2.1) [26, 27], agglutinates may even be visible at the wall of the EDTA-tubes. The following results may indicate the presence of cold agglutinins:

- very low erythrocyte count at normal hemoglobin concentration
- Grossly enhanced MCV-values
- very low values of calculated hematocrit and resulting in unplausibly high MCH- and MCHC-values.
- Leucocytes- and thrombocyte counts are falsely elevated, because the agglutinates, depending on their size, are either counted in the leucocyte or thrombocyte channel.
- The blood smear shows agglutination of erythrocytes.

Further interferences:

- The antibody causing panagglutination can influence the correct assignment of blood group antigens and cross matching procedure. In addition, the presence of cold antibodies may mask other types of antibodies which may affect analytical procedures in a different manner.
- At high cold agglutinins increased blood sedimentation rate may occur.

Recommendations:

- EDTA-blood sample may be transported at 37 °C (in a prewarmed sand bath) to the laboratory and measured instantly.
- Whenever agglutinates are observed at the wall of the sample tube by the laboratory, sample is to be warmed up at 37 °C for 5–10 min. A correct result is to be expected when reversible binding is the cause of agglutination.

EDTA-dependent antibodies

Thrombocyte antibodies

Antibody induced falsely decreased platelet counts (without hemorrhagic diathesis) can be due to cold agglutinins, or antibodies active in the presence of EDTA.

In both cases, measurable or visible agglutination takes some time. Thus, a prolonged delay between obtaining the sample and platelet counting results in a more pronounced pseudothrombocytopenia (too low platelet count with missing hemorhagic diathesis).

Platelet aggregates may be counted as falsely elevated leucocyte counts.

Platelets of patients with thrombasthenia, which lack the membrane glycoproteins IIb and IIIa, do not react with EDTA-dependent antibodies. This observation suggests that these membrane glycoproteins are actively involved in binding the antibody.

Detection:
- Particles of the size of lymphocytes in the white cell histogram. Detection of aggregates of platelets or satellism of platelets on granulocytes in the stained peripheral blood smear.

Recommendation:
In addition to EDTA-blood sample, citrated or heparinized blood samples should be obtained, because no agglutination appears in these samples.

Other causes of falsely low platelet counts can be due to:
- Giant platelets
- Partially coagulated blood sample caused by wrong sampling technique or insufficient mixing of anticoagulant.

Leucocyte antibodies

EDTA-dependent leukoagglutination has also been described [13] characterized by:
- normal leukocyte count in citrated or heparinized samples and
- Leukocyte aggregates in peripheral blood smear.

Recommendation: Add 0.01 Mol/L dithioerythrol to patient samples. This prevents the disturbing agglutination of leucocytes in patient sample.

Cryoglobulins [15]

Cryoglobulins cristallize in samples kept at room temperature. The resulting particles are of varying shape and may mimic leucocytes, resulting in a falsely elevated leucocyte count. The degree of pseudoleucocytosis depends on the time of exposition, temperature, cryoglobuline concentration and the interaction of cryoglobulins with other plasma proteins [28].

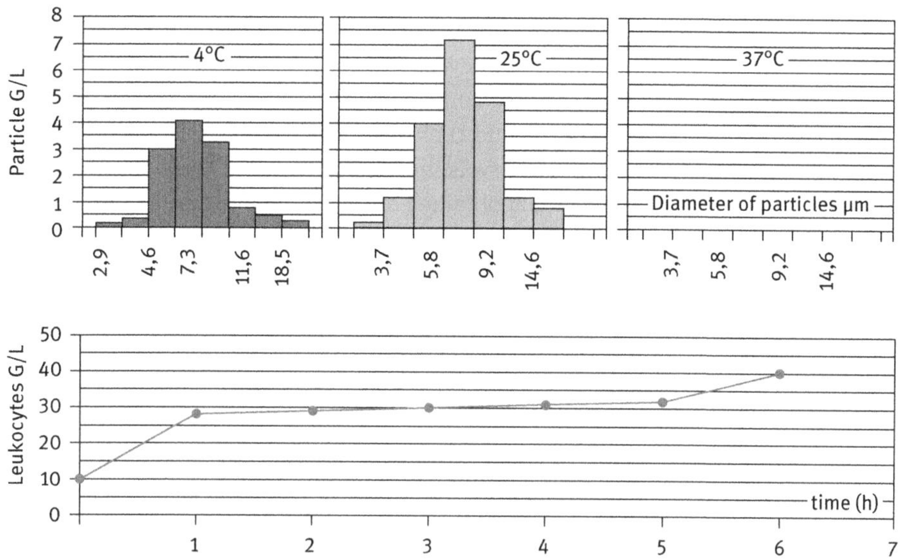

Fig. 7.2.2: Distribution of cryoglobuline particles at different tempartures and the corresponding increase in leukocyte count at different storage times at room temperature [28].

False determinations due to cryoglobulinemia:
- Pseudoleucocytosis
- Pseudopolycythemia
- Pseudothrombocytosis
- Falsely high hemoglobin concentration

The following signs can point to the presence of cryoglobulins:
- Very different cell counts in different investigations
- Blue colored sediments in differential count samples [29]
- In a sample warmed up to 37 °C significantly lower cell counts are observed.

The blood smear shows flocculated dark blue crystals.

The immunological platelet asssessment using fluorescent monoclonal antibodies against the Gp IIb/IIIa receptor (CD 61/Celldyn 4000) is not disturbed by the presence of cryoglobulins [30].

7.2.5 Special considerations for platelet content measurement

To assess in vivo activation of platelets by measurement of constituents in the platelet α – granules such as platelet factor 4 (PF4), β–thromboglobulin, fibronectin and platelet-derived growth factor, activation of platelets after blood collection should be minimized. This can be accomplished either by preventing the formation of thromboxane A_2 or by maintaining high levels of cyclic AMP within the platelets. An additive mixture used for this purpose consists of 0.11 mol/L citric acid, 15 mmol/L theophylline, 3.7 mmol/L adenosine and 0.198 mmol/L dipyridamole with the final pH being adjusted to 5.0 [11, 31]. PF4 levels in blood collected in this additive called CTAD is almost 10 times less than that obtained in a conventional citrated tube [32].

7.2.6 Flow cytometric immunotyping [33, 34, 35]

Sampling

- Sampling should be performed in the early morning, since flow cytometric analysis is a time consuming procedure and may need special transport conditions. Sampling is recommended after 12 hours fasting (because of a postprandial change of cell types), and if possible in the morning (circadian rhythm).
- Use polypropylene tubes for sampling and transport of samples. No polystyrene- or untreated glas tubes ahould be used to prevent adsorbing effects of the tube walls.
- Polypropylene tubes should not contain solid particles like beads to prevent cell damage during transport.
- When posting samples: prepare and add unstained smears.
- Anticoagulants: EDTA versus heparin

Use **K_2- EDTA** for immuno phenotyping
Advantage:

- Diminished adsorption of cells of the myeloid lineage on the tube wall [33] compared to heparin blood, decreased platelet aggregation. Particle count and morphological investigation by microscopy possible with the same sample.

Disadvantage:

- Cellular light scattering characteristics are more rapidly changed in EDTA-blood compared to heparin-blood. Function analysis is not possible in EDTA-blood samples, since calcium is irreversibly extracted from cells. Neutrophils are damaged heaviest.
- After cooling the blood sample a selective dysenrichment of subpopulations can occur [36].

Heparin

The most preferred anticoagulant is heparin (50 U additive free heparin/ml).

Disadvantages:

- Preparation of smears no longer possible (heparin screen)
- Preactivation of cells, especially monocytes, and aggregation of thrombocytes and partially leukocytes by endotoxin contents of heparin. This effect is stronger, the longer the time interval is between sampling and analytical assessment.
- Best for use are pharmaceutical heparin einsetzen and propylene tubes.

Advantage:

- Can be used for function test and transport of samples.

Recommendations: Instant working up would be ideal. Storage time exceeding 4 h should be performed at 17 °C [37].

Transport temperature: room temperature (18–22 °C)

Lymphocytes can be purified using the Ficoll-Hypaque centrifugation procedure [5]. Density gradient cell preparation of mononuclear cells, prepared 6 h after blood sampling, can be contaminated by neutrophils and erythrocytes.

References

[1] Banfi G, Salvagno GL, Lippi G. The role of ethylenediamine tetraacetic acid (EDTA) as in vitro anticoagulant for diagnostic purposes. Clin Chem Lab Med 2007; 45:565–76.

[2] International Council for Standardization in Haematology (ICSH). Recommendations of the I C S H for ethylene diamine tetraacetic acid anticoagulation of blood for blood cell counting and sizing: expert panel on cytometry. Am J Clin Pathol 1993; 100:371–2.

[3] Clinical Laboratory Standards Institute (CLSI). Additives to blood collection devices: EDTA. Approved standard. Wayne PA: Document H35-P 2010.

[4] Guder WG, Hagemann P, Wisser H, Zawta B. Fokus Patientenprobe; Kompendium Präanaltik. Heidelberg BD 2007.

[5] Guder WG, Narayanan S, Wisser H, Zawta B. Diagnostic Samples: From the Patient to the Laboratory. 4th ed. Weinheim Whiley-Blackell 2009. pp 56–57.

[6] Goosens W, van Duppen V, Verwilghen RL. K_2 – or K_3-EDTA: The anticoagulant of choice in routine haematology? Clin Lab Haemat 1991; 13:291–5.

[7] Sears D, Charache S, Perlstein M. Electronic blood cell counters. Faulty calibration due to type and amount of anticoagulant in collection tubes. Arch Pathol Lab Med 1985; 109:247–9.

[8] Lippi U, Schinella M, Modena M, Nicoli M. A new anticoagulant solution to stabilize platelet indices. Prog Med Lab 1989; 3: 111–15.

[9] Thompson CB, Diaz DD, Quinn PG, Lapins M, Kurtz SR, Valeri CR. The role of anticoagulation in the measurement of platelet volumes. Am J Clin Pathol 1983; 80:327–32.

[10] Trowbridge EA, Reardson DM, Hutchinson D, Pickering C. The routine measurement of platelet volume: a comparison of light scattering and aperture-impedance technologies. Clin Phys Physiol Meas 1985; 6:221–38.

[11] Contant G, Gouault-Heilmann M, Martiolini JL. Heparin inactivation during blood storage: its prevention by blood collection in citric acid, theophilline, adenosine, dipyramidole (CTAD) mixture. Thromb Res 1983; 31:365–74.

[12] Epstein HD, Kruskall MS. Spurious leukopenia due to in vitro granulocyte aggregation. Am J Clin Pathol 1988; 89:652–5.

[13] Hillyer ChD, Knopf AN, Berkman EM. EDTA-dependent leukoagglutination. Am J Clin Pathol 1990; 94:458–61.

[14] Banfi G. La pseudocytopenia da EDTA. Biochim Clin 1987; 11:957–62.

[15] Patel KJ, Hughes CG, Parapia LA. Pseudoleucocytosis and pseudothrombocytosis due to cryoglobulinemia. J Clin Pathol 1987; 40:120–1.

[16] Sakurai S, Shiojima I, Tanigawa T, Nakahara K. Aminoglycosides prevent and dissociate the aggregation of platelets in patients with EDTA dependent pseudothrombopenia. Br J Haematol 1997; 99:817–23.

[17] Narayanan S. Preanalytical issues in hematology. J Lab Med 2003; 27:243–8.

[18] Pewarchuk W, Vanderboom J, Blajchman MA. Pseudopolycythemia, pseudothrombocytopenia, and pseudoleukopenia due to overfilling of blood collection vacuum tubes. Arch Pathol Lab Med 1992; 116:90–2.

[19] Solanki D:, Blackburn BC. Spurious leucocytosis and thrombocytopenia. A dual phenomenon caused by clumping platelets in vitro. J Am Med Ass 1983; 250:2514–15.

[20] Sacker IS, Specimen Collection. In: Lewis SM, Coster IF, editors. Quality Control in Haematology. Symposium of the International Committee for Standardization in Haematology. New York: Academic Press 1975: 211–29.

[21] Guder WG, Fiedler M, daFonseca Wollheim F, Schmitt Y, Töpfer G, Wisser H, Zawta B. The Quality of Diagnostic Samples. 4th completely revised ed. 2015 Oxford: BD Diagnostics.

[22] Heins M, Schossow B, Greiner L, Heil W. Influence of storage time and temperature on haematological measurements using a Sysmex NE 8000 analyser. J Lab Med 2000; 24:236–42.

[23] Kohse KP, Wisser H. Antibodies as a source of analytical errors. Eur J Clin Chem Clin Biochem 1990; 28:881–92.

[24] Caville I, Kraaijenhagen R, Pradella R, D'Onofrio G, Herkner K, Rowan RM, et al. In vitro stability of the reticulocyte count. Clin Lab Haematol 1996; 18:9–11.

[25] Bessman JD, Banks D. Spurious macrocytosis, a common clue to erythrocyte cold agglutinins. Am J Clin Pathol 1980; 74:797–800.

[26] Wisser H. Störungen der Messgröße des kleinen Blutbildes durch Antikörper. GIT Lab Med 1995; 18:83–5.

[27] Hatterley PG, Gerard PW, Cacciano V, Nash DR. Erroneous values on the Model S Coulter Counter due to high titer cold agglutinins. Am J Clin Pathol 1971; 55:442–6.

[28] Abela M, Mc Ardle B, Qureshi M. Pseudoleucocytosis due to cryoglobulinaemia. J Clin Pathol 1980; 33:796.

[29] Nebe Th. Stolpersteine im Laboratorium und diagnostischer Wink. Abbott Times 2002: Issue 1:24–6.

[30] Von Ahsen N, Ehrlich B, Scott C, Riggert J, Oellerich M. Cryoglobulines interference with platelet counts and impedance method, but not with the CD 61 count. Clin Chem 2001; 47:1858–60.

[31] Narayanan S. Inhibition of in vitro platelet aggregation and release and fibrinolysis. Ann Clin Lab Sci 1989; 19:260–5.

[32] Van den Besselaar AMH, Meeuwisse-Braun J, Jansen-Gruter R, Bertina RM. Monitoring heparin therapy by the activated thromboplastin time. The effect of preanalytical conditions. Thromb Haemostasiol 1987; 58:226–31.

[33] Rothe G, Schmitz G for the Working Group on Flow Cytometry and Image Analysis. Consensus protocol for the flow cytometric imunophenotyping of hematopoietic malignancies. Leukemia 1996; 10:877–95.

[34] Nebe Th. Immunphänotypisierung von Blutzellen In: Thomas L, Herausgeber. Labor und Diagnose. Frankfurt/Main: TH-Books, 5. Auflage 1998:546–51.
[35] Sack U, Rothe G, Barlage S, Gruber R, Kabelitz D, Kleine TO, Lun A, Renz H, Ruf A, Schmitz G. Durchflußzytometrie in der Klinischen Diagnostik. J Clin Lab Med 2000; 24:277–97.
[36] Thornthweite JT, Rosenthal PK, Vazquez DA, Seckinger D. The effects of anticoagulant and temperature on the measurements of helper and suppressor cells. Diagn Immunol 1984; 2:167–74.
[37] Ekong T, Küpek E, Hill A, Clark C, Davies A, Pinching A. Technical influences on immunophenotyping by flow cytometry. The effect of time and temperature of storage on the viability of lymphocyte subsets. J Immunol Meth 1993; 3:296–300.

Ana-Maria Simundic

7.3 Blood Gases, Ions and Electrolytes

7.3.1 Introduction

Blood gas testing is unique in so many ways. First of all, blood gas testing is most commonly requested for patients in a critical, life threatening condition. Such patients may have some serious metabolic or respiratory disorder and need immediate medical intervention. Another prominent feature of the blood gas testing is the invasiveness of the blood collection procedure. Finally, for many of the blood gas parameters stability of the sample is extremely limited. Due to the low biological variability of many blood gas parameters, allowable total error is quite low and even small differences in serial measurements can be clinically meaningful. Blood gas testing therefore requires special attention and continuous quality management [1]. Standardization of all procedures and unquestionable compliance to the recommended practice is the key to the proper quality management [2]. International standards for blood gas and pH analysis [3, 4] as well as for ionized calcium [5, 6] are available and may serve as a basis for such standardization.

7.3.2 Patient identification

The frequency of identification (ID) errors in laboratory medicine is still unacceptably high. It has been estimated that ID error occurs at the rate of 0.1–1 % in laboratory medicine, whereas they are much less frequent in transfusion medicine [0.05 %] [7]. Nevertheless, actual ID error rate is probably even higher, since most of them go undetected. ID errors are a major healthcare issue since they are potentially associated with serious adverse consequences and hence there should be a zero tolerance to ID errors.

According to the international standards [8], proper patient identification should be done by using at least two acceptable unique patient identifiers (full name, assigned ID number, date of birth, photo ID on government issued ID card or a driver's licence or any other person specific identifier). Patient should be identified in an active way, by an open ended question (e. g. what is your name? what is your date of birth?).

In case some discrepancies are identified, sample collections should be delayed until issues are resolved. Bar-coded ID bracelets and sample identifiers or radiofrequency identifier devices (RFID) can be used to minimize the error risk. Where advanced technological equipment is not available, there is still a lot that can be done just by following a simple recommended standardized procedure. Labelling needs to be done at the time and site of collection and most importantly, in the presence of the patient.

7.3.3 Patient condition

Patient condition may substantially affect the blood gas test results and should therefore be carefully considered. All important variables with potential effect on the test result should be recorded on the test report. Blood sampling should be done when a patient is in a stable resting state. It is extremely important to assess carefully the patient status and record any deviations from the steady state, such as if a patient is exercising, crying, being anxious, etc. Moreover, a note should be taken about all changes in the patient's ventilatory setting (spontaneous breathing or assisted mechanical ventilation) and mode of the oxygen delivery (fraction of inspired oxygen (Fi02) through nasal cannula or Ventouri mask). Respiratory rate (hyperventilation, hypoventilation) and body temperature at the time of blood collection should also be assessed. Hypoventilation and increasing body temperature are associated with increase of ionized calcium, pCO_2 and decrease of pH [9].

If patient condition is changing, a sufficient time should be allowed for patient to stabilize. For patients without pulmonary disease, a period of 3–5 min is usually enough to stabilize. However, in patients with pulmonary disease, this period is significantly longer. Although some recent evidence shows that oxygen equilibration relevant for clinical interpretation in patients with chronic obstructive pulmonary disease (COPD), receiving long-term oxygen therapy requires only 10 min following an increase in oxygen delivery and 16 min following a decrease in oxygen delivery [10], CLSI 46-A2 guidelines [4] recommend that 20–30 min are adequate for most patients to reach a stable state, following ventilatory changes.

For a proper interpretation of a test result, the exact time of the blood collection and the site of sampling should always be recorded and reported with a test result.

7.3.4 Sample type

Arterial blood collected under anaerobic conditions is the only acceptable sample type for an accurate evaluation of the gas exchange function of the lungs (pO_2 and pCO_2).

There has been a large debate over the years as to whether the capillary blood could be acceptable alternative to arterial blood. In a large metaanalysis, Zavorsky et al. have showed that capillary sampling accurately reflects arterial pCO_2 and pH over a wide range of values but is not an adequate substitute for arterial blood for accurate pO_2 measurement [11]. Many subsequent recent studies have confirmed these findings. So, there is really no adequate substitute for arterial blood if accuracy of pO_2 measurement is important [12]. The difference in oxygen content is even more pronounced with increasing actual arterial pO_2 [13].

Therefore, capillary sample may be considered as acceptable alternative only when arterial sample is not available (neonates, paediatric population and some adults) and during medical transport and pre-hospital critical care. Results obtained from capillary sample should be interpreted with extra caution.

Capillary blood should be collected using an arterialisation technique by warming the skin to 40–45 °C with a warm towel or by using the vasodilating cream. Earlobe is better sampling site than a fingertip, because the blood sampled from an arterialized earlobe better reflects arterial blood values.

7.3.5 Anticoagulant

The recommended anticoagulant for arterial blood gas and ionized calcium testing is lyophilized balanced Li-heparin. Heparin binds cations (Ca^{++}, Na^{+}, K^{+}) and that is the reason why only balanced heparin (i. e. heparin which is pre-saturated with cations) should be used [3]. CLSI 46-A2 [4] states that final sample heparin concentration should be 20 [12–30] IU/mL blood.

7.3.6 Sample contamination

Contaminated sample is not rare. Sample contamination may substantially affect blood sample quality and cause significant bias. Arterial blood gas samples are most commonly contaminated with flush solution, liquid heparin, venous blood or air bubbles. Flushing the sampling device (syringe) with liquid heparin causes contamination with heparin and sample dilution resulting in significant differences in several blood gas parameters. To avoid contamination with liquid heparin, arterial blood sampling should preferably be performed using the dedicated blood gas syringes, heparinised with dry electrolyte balanced heparin. Noteworthy, flushing with therapeutic heparin is to be avoided, since therapeutic heparin contains high heparin concentration and thus affects sample pH and electrolyte concentration. Moreover, liquid heparin has high atmospheric pO_2 and thus also affects pO_2 results. Dilution with liquid heparin will therefore cause increase in pO_2. Moreover, it will also lead to the increase in sodium concentration (if sodium heparin is used) and decrease in pH, pCO_2, potassium, calcium and magnesium concentrations [14].

Contamination of arterial sample with venous blood occurs if, during the arterial puncture, needle is not correctly positioned and vein is accidentally punctured. The recommended way to make a successful arterial puncture is to position the needle at an angle of 45 °. Moreover, it is also recommended that short-bevelled needles are used, since they are much easier to position inside the artery.

In an arterial blood sample contaminated with venous blood, there is a decrease of pO_2 and sO_2 and increase of pCO_2. To avoid such preanalytical error, it is highly recommended that self-filling syringes are used. These syringes fill more quickly and much easier when a needle is puncturing an artery instead of a vein, as a result of a difference in blood pressure between the vein and the artery [15].

The aspiration of air during the arterial blood sampling, or bubble formation can result in significant changes in the concentration of some blood gas parameters (↑pH, ↑pO_2 ↑sO_2, ↓pCO_2). It should be noted that even a bubble as small as 1% of the total sample volume may cause significant changes in the sample oxygen content. Contamination with air bubbles can be prevented by visual inspection of the sample immediately after the blood sampling. If air bubbles are present in the sample, they should be expelled as soon as possible and certainly prior to the sample mixing.

7.3.7 Hemolysis

Sample hemolysis is another big concern related to both arterial and capillary blood gas testing. It has been shown that pO_2 decreases and pCO_2 increases in hemolyzed sample. The observed bias for pO_2 and pCO_2 in hemolyzed sample is greater than allowable bias (desirable specifications) based on biological variation [16]. Electrolyte concentration (potassium, calcium) is also dramatically affected by hemolysis. Hemolysis can occur during sampling and due to the inadequate transport and storage conditions. To avoid sample hemolysis samples should not be stored directly on ice cubes and vigorous mixing should be avoided. Since sample turbulence is associated with tendency of hemolysis, special care should be taken to avoid or minimize turbulence due to the use of narrow needles, high vacuum and older pneumatic tube systems. Possible difficulties during the blood sampling should therefore always be recorded, to enable proper interpretation of test results.

7.3.8 Sample transport and handling

Once blood sampling is done, blood will coagulate if not mixed properly immediately after the sampling. Mixing should be done gently (to avoid hemolysis) by inverting the syringe several times and rolling it between the palms. Clotted sample will cause analyzer malfunction. Moreover, if such sample is analyzed, potassium concentration will be increased, due to the efflux of the potassium from the platelets during blood clotting.

It is highly recommended to analyse the sample as soon as possible. Time is a critical modifiable variable in blood gas testing and may be considered as the key to the sample quality. Prolonged storage prior to analysis introduces significant bias due to the cell metabolism and oxygen exchange between the sample and atmosphere. Moreover, prolonged storage can also cause spurious results due to the blood sedimentation. To avoid that, samples should be visually inspected and properly mixed to homogenize the blood inside the syringe.

Samples should be transported by hand, in a plastic syringe, at room temperature and analyzed within 30 min of collection. In cases when expected delivery time is longer than 30 min, glass syringes should be used and sample should be transported on ice. As already mentioned before, vigorous movement during sample transport should be avoided.

Once sample is delivered to the laboratory, it should again be properly mixed prior to analysis in order to obtain a homogeneous sample.

Points to remember

Blood gas testing is of vital importance for a critically ill patient. The role of the laboratory is to ensure that the right sample is taken from the right patient, at the right time, and that correct test result is provided to the ordering physician without delays. If the quality of the sample is compromised to the degree when the expected effect is larger than allowable error, sample should be rejected for analysis. No result is always better than a bad result. Erroneous test result can cause some kind of harm to the patient to a varying degree, resulting in delayed, missed diagnosis or misdiagnosis.

References:

[1] Baird G. Preanalytical considerations in blood gas analysis. Biochem Med 2013; 23:19–27. http://dx.doi.org/10.11613/BM.2013.005

[2] Dukic L, Simundic AM. Institutional practices and policies in acid-base testing: a self reported Croatian survey study on behalf of the Croatian society of medical biochemistry and laboratory medicine Working Group for acid-base balance. Biochem Med 2014; 24:281–92.

[3] Burnett RW, Covington AK, Fogh-Andersen N, Külpmann WR, Maas AHJ, Müller-Plathe O et al. Approved IFCC recommendations on whole blood sampling, transport and storage for simultaneous determination of pH, blood gases, and electrolytes. Eur J Clin Chem Clin Biochem 1995; 33:247–53.

[4] Clinical and Laboratory Standards Institute (CLSI) C46-A2 - Blood gas and ph analysis and related measurements; approved guideline. Wayne, PA: Document C46-A2, 2010.

[5] Boink ABTJ, Buckley BM, Christiansen TF, Cavington AK, Maas AHJ, Müller-Plathe O et al. IFCC – Recommendations on sampling, transport and storage for the determination of concentration of ionized calcium in whole blood, plasma and serum. Europ J Clin Chem Clin Biochem 1991; 29:767–72.

[6] Clinical and Laboratory Standards Institute (CLSI) C31-A2 – Ionized Calcium Determinations: Precollection Variables, Specimen Choice, Collection, and Handling; Approved Guideline. 2nd Edition. Wayne, PA: Clinical and Laboratory Standards Institute, Document C31-A2, 2001.

[7] Lippi G, Chance JJ, Church S, Dazzi P, Fontana R, Giavarina D et al. Preanalytical quality improvement: from dream to reality. Clin Chem Lab Med 2011; 49:1113–26.

[8] Clinical and Laboratory Standards Institute (CLSI) GP33-A - Accuracy in Patient and Sample Identification; approved guideline. Wayne, PA: Clinical and Laboratory Standards Institute,- Document GP33-A, 2010.

[9] Groenendaal F, De Vooght KM, van Bel F. Blood gas values during hypothermia in asphyxiated term neonates. Pediatrics 2009; 123:170–2.

[10] Weinreich UM, Thomsen LP, Hansen A, Kjærgaard S, Wagner PD, Rees SE. Time to steady state after changes in FIO[2] in patients with chronic obstructive pulmonary disease (COPD). COPD 2013; 10:405–10
[11] Zavorsky GS, Cao J, Mayo NE, Gabbay R, Murias JM. Arterial versus capillary blood gases: a meta-analysis. Respir Physiol Neurobiol 2007; 155:268–79.
[12] Higgins C. Capillary-blood gases: To arterialize or not. Med Lab Obs 2008; 40:42–7.
[13] Vaquer S, Masip J, Gili G, Gomà G, Oliva JC, Frechette A, et al Earlobe arterialized capillary blood gas analysis in the intensive care unit: a pilot study. Ann Intensive Care 2014; 4:11.
[14] Küme T, Şişman AR, Solak A, Tuğlu B, Çinkooğlu B, Çoker C. The effects of different syringe volume, needle size and sample volume on blood gas analysis in syringes washed with heparin. Biochem Med 2012; 22:189–201. http://dx.doi.org/10.11613/BM.2012.022
[15] Bender JJ, Allison JR, Goehring JJ, Patel MD, Niederst SM, Douce FH. Arterial sampler filling time during arterial and venous punctures, and its relationship with mean arterial pressure in human subjects. Respir Care 2012; 57:1945–8.
[16] Lippi G, Fontana R, Avanzini P, Sandei F, Ippolito L. Influence of spurious hemolysis on blood gas analysis. Clin Chem Lab Med 2013; 51:1651–4.

Sheshadri Narayanan, Walter G. Guder

7.4 Clinical Chemistry including Metabolites, Enzymes, Hormones and Proteins

In clinical chemistry, a number of reagent- and analyzer-specific problems have to be considered. Thus, many interference factors act in a reagent-specific way. In addition, there are differences between analyzers and analytical principles (i. e."dry" and "wet" chemistry, direct and indirect potentiometry). In this chapter, only a few examples are demonstrated [1, 2], as summarized recently [3].

7.4.1 Sampling for analysis by clinical chemical and immunochemical methods

Sensitive immunochemical methods lend themselves to the measurement of trace quantities of labile hormones, proteins and other analytes in blood. Because of the wide range of analytes measured by immunoassays, this chapter will attempt to focus on a few representative analytes.

7.4.1.1 Collection of blood in an appropriate anticoagulant

Most analytes determined by chemical or immunochemical methods can be measured in serum and/or heparinized plasma (for detailed information see Annex and respective updated publication [4].

Collection of blood in EDTA and promptly freezing the plasma has been found to be adequate for preserving labile polypeptide hormones such as endorphin, vasoactive intestinal peptide, substance P and pancreatic peptide. Because of its inhibitory effect on metalloproteinases, EDTA plasma is also recommended for corticotropin (ACTH), parathyrin (PTH) and glucagon.

7.4.1.2 Collection of blood with proteolytic enzyme inhibitor

A proteinase inhibitor called aprotinin (also known by its trade name Trasylol) added to an anticoagulant such as EDTA or heparin has found application in the stabilization of labile polypeptide hormones and enzymes [5]. Since aprotinin inhibits kallikrein, its potency is expressed in terms of kallikrein inhibitory units (KIU). The concentration of aprotinin used for the preservation of labile hormones and enzymes ranges from 500 to 2000 KIU/mL. Thus, a mixture of EDTA – aprotinin has been used to stabilize glucagon, ACTH, renin and certain gastrointestinal hormones such as β-endorphin, secretin, neurotensin, gut glucagon, somatostatin and vasoactive intestinal peptide [5]. In one study it was demonstrated that glucagon levels measured by RIA in EDTA plasma were approximately 26 % higher than in plasma obtained from blood collected in a mixture of 1.5 mg EDTA and 2000 KIU of aprotinin per mL of blood. The high glucagon level in samples collected without aprotinin was due to the fragments of hormone produced by proteolytic enzyme degradation being apparently recognized as an intact molecule by the antibody used in the assay. In addition, some of the radio-labeled hormone underwent proteolytic enzyme degradation, thus limiting the amount of radio-label available to compete with the hormone in plasma for binding sites on the antibody [6].

A mixture of lithium heparin and aprotinin has also been used to stabilize immunoreactive somatostatin, secretin, glucagon, C-peptide and cholecystokinin-pancreozymin.

There are instances where EDTA alone is not sufficient for stabilizing an analyte. Thus, even when blood is collected in EDTA and the plasma is separated promptly and stored under ideal conditions, there is activation of complement. However, when a synthetic protease inhibitor such as nafamostat mesylate is added to EDTA, the stability of complement components (C_{3a}, C_{4a} and C_{5a}) is significantly improved. Furthermore, while the activity of complement components doubles with each freeze–thaw cycle when blood is collected with EDTA alone, no such discrepancy is seen in samples collected in EDTA supplemented with nafamostat mesylate [7].

Traditional anticoagulants such as EDTA and heparin have been ineffective in stabilizing a labile constituent such as parathyroid hormone related protein (PTH-RP). This tumor marker is so unstable that less than 10 % of the original activity remains after 16 hours of storage at room temperature in blood specimens collected in either heparin or EDTA [8]. The optimum mixture for blood collection is EDTA (1.5 g/L of blood), aprotinin (500 KIU/mL), leupeptin and pepstatin (each 2.5 mg/mL). With this additive mixture, PTH-RP is stable for up to 1 h at room temperature and for 24 h if maintained refrigerated at 4 °C [8].

For the measurement of catecholamines in plasma, blood should be collected in an anticoagulant mixture such as EGTA (90 g/L) supplemented with sodium metabisulfite or glutathione (60 g/L). Even with this additive mixture, the blood has to be kept cold in an ice bath and centrifuged in a refrigerated centrifuge. The plasma should then be transferred to a plastic vial, capped and stored in a freezer at a temperature below –20 °C until ready for analysis. Frozen plasma prepared under the above-mentioned conditions preserves catecholamines for at least 3 months [9].

7.4.2 Posture and timing of sampling

Variables such as posture during blood sampling and diurnal changes have to be taken into consideration (see Chapters 2.3 and 3.2).

Postural variations have a significant impact on renin, elevation of enzyme activity being observed when moving from the recumbent to the erect position.

Cortisol has a diurnal rhythm with a peak value between 4 a. m. and 6 a.m (see Fig. 3.2.3). Hormones such as growth hormone, lutropin (LH) and follitropin (FSH) are released in bursts, and as such, several blood specimens taken within closely timed intervals are needed to establish a median value.

7.4.3 Refrigeration and freezing of samples for hormone assays

Some hormones such as insulin, proinsulin and C-peptide can be stabilized by merely placing blood specimen containers on ice immediately after collection. Such specimens should be promptly centrifuged, preferably in a refrigerated centrifuge at 4 °C, and the serum kept frozen until assayed. Of course, the frozen specimen should be completely thawed prior to assay. It is also important that hemolysis be avoided, since this will decrease both insulin and proinsulin values.

Processing of blood promptly upon clotting and freezing of serum at –70 °C provides long-term stability for analytes such as gastrin, pepsinogen-1, and insulin.

7.4.4 Special aspects when using the so-called "dry chemistry" [10, 11]

When a carrier-bound reagent analyzer for capillary blood is used, the correct sampling of capillary blood samples is of decisive importance for the reliability of the results. The producer of the Reflotron® System (Roche Diagnostics, Germany) recommends:

> When a large free-hanging droplet has formed, it should be applied to the sample application field of the test carrier without directly touching the carrier with the finger. For an additional measurement it is necessary to prick the finger at a different site.

Results in a whole blood analyzer may depend on sample haematocrit, because the amount of plasma available to the reactive zone of the test strip varies at different packed cell volumes.

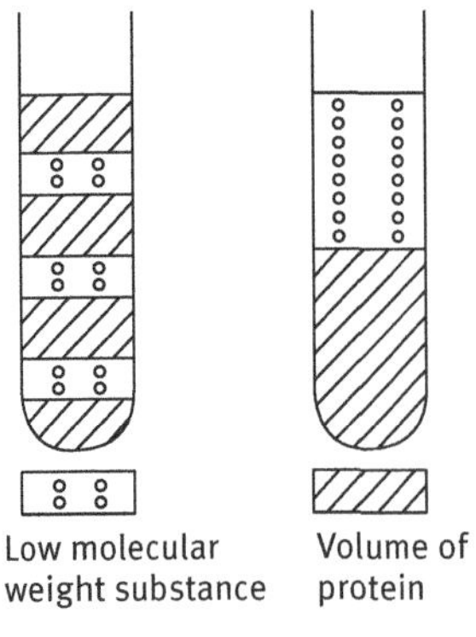

Fig. 7.4.1: Change of concentration of low molecular weight substances after deproteinization (from [12]).

Dry chemistry methodologies on the one hand offer the possibility of separating the analyte from many disturbing components in the matrix. On the other hand, the matrix is more diluted in wet chemistry procedures, resulting in lower analyte and possible interfering concentrations.

7.4.5 Different results using methods with and without deproteinization?

Protein molecules in serum/plasma occupy a defined volume, dependent on the concentration and the size of the protein. As a result, the concentration of low molecular solutes (e. g. glucose, electrolytes, etc.) in protein-free filtrates are found to be approximately 5 % higher than in untreated serum/plasma samples without deproteinization. The influence of deproteinization on the determination of low molecular weight substances which are dissolved in serum-/plasma water, is shown in Fig. 7.4.1 [12]. The volume displacement effect of proteins is especially important for the determination of electrolytes using direct potentiometric in comparison to indirect measuring procedures [13] as well as for determination in deproteinised whole blood (e. g. glucose).

7.4.6 The volume displacement effect of lipids

A similar displacement effect is observed with triglycerides in blood. At 5000 mg/dL (57 mmol/L) triglyceride concentration, direct potentiometry gives about 5 % higher (and more true) results of electrolytes compared to flame photometry or indirect (after dilution) potentiometry [13].

7.4.7 Whole blood versus plasma glucose

When comparing glucose in whole blood with that in plasma, similar, but larger differences are observed. Since blood cells have a higher protein and lipid content and glucose is not equally distributed between the intracellular and the extracellular space, results obtained in plasma are approximately 15 % higher compared to whole blood, when related to the same volume. The WHO [14] and the ADA criteria [15] for the diagnosis of diabetes mellitus are therefore different for plasma and whole blood. This has recently been considered in the recommendations of the national diabetes society suggesting plasma glucose as gold standard for diagnosing diabetes mellitus [16]. In addition, the recent change to acidified sample has increased the stability of glucose in blood samples and allows plasma correct measurements even after hours of sampling [17].

7.4.8 Electrolytes (Na^+, K^+, Cl, HCO_3)

If during transport and storage of whole blood, the glucose concentration falls below a critical concentration, the cells lose their intracellular potassium and take up sodium in its place. If CO_2 escapes from a blood sample, the cells lose bicarbonate and replace it with chloride from the surrounding plasma (chloride shift) (Fig.7.4.2). This limits the stability of whole blood for the plasma/serum analysis of electrolytes. Interestingly, the increase in plasma potassium is higher in refrigerated blood samples compared to samples stored at room temperature [18]. This is caused by an inhibition of the cellular Na^+,K^+-ATPase activity brought about by cold.

7.4.9 Trace elements

Contamination plays a significant role in trace element analysis. Blood should therefore be obtained in a suitable sampling system that has been declared trace element-free by the laboratory.

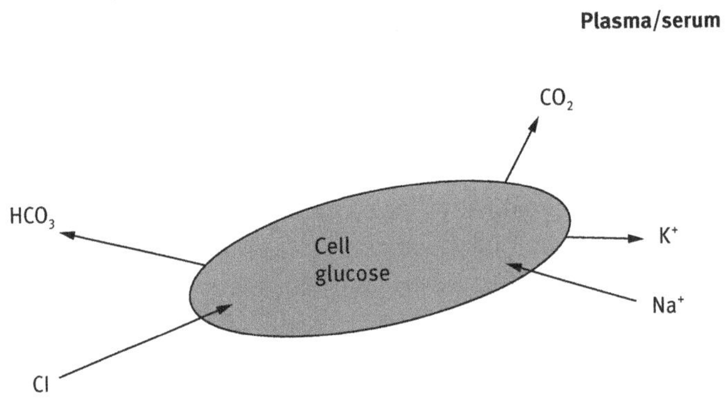

Fig. 7.4.2: Electrolyte fluxes between blood cells and plasma/serum during storage of whole blood.

7.4.10 Lipids

Storage alters triglyceride concentration due to the action of endogenous lipases. The triglyceride concentration in the sample falls while that of free glycerol rises. The extent of this effect varies from person to person and does not correlate with the initial triglyceride concentration. In addition, the composition of plasma lipids determines their behavior during centrifugation. Chylomicrons and their remnants tend to float to the top layer of plasma, whereas other lipoproteins remain equally distributed. This has to be considered when primary tubes are used in analyzers [19].

7.4.11 Creatinine

The concentration of non-creatinine chromogens increases at room temperature, this effect being more pronounced in whole blood than in serum/plasma (the higher the temperature, the greater the effect). This increase – which varies from person to person and is independent of the initial creatinine concentration – occurs when creatinine is determined by the Jaffé method [20].

Recommendation: In clinical chemical analysis, method and analyzer-specific prenalytical influences and interferences have to be taken into consideration. Results obtained with one system are not transferable to others without experimental proof.

Detailed information may be obtained from reagent and analyzer manufacturers.

References

[1] Thomas L. (ed). Clinical Laboratory Diagnostics. Frankfurt/Main: TH-Books 1998.

[2] Young DS. Effect of preanalytical Variables on Clinical Laboratory tests. 3rd ed. Washington: AACC-Press 2007.

[3] Guder WG, Narayanan S, Wisser H, Zawta B. Diagnostic Samples: From the Patient to the Laboratory. 4th updated ed. Weinheim: Wiley-Blackwell 2009.

[4] Guder WG, Fiedler M, da Fonseca Wollheim F, Schmitt Y, Töpfer G, Wisser H, Zawta B. Quality of Diagnostic Samples. 4th completely revised ed. Oxford: BD Diagnostic 2015.

[5] Narayanan S. Protection of peptidic substrates by protease inhibitors. Biochim Clin 1987; 11:954–6.

[6] Echemendia E, Van Heertum RL, Guzman L, Narayanan S, Ince G. The stabilizing effect of aprotinin on selected hormone assays. (Abstract) Clin Chem 1984; 30:1051.

[7] Watkins J, Wild G, Smith S. Nafomostat to stabilize plasma sample staken for complement measurement. Lancet 1989; i:896–7.

[8] Pandian MR, Morgan CH, Carlton E, Segre GV. Modified immunoradiometric assay of parathyroid hormone-related protein: clinical application in the differential diagnosis of hypercalcemia. Clin Chem 1992; 38:282–8.

[9] Boomsma F, Alberts G, van Eijk L, Man in't Field AJ, Schalekamp ADH. Optimal collection and storage conditions for catecholamine measurement in human plasma and urine. Clin Chem 1993; 39:2503–8.

[10] Junker R, Schlebusch H, Luppa PB. Point of care testing in hospitals and primary care. Dtsch. Ärztebl. 2010; 107:561–7.

[11] Sonntag O. Trockenchemie , Analytik mit trägergebundenen Reagenzien. Stuttgart: Thieme,1988.

[12] Bürgi W, Richterich R, Mittelholzer ML. Der Einfluss der Enteiweißung auf die Resultate von Serum– und Plasmaanalysen. Klin Wschr 1967; 45:83–6.

[13] Külpmann WR. Determination of electrolytes in serum and serum water. Wiener Klein Wschr 1992; 104 (Suppl 192):34–8.

[14] Report of the WHO consultation. Diagnosis and Classification of Diabetes mellitus Geneva: WHO 1999.

[15] American Diabetes Association. Report of the expert committee on the diagnosis and classification of diabetes mellitus. Diabetes Care 1997; 20:1183–97.

[16] Kerner W, Brückel J. Definition, Klassifikation und Diagnostik des Diabetes Mellitus. Diabetologie und Stoffwechsel 2013; 8:104–7.
[17] Gambino R, Piscitelli J, Achattupathil TA, Theriault JL, Andrich RD, Sanfilippo ML, Etienne M. Acidification of blood is superior to sodium fluoride alone as an inhibitor of glycolysis. Clin Chem 2009; 55:1019–21.
[18] Heins M, Heil W, Withold W. Storage of serum or whole blood samples? Effect of time and temperature on 22 serum analytes. Eur J Clin Chem Clin Biochem 1995; 33:231–8.
[19] Miller M, Bachorik PS, Cloey TA. Nrmal variation n plasma lipoproteins: postural effects on plasma concentration of lipids, lipoproteins and apolipoproteins. Clin Chem 1992; 38:569–74.
[20] Narayanan S, Appleton HD. Creatinine: a review. Clin Chem 1980; 26:1119–26.

Sheshadri Narayanan, Yashpal Agrawal

7.5 Preanalytical Variables in Therapeutic Drug Monitoring

7.5.1 Introduction

Basic information on the pharmacokinetics of the drug to be monitored is a prerequisite to the question of when to collect a sample for therapeutic drug monitoring (TDM).

- Where the metabolite of the parent drug is active it may be necessary to measure both the parent drug and its active metabolite.
- Pharmacogenetic variability due to mutations in enzymes involved in the metabolism of drugs can affect TDM by either lack of therapeutic effect or toxicity unless the dosage of the drug is adjusted either upward or downward respectively.
- Drug-drug and drug-herb interactions can influence the results obtained by TDM.
- Interaction of sample collection materials (e. g. gel in serum separator tube) with the drug to be measured.
- The effect of anticoagulants used for specimen collection are preanalytical variables that can affect results obtained by TDM.

We will attempt to explore these issues in this chapter.

7.5.2 General considerations [1, 2]

Drugs that have a narrow and well-defined therapeutic range are good candidates for TDM. The therapeutic range or therapeutic window spans the minimum effective concentration (MEC) and minimum toxic concentration (MTC). The trough concentration obtained from a blood specimen collected within one hour before the administration of the next dose should not fall below the MEC while the peak concentration should not exceed the MTC. The knowledge of therapeutic index which is the ratio of minimum concentration in blood or the dose that is toxic divided by the minimum concentration in blood or dose that gives the therapeutic effect is valuable. If the therapeutic index is 2 or less, toxicity can be encountered. Drugs that have a narrow therapeutic index require close monitoring by TDM. In this class are aminoglycoside antibiotics such as gentamicin and tobramycin which require monitoring by measuring both peak and trough levels since with these drugs the peak level corresponds to efficacy and toxicity is a cumulative effect governed by the duration of treatment.

Drugs circulate in plasma bound to proteins and in this form they are unable to cross cell membranes. The fraction of the drug that is free (not bound to protein) is able to cross the cell membrane and exert its action by interacting with specific receptors. Thus, it is the free drug that is biologically active. Hence, the rationale for the measurement of the free drug level especially if either toxicity or lack of therapeutic effect is encountered or if the metabolite of the parent drug is also active. Generally, acidic drugs are bound to albumin and basic drugs are bound to α_1-acid glycoprotein. Examples of basic drugs that are bound to α_1-acid glycoprotein are quinidine, propanolol, lidocaine, dysopyramide and meperidine. Binding equilibrium of drugs can be altered in presence of other drugs or by an increase in fatty acid level. Displacement of a drug from its binding protein by a competing drug or fatty acid is readily accomplished with drugs that are weakly bound as compared to those that are tightly bound. A clinical consequence of an increased free drug level caused by displacement from binding protein (unless the free drug is rapidly metabolized) is to cause toxicity at a dose that would otherwise be considered therapeutic.

Protein binding is affected in renal disease and in liver disease due to reduced amounts of albumin available for drug binding and also when two or more drugs compete for binding sites on a protein. Thus, for example valproic acid competes with phenytoin for binding sites on protein resulting in a decrease in phenytoin level since the displaced free phenytoin is rapidly converted to an inactive metabolite. The concentration of phenobarbital can increase by 10–20 % due to co-administration of salicylate or valproic acid, thus requiring lowering of the dose of phenobarbital to prevent toxicity.

Protein binding of drug is affected in disease. Thus while phenytoin in healthy subjects is 90 % protein-bound and 10 % free, in uremic subjects, however, protein binding of drug can range from virtually zero to as much as 70 %. Hence for the same total phenytoin concentration that is seen in healthy subjects, the free drug concentration in uremic subjects, in the best case scenario, can amount to as much as 30 % causing toxicity unless the dose is adjusted downward. The reduction of protein binding is apparently due to the endogenous substances accumulating in uremia competing with phenytoin for binding sites on protein.

Some drugs exhibit saturation of protein binding sites at optimal total drug concentrations. Hence a slight increase over the optimal drug concentration can significantly increase the free drug concentration thus causing toxicity. For instance, the optimal therapeutic trough concentration of valproic acid is approximately 100 µg/mL (mg/L) and the upper limit of the therapeutic range is between 120–125 µg/mL(mg/L). A slight increase over the upper limit of the therapeutic range will result in an increase in free drug concentration and a potential for toxicity. A decrease in albumin concentration can also increase the free drug level and cause toxicity.

7.5.3 Timing of specimen collection

Before collection of a blood specimen it is important that steady state therapeutic concentrations of drug in serum have been obtained. Steady state conditions are reached when the amount of drug in circulation is equal to the amount that is being eliminated. The time taken to reach steady state can be estimated from the elimination half-life of the drug. The elimination half-life ($t_{1/2}$), is the time needed for one-half of the dose (50 %) of the drug to be eliminated from the body. Thus, it takes one half-life to reach 50 %, two half-lives to reach 75 %, three half-lives to reach 87.5 %, four half-lives to reach 93.75 % and five half-lives to reach 95 % of steady state levels. In practice blood specimens are collected after the completion of 4 or even 5 half-lives, depending on the clinical situation. It is also possible to reach steady state condition by injecting a single loading dose equal to the amount found at steady state. The dose needed to maintain a mean steady state concentration is called the maintenance dose. A blood sample collected just before administration of the next drug dose will provide a trough level of the drug which represents the lowest concentration of the drug to be found with the established drug dose. Blood samples collected when the drug concentration is at its highest level in serum or plasma is referred to as the peak level. Specimens representing peak level are usually drawn between 1 and 5 h after oral administration, the exact timing being dependent on the rate of distribution of the drug. Peak level specimens should be drawn within 1–2 h after intramuscular administration and 15–30 min after intravenous administration. Blood specimens drawn during an infusion should never be drawn from the arm that is receiving the infusion, but instead should be drawn from the opposite arm. In general, blood specimens collected for TDM are drawn to provide trough levels.

It is difficult to accurately ascertain peak levels after an oral dose since the rate of absorption varies from individual to individual. However, there are occasions when it may be necessary to know the peak levels after intravenous administration, and it may be of value in the monitoring of drugs that can be toxic if the upper limit of the therapeutic range is slightly exceeded. For some drugs, that have a long distribution phase, such as digoxin, although the peak concentration is reached 2–3 h after an oral dose, the peak tissue concentration is reached only after 6–10 h after administering the oral dose. Hence for the monitoring of digoxin, blood sample should be collected at least 8 h after the last oral dose (trough specimen). Likewise, for the monitoring of lithium, blood specimens should be collected 12 h after the last dose [1, 2, 3].

While most drugs are measured on either serum or plasma for the TDM of immunosuppressants such as cyclosporine, tacrolimus and sirolimus, however, whole blood collected in EDTA is the specimen of choice.

Of the various anticoagulants used for specimen collection, EDTA, by its ability to chelate divalent cations and thus protecting some drugs from metal oxidation and also by preventing volume shifts between cellular and plasma compartments has a stabilizing influence on basic drugs, such as tricyclic antidepressants.

Heparin has been reported to alter protein binding of some drugs [4]. Heparin is also reported to form an inactive complex with gentamicin. This complex could cross react with antibodies used to measure gentamicin.

Although theoretically you would expect the drug concentration in heparinized plasma to be higher than in serum, assuming that the drug would also be bound to fibrinogen, in reality the serum concentrations of amiodarone, desethylamiodarone, chloroquine and desethylchloroquine were higher than the plasma values, apparently due to the release of cell-bound drug during the process of clotting of blood [5]. Thus, results obtained for amiodarone were 11.5–13.5 % lower in heparinized plasma than in serum, while results for desethylamiodarone were 4.4–8.5 % lower in plasma than in serum [6]. In the case of chloroquine and its metabolite desethylchloroquine, both of which are mainly present in the platelets and granulocytes, the release of these two drugs during the coagulation process into serum results in their concentration in serum to be higher than in heparinized plasma. The concentration of chloroquine was approximately 2-fold higher in serum than in plasma and the concentration of desethylchloroquine, was approximately four times higher in serum than in plasma [7].

Heparin is not recommended as an anticoagulant for the measurement of cyclosporine and tacrolimus. Citrate as an anticoagulant is not suitable for the measurement of tacrolimus. Neither EDTA nor Citrate is suitable as anticoagulants for the measurement of Digitoxin [8].

Table 7.5.1 provides information on the half-life, therapeutic range, and information on specimen collection of selected drugs.

Table 7.5.1: Biological half-life, therapeutic range, and information on specimen collection of selected drugs (Adapted from 2, 9, 10)

Abbreviations: S = serum, HP = heparin-plasma, H= heparin, h = hour(s), d = day(s), w = week(s), y = year(s).

Drug	Half-life h	Therapeutic range mg/L (μg/mL)	Specimen	Notes on specimen collection	Stability in blood	Stability in plasma/ serum at RT
Anticonvulsants						
Carbamazepine	10–25	4–12	S, HP, EDTA-Plasma	Trough	2 d	5 d
Ethosuximide	10–60	40–100	S, HP, EDTA-Plasma	Trough		
Phenobarbital	50–120	15–40	S, HP, EDTA-Plasma	Trough or mid interval point	2 d	1 d
Phenytoin	25–200	10–20	S, HP, EDTA-Plasma	Toxic >20; trough, free phenytoin useful	2 d	2 d
Primidone	6–8 h	5–12	S, HP, EDTA-Plasma	Trough; metabolized to Phenobarbital	1 y	
Valproic acid	8–15	50–125	S, HP, EDTA-Plasma	Trough; free drug useful	2 d	2 d

Table 7.5.1: (continued)

Drug	Half-life h	Therapeutic range mg/L (µg/mL)	Specimen	Notes on specimen collection	Stability in blood	Stability in plasma/ serum at RT
Anti-arrhythmic agents						
Amiodarone	4 h–25d	0.5–2	S, H- and EDTA-Plasma	Trough; >2.5 µg/ mL toxic	< 4 h	1 d
Flecainide	14	0.2–1.0	S, H- and EDTA-Plasma	Trough or midpoint		
Lidocaine	70–200 min	1.5–5	S, HP, EDTA-Plasma	during infusion; metabolite active		
N-acetyl-Procainamide (NAPA)	6–10 h	10–20	S, H- and EDTA-Plasma	PA & NAPA measured on same sample		
Procainamide (PA)	3–5	4–8	S, H- and EDTA-Plasma	Trough or mid-point		
Quinidine	6–7	2–5	S, H - and EDTA-Plasma	Trough or mid-point		
Antibiotics						
Amikacin	1.5–15	Peak: 15–30 Trough:< 2	S, H- and EDTA-Plasma	peak: 30 min after infusion		2 h
Gentamicin	1.5–6	Peak: 5–10 Trough: < 2	S, H- and EDTA-Plasma	peak:1 h after infusion, trough: before next dose	4 h	4 h
Tobramycin	0.5–3 h	Peak: 4–8 Trough: < 1	S, H- and EDTA-Plasma	peak: 1h after infusion, trough: before next dose		2 d
Vancomycin	4–10	Trough: 5–15	S, H- and EDTA-Plasma	trough: before next dose, monitoring peak not recommended		2 d
Cardiac Glycosides	**days**	**µg/ L (ng/mL)**				
Digitoxin	6–8 d	13–25	S, H- and EDTA-Plasma	Trough: before next dose	1–2 d	2 w
Digoxin	1–2	0.5–2.0	S, H- and EDTA-Plasma	Collect > 8 h after last dose	1–2 d	2 w
Immunosuppressants						
Cyclosporine	10–27	100–400	whole EDTA blood;	Trough, drug adheres to plastic	13 d	3 w in hemolysate
Sirolimus	59 ± 19	4–20	whole EDTA blood;	trough	1 d	8 h
Tacrolimus	6–21	5–20	whole EDTA blood;	trough 12 h after dose	7 d	7 d

Table 7.5.1: (continued)

Drug	Half-life h	Therapeutic range mg/L (µg/mL)	Specimen	Notes on specimen collection	Stability in blood	Stability in plasma/ serum at RT
Psychopharmaceutic agents	**hours**	**µg/ L (ng/mL)**				
Amitryptyline	17–40	125–250	S, H- and EDTA-Plasma	Trough; metabolite NT		1 d
Desipramine (DE)	12–54	115–250	S, H- and EDTA-Plasma			
Doxepin	8–25	50–250	S, H- and EDTA-Plasma	Trough	2 h	1 d
Imipramine (IM)	9–24	180–240 ng/ mL Level for IM includes DE its active metabolite	S, H- and EDTA-Plasma	Trough		
Lithium	14–33	Acute: 0.6–1.2 [1.6] mmol/L, Chronic: 0.6–0.8 [1.2] mmol / L	Serum or ammonium – H - plasma; do not use Li-H	Collect 12 h after dose, (may increase in hyponatremia)	1 h	1 d
Nortryptyline (NT)	18–56	50–150 ng/ mL	S, H- and EDTA-Plasma	Trough		1 d
Bronchodilator						
Theophylline	3–12 h	5–20 µg/mL (Less toxic and effect-tive at 5–15 µg/mL) (formulation dependent)	S, H- and EDTA-Plasma	Trough or mid-interval		1 d

Note: Therapeutic ranges given in table are approximate and are method dependent.

7.5.4 Importance of measuring active metabolites

Many of the parent drugs are converted to active metabolites by Cytochrome 450 (CYP) family of enzymes. If the half-lives of these metabolites are longer than that of the parent drug, it can cause toxicity if they accumulate. Variants of CYP isoforms can either delay the formation of the active metabolite or lead to rapid depletion of the active metabolite in ultra-rapid metabolizers. Thus the measurement of both the parent drug and its active metabolite ensures that the parent drug is properly adjusted to maintain both efficacy and the prevention of adverse toxic reactions.

Induction or inhibition of CYP isoforms by herbs can also alter the parent drug to active metabolite ratio. We'll address the interaction of herbs with drugs later in this chapter.

Table 7.5.2: is a selective listing of drugs with active metabolites.

Drug	Active metabolite
Amitriptyline	Nortriptyline
Imipramine	Desipramine
Primidone	Phenobarbital
Procainamide	N-Acetylprocainamide
Carbamazepine	10, 11-Carabamazepine epoxide
Codeine	Morphine
Methamphetamine	Amphetamine
Lidocaine	Monoethylglycinexylidide (MEGX)

We'll illustrate the importance of measuring both the parent drug and the active metabolite with some examples.

Phenobarbital, the active metabolite of primidone has a much longer half-life (50–120 h) as compared to primidone (6–8 h). If primidone is rapidly metabolized, the concentration of phenobarbital can increase, causing toxicity unless the dose of primidone is adjusted downward.

In patients who are fast acetylators (50 % of African-Americans and Caucasian-Americans and 60–70 % of Northern Europeans) the concentration of N-acetyl-procainamide, the active metabolite of procainamide can increase necessitating the measurement of both these drugs in the same sample.

Amitriptyline is metabolized by CYP2C19 isoform to the active metabolite, nortriptyline.

Nortriptyline in turn is inactivated by CYP2D6 isoform. Variants of CYP2D6 that slow the inactivation and clearance will cause accumulation of nortriptyline to toxic levels unless the dosage of amitryptyline is reduced. In contrast, variants of CYP2D6 isoform which result in fast metabolizers would require a larger dose of amitryptyline to maintain the therapeutic level of both the parent drug and its active metabolite [11].

7.5.5 Pharamacogenetic variability

Variants of genes that code for enzymes involved in the metabolism of drugs can affect the therapeutic management of the patient unless the dose of the drug is appropriately adjusted. For instance, therapy with the anticoagulant drug warfarin is influenced by variations in genes involved in its metabolism. Warfarin exists in two enantiomeric forms (R-and S-warfarin). R-warfarin is metabolized by CYP1A2 and CYP3A4 isoforms.

S-warfarin which is two to five times more potent than the R-enantiomer is metabolized by the hepatic microsomal CYP2C9 isoform to the inactive S-7-hydroxywarfarin [12]. Carriers of CYP2C9*2 and CYP2C9*3 variant alleles had a 30 % and 80 % decrease in enzymatic activity, respectively, subjecting them to an increased risk for overanticoagulation and bleeding unless the warfarin dose was reduced [13].

Variations in vitamin K epoxide reductase complex subunit 1 (VKORC1) gene can also affect the efficacy of warfarin. The efficacy of warfarin is dependent on its inhibiting vitamin K epoxide reductase enzyme. This enzyme is involved in the pathway of the production of the active form of vitamin K which is required to add gamma carboxyl groups to vitamin K dependent clotting factors II, VII, IX and X and thus facilitate the process of clotting [14]. Variations in the VKORC1 gene dictated the warfarin dose required to maintain stable anticoagulation. Compared to wild type, the two variants of the VKORC1 gene (the CT and TT genotypes) required 27 % and 47 % reduction in warfarin dosage, respectively, to maintain stable anticoagulation [15]. This inter-individual genetic variability makes it imperative for warfarin dosage to be determined by monitoring the patient's INR (International normalized ratio) derived from prothrombin time measurements.

Another example of pharmacogenetic variability is illustrated with therapy utilizing Clopidogrel which is a platelet aggregation inhibitor. This drug belongs to a class of molecules called thienopyridines that irreversibly inhibit the platelet ADP $P2Y_{12}$ receptor. Binding of ADP to this receptor activates the GPIIb-IIIa receptor on the platelet and the formation of a stable platelet aggregate [12]. Clopidogrel is a pro-drug which is converted by cytochrome P450 (CYP2C19) isoform in the liver to its active form that inhibits ADP from binding to platelet $P2Y_{12}$ receptor thus preventing platelets from aggregating. Mutations in the alleles *2 to *5 of CYP2C19 are poor metabolizers of clopidogrel and present a risk of thrombosis as compared to the wild type *1 allele in patients who are normal metabolizers. In contrast, persons with mutation in allele *17 of CYP2C19 are ultra rapid metabolizers in whom a smaller dose of drug is required. For this reason an automated assay to detect CYP2C19 mutations in alleles *2, *3 and *17 has commercially been introduced in order that patients about to undergo clopidogrel therapy could be screened to determine if they have these mutations that would require adjustment of the dose of the drug. In a meta-analysis of 9 studies of patients who had coronary artery stents and were on clopidogrel therapy, carriers who had just one reduced-function CYP2C19 allele had a 167 % increased risk for stent thrombosis compared to those who had wild type allele. The risk increases even more dramatically in carriers of 2 reduced-function alleles [16].

It is sobering to note that codeine which is a component of many over-the-counter medicines admixed with other drugs like acetaminophen may have potential for adverse effects in babies and even adults considering that the active metabolite of codeine is morphine. Subjects who have a mutation in the CYP2D6 isoform that converts codeine to morphine (specifically CYP2D6*2 × 2 genotype) are ultra-rapid metabolizers and would potentially have higher than expected levels of morphine in serum.

The above examples serve to highlight the effect of pharmacogenetic variability due to mutations in genes that code for enzymes involved in the metabolism of drugs thus affecting therapy.

7.5.7 Drug-drug interactions

Literature on drugs and their interactions with other drugs is so voluminous it has been the subject of a weighty compendium [17]. Thus we will highlight the importance of this subject with a few examples.

The metabolism of phenytoin is increased in the presence of alcohol, carbamazepine and barbiturates which induce oxidative enzymes. Drugs which compete with phenytoin metabolism such as dicumarol, chloramphenicol, isoniazid and disulfiram cause an increase in both total and free phenytoin concentration in serum.

Carbamazepine and phenytoin can increase the rate of conversion of primidone to phenobarbital.

Salicylates will compete with methotrexate and thus reduce its clearance.

The hepatic clearance of tricyclic antidepressants is increased in subjects taking drugs that induce microsomal enzymes, such as carabamazepine, barbiturates and phenytoin. The clearance will be decreased in subjects taking drugs such as cimetidine which inhibit microsomal enzymes.

Digoxin concentration in serum is increased in presence of quinidine which competes for receptors for digoxin and may also reduce its clearance.

In general, since the majority of drugs are metabolized by the cytochrome P450 3A (CYP3A4) isoform found in the liver and the gut, competing drugs which induce or inhibit the CYP3A4 isoform or for that matter other CYP isoforms such as CYP2C9, CYP2C19, etc. can affect the therapeutic concentration of the measured drug, and in turn will have a bearing on its efficacy.

7.5.8 Herb-drug interactions

We refer the reader to the Chapter 3.4 "Effect of Herbs" in this book where this subject has been discussed in detail. Hence we would just like to recapitulate the mechanism by which herbs can increase or decrease the concentration of a drug in blood and provide a few examples. As we mentioned in the previous section on drug-drug interactions a majority of drugs are metabolized by the cytochrome p 450 (CYP3A4) isoform found in the liver and the gut. Other CYP isoforms are also responsible for the metabolism of some drugs. The alteration in P-glycoprotein, an efflux transporter found in the gut and kidney could also be a contributing factor if the drug is a substrate for this efflux transporter. Furthermore if a drug is dependent on its uptake by organic anion-transporting polypeptide A (OATAP1A2) its concentration could be affected if the herb influences the uptake. The herb St. John's wort which is used widely to treat depression is known to decrease the concentration of drugs such as cyclosporine, tacrolimus, sirolimus, theophylline, amitryptyline and nortriptyline to list just a few. St. John's wort by inducing the CYP3A4, CYP1A2 and CYP2C9 isoforms accelerates the metabolism of warfarin causing a decrease in the International Normalized ratio (INR) and its loss of efficacy as an anticoagulant. The herb Ginseng, which is used as a restorative tonic affects warfarin therapy by decreasing the INR. Ginseng, however, increases the concentration of digoxin in serum [18]. Again, examples of other herb-drug intractions can be found in the Chapter 3.4 on "Effect of Herbs".

7.5.9 Effect of Storage and transport of specimens and endogenous interferences

In general, it is imperative to ensure that the integrity of the constituents in the sample is unaltered during storage and transport of specimens. The prime requirement is that the contents of the specimen container are not leached into the specimen and thus affect the quality of the samples. We'll discuss the interaction of specimen collection devices on the integrity of the sample in the next section. Here, however, we'll provide a few specific examples of the effect of the handling of specimens.

Nitrazepam, cocaine and chlorazepam are unstable and decompose during storage of blood samples at 4 °C and storage of samples intended for the measurement of these analytes at 4 °C should be avoided. Cyclosporine is sensitive to freezing and hence freezing of samples intended for the determination of cyclosporine should be avoided. Furthermore, cyclosporine partitions into the erythrocyte rapidly when blood cools from 37 °C (body temperature) to 20 °C (room temperature) after blood collection making it imperative to use whole blood, preferably anticoagulated with EDTA as the specimen of choice (Table 7.5.1) [2, 10]. Some analytes are sensitive to temperature. Thus for instance samples intended for the analysis of fluorouracil should be collected on ice since this analyte is unstable at room temperature. As to the thawing of the frozen samples this should be done at room temperature since too rapid thawing of the sample by warming the sample can by overheating degrade the sample. After thawing of the sample the sample should be mixed thoroughly to ensure homogeneity prior to analysis.

We would like to illustrate the effect of endogenous interferences by mentioning the effect of digoxin-like immuno-reactive factor or substance (DLIF/DLIS) present in serum on the measurement of digoxin and interference in general, in immunoassays of the presence of human anti-mouse antibodies (HAMA).

DLIF/DLIS may be present in the sera of neonates, pregnant women especially in the third semester and in patients with hepatic or renal failure [8]. DLIF/DLIS and digoxin antidotes such as, Digibind and DigiFab which also interfere in digoxin assays can however, be removed by ultrafiltration techniques.

A major problem with immunoassays using mouse (murine) monoclonal antibodies, including those used in TDM is the presence of potential human anti-mouse antibodies (HAMA) in patient's sera. Patients who either are receiving immunotherapy with mouse monoclonal antibodies or are exposed to imaging techniques or handle experimental animals such as mice are apt to develop HAMA, which in turn can result in spurious analytical results [19].

7.5.10 Interaction with specimen collection devices

Back in the days in the nineteen-forties to the early fifties to the present time interaction of blood specimen with specimen container devices has introduced interferences in TDM. Interferences could come from needles used for collecting blood specimens; the lubricant used to coat the needle, the stopper used in blood collection tubes, the serum or plasma separation devices, to list a few [20].

Of historical interest is the case of a plasticizer, tris (2-butoxyethyl) phosphate (TBEP) which was a component of rubber stoppers used in blood collection tubes in the nineteen seventies, interfering with TDM. TBEP displaced basic drugs bound to α_1-acid glycoprotein. The displaced drug redistributed into the cellular (erythrocyte) fraction thus artifactually lowering the concentration of drug measured in plasma [21]. This cellular redistribution introduced by TBEP affected a wide range of drugs from propranolol, tricyclic antidepressants, lidocaine, quinidine, chlorpromazine to fluphenazine resulting in apparently lower measured concentrations of these drugs [22]. TBEP was subsequently eliminated from the stopper formulations in blood collection tubes.

While TBEP displaced drugs bound to α_1 acid glycoprotein, blood collection tubes containing serum separator gels introduced in the nineteen-seventies and popular to this day, albeit with modifications had a different mechanism of interference in TDM. The gel, because of its hydrophobic nature had an adsorptive effect on certain hydrophobic drugs such as lidocaine, quinidine, phenobarbital, phenytoin and carbamazepine artifactually lowering their measured concentration in serum. The magnitude of the decrease was related to the length of time the serum was in contact with the separator gel even when stored at 4 °C [23, 24].

Manufacturers are constantly striving to minimize absorption of drugs to serum separator gels and have in recent years introduced newer polymeric gels, the composition of which has been described elsewhere [25], (see also Chapters 4.5 and 4.6 of this book). However, each laboratory should evaluate any claim made by the manufacturer as to the lack of interference of their blood collection device for TDM. In the event one has to use gel separation devices for the measurement of drugs that were subject to interferences previously it is best to minimize the contact time of serum or plasma with such devices by promptly separating the serum or plasma from the top of the gel barrier.

Conclusion

In this chapter, we have attempted to provide an overview of the variables that affect the results obtained in the assay of therapeutic drugs. As the repertoire of drugs used in therapy increase and novel assays are introduced and improvements are made in specimen collection devices the clinical laboratory will have the task and responsibility of uncovering and resolving any preanalytical and analytical issues and ensuring the quality of laboratory results that will be used by the clinician to guide therapy.

References

[1] Narayanan S. Considerations in therapeutic drug monitoring. Italian J Clin Patol 1984; 1:73–5.

[2] Guder WG, Narayanan S, Wisser H, Zawta B. The right time of drugs. Special aspects in therapeutic drug monitoring (TDM). In Diagnostic Samples: From the Patient to the Laboratory. 4th ed. Weinheim , Germany: Wiley-Blackwell, 2009, pp. 70–73.

[3] Mehta AC. Preanalytical considerations in drug assays in clinical pharmacokinetic studies. J Clin Pharm Ther 1989; 14:285–95.

[4] Wood M, Shand DG, Wood AJJ. Altered drug binding due to the use of indwelling heparinized cannulas (heparin lock) for sampling. Clin Pharm Ther 1979; 25:103–7.

[5] Narayanan S. The preanalytic phase. An important component of laboratory medicine. Am J Clin Pathol 2000; 113:429–52.

[6] Siebers RW, Chen CT, Fergusson RI, Maling TJ. Effect of blood sample tubes on amiodarone and desethylamiodarone concentrations. Ther Drug Monit 1988; 10:349–51.

[7] Berquist Y, Domeji-Nyberg B. Distribution of chloroquine and its metabolite desethylchloroquine in human blood cells and its implication for the quantitative determination of these compounds in serum and plasma. J Chromatogr 1983; 272:137–48.

[8] Young DS. Effects of Preanalytical Variables on Clinical Laboratory tests. 3rd ed. Washington, DC, USA: AACC Press, 2007.

[9] Agrawal Y. Critical values for therapeutic drug levels. Supplement to the Medical laboratory observer (MLO), Clinical laboratory reference (CLR). 2014; 46:4.

[10] Guder WG, Fiedler M, da Fonseca Wollheim F, Schmitt Y, Töpfer G, Wisser H, Zawta B. Quality of Diagnostic Samples, 4th ed. Oxford GB: BD Diagnostics Preanalytical Systems, 2015.

[11] Brunton L, Chabner B, Knollman B. Goodman and Gilman's The Pharmaceutical Basis of Therapeutics, 12th ed. New York: McGraw-Hill Professional, 2010.

[12] Narayanan S. Expanding frontiers of coagulation: a window on therapeutic advances. APFCB (Asian Pacific Federation of Clinical Biochemistry) News 2010; 1:85–92.

[13] Higashi MK, Veenstra DL, Kondo LM, Wittkowski AK, Srinouanprachanh SL et al. Association between CYP2C9 genetic variants and anticoagulation-related outcomes during Warfarin therapy. J Am Med Assoc 2002; 287:1690–8.

[14] Narayanan S, Hamasaki N. Current concepts of coagulation and fibrinolysis. Adv Clin Chem 1998; 33:133–68.
[15] Carlquist JF, Horne BD, Muhlestein JB, Lappe DL, Whiting BM, et al. Genotypes of the cytochrome p-450 isoform, CYP2C9, and the vitamin K epoxide reductase complex subunit I conjointly determine stable warfarin dose: a prospective study. J Thromb Thrombolysis 2006; 3:191–7.
[16] Mega JL, Simon T, Collet J-P, Anderson JL, Antman EM et al. Reduced function CYP2C19 genotype and risk of adverse clinical outcomes among patients treated with clopidogrel predominantly for PCI: A meta analysis. J Am Med Assoc 2010; 304:1821–30.
[17] Young DS. Effects of Drugs on Clinical Laboratory Tests. 6th ed. Washington DC USA: AACC Press, 2007.
[18] Narayanan S, Young DS. Effect of Herbs and Natural Products on Clinical Laboratory Tests. Washington DC, USA: AACC Press, 2007.
[19] Narayanan S, Schuetz AN. Current trends in instrumentation and technology: Outlook for the future. In: Garcia LS, editor. Clinical Laboratory Management, 2nd ed. Washington, DC USA: ASM Press 2014, pp. 933–65.
[20] Narayanan S, Lin FC. Sampling technique. In: Wong SHY, editor. Therapeutic Drug Monitoring and Toxicology by Liquid Chromatography. New York: Dekker, 1985 79–88.
[21] Borga O, Piafsky KM, Nilsen OG. Plasma protein binding of basic drugs. I. Selective displacement from α-1 acid glycoprotein by Tris (2-butoxyethyl) phosphate. Clin Pharmacol Ther 1977; 22:539–44.
[22] Cotham RS, Shand D. Spuriously low plasma propranolol concentrations resulting from blood collection methods. Clin Pharmacol Ther 1975; 18:535–8.
[23] Bergquist Y, Eckerbom S, Funding L. Effect of use of gel-barrier sampling tubes on determination of some antiepileptic drugs in serum. Clin Chem 1984; 30:465–6.
[24] Koch TR, Platoff G. Suitability of collection tubes with separator gels for therapeutic drug monitoring. Ther Drug Monit 1990; 12:277–80.
[25] Bowen RAR, Remaley AT. Interferences from blood collection tube components on clinical chemistry assays. Biochemia Medica 2014; 24:31–44.

Sheshadri Narayanan, Audrey N. Schuetz

7.6 Preanalytical Variables in Microbiology

Introduction

A key to good laboratory practice is appropriate specimen collection, such as adequate specimen quantity, proper site and timing of collection, proper collection technique and prompt transport of the specimen to the laboratory. Common specimens submitted for bacteriologic examination include blood for culture, respiratory, urine and wound specimens and specimens collected specifically for virology, parasitology and mycology studies. Transport systems and transport times of each vary depending on the type of specimen. In this chapter, we examine some of the more common types of specimen submitted for microbiological analysis and the appropriate transport of such specimens.

It is imperative that the sample submitted to the laboratory is adequate in volume. Often, multiple test requests are made on a single specimen, and the laboratory occasionally cannot perform all requested testing on the single specimen and so must prioritize testing with the help of the submitting healthcare provider.

Microbiological culture typically is more adversely affected by nonconformity to pre-analytical requirements than are molecular studies. However, specific transport and collection requirements for molecular studies also exist and should be investigated before submitting a specimen. Often, specimen submission requirements for molecular studies vary from institution to institution and between assays. Aseptic collection of tissue for molecular analysis is critical.

For mailing microbiological specimens through the postal service, special containers and standard procedures for the handling and transport of diagnostic specimens and etiological agents as proposed by the Clinical and Laboratory Standards Institute (CLSI, formerly NCCLS) should be followed [1].

7.6.1 Blood cultures

Volume of blood collected

The volume of blood collected influences the yield of positive cultures. Ideally, 10 mL of blood should be injected into each blood culture bottle for adults. Therefore, 20 mL per blood culture set should be ideal in adults, as one set usually refers to one aerobic and one anaerobic bottle. A direct relationship exists between the volume of blood collected and the yield of culture results. In one study, the yield of 20 mL of cultured blood was approximately 30 % greater than the yield of only 10 mL, and when 30 mL of blood was cultured the yield was nearly 47 % greater as compared to 10 mL [2]. In infants and children, the amount of blood collected should be between 0.5 mL to 5 mL per bottle when pediatric blood bottles are used and should be guided by patient weight. The blood to broth media ratio of each bottle should be between 1:5 and 1:10 in order to minimize the activity of antibiotics in blood and the effect of other antimicrobial substances or circulating antibodies.

Preparation of patient for blood collection

The preparation of patient for venipuncture involves multiple steps. First, the venipuncture site should be prepared to ensure elimination of colonizing or contaminating organisms from the skin. If the site of venipuncture is grossly contaminated with dirt, it should be cleansed with soap followed by rinse with water. Subsequently, a skin cleansing solution such as alcohol, chlorhexidine, iodine tincture or povidone-iodine (or a combination of the above) should be applied and allowed to dry for between 30 sec and 3 min, depending upon the manufacturer's instructions. If iodine tincture is used, excess iodine should be removed with a 70 % alcohol wash. Use of povidone-iodine does not require an alcohol wash. Prior to performing the venipuncture, it is important to ensure that the skin is dry after use of the cleansing solution. Proper collection of blood for blood culture is a prerequisite to minimizing contamination and avoiding unnecessary antibiotic exposure [3]. In one study of collection technique, only 115 of 219 total positive blood culture episodes were considered truly positive, while an almost equal number (104) were considered falsely positive due in part to contamination from improper collection [4]. It is preferential to draw blood from a peripheral venous site rather than from an indwelling intravenous or arterial catheter. If a central line infection is suspected, blood may be drawn from both a peripheral vein and through the central line and the cultures compared for time to positivity or compared for the relative numbers of organisms growing from each site [5]. It is helpful in interpretation of positive cultures to obtain blood cultures from two different sites, and it is preferable that at least one site is peripheral [6].

Timing of blood collection

Although it has been described in the past that the optimum time to draw blood for culture is before the onset of fever and chills, it is now thought that timing of blood culture collection is less important than the total volume of blood collected [5]. A multicenter study demonstrated that the probability of diagnosing a positive bloodstream infection was not significantly enhanced by collecting the blood when the patients' temperatures were highest [7]. Further information on the timing of blood collections for other conditions is included as references [8–10].

Other variables influencing yield of blood culture

Inhibition of phagocytosis by complement and lysozymes is accomplished by the anticoagulant sodium polyanethole sulfonate (SPS) which is present in blood culture media usually at a concentration of 0.025–0.05 %. As an anticoagulant, SPS protects the bacteria against killing by innate cellular and humoral factors. In addition to its anticomplement activity and its ability to inactivate neutrophils, SPS inactivates antibiotics such as gentamicin, kanamycin, streptomycin and polymyxin B. However, SPS also inhibits the growth of some bacteria to different degrees, including *Neisseria gonorrhoeae*, *Neisseria meningitidis*, *Gardnerella vaginalis*, *Streptobacillus moniliformis* and *Peptostreptococcus anaerobius*. Some of these bacteria may still grow in the presence of SPS, but an anticoagulant-free culture medium, such as a lysis-centrifugation system, may be used if the culture is negative and these organisms are high on the differential. The lysis-centrifugation blood culture system involves the use of a special tube containing saponin that lyses the leukocytes upon processing, thus allowing growth of certain intracellular organisms, such as *Bartonella* spp. and *Histoplasma capsulatum*. Many commercial blood culture systems also have synthetic antibiotic-binding resins within the blood culture media, which improves yield. Refer to Table 7.6.1 for a summary of variables which affect the yield of blood cultures.

Table 7.6.1: Variables affecting blood culture yield.

Volume of blood collected for culture
Timing of blood collection
Decontamination of skin before blood collection
Site of blood collection (peripheral versus central line)
Blood to media broth ratio (ratio between 1:5 and 1:10 minimizes antimicrobial activity of blood and dilutes other interfering substances)
Sodium polyanethole sulfonate (SPS) (0.025–0.05 % SPS inhibits phagocytosis and inactivates several antibiotics; SPS also inhibits certain bacteria)
Lysis-centrifugation system (increases yield of certain organisms)

7.6.2 Specimens submitted for bacteriological studies

Swabs should not be used if more than one type of culture is ordered (such as several bacterial, fungal and mycobacterial cultures) due to the limitation of culture sensitivity since little actual specimen is received on a swab. Additionally, swabs generally should not be submitted for anaerobic culture [5]. Flocked swabs have been proven to improve recovery of cells over nonflocked swabs and are generally preferable for bacteriological studies [11]. Other factors which affect the yield of specimens sent for bacteriological analysis include refrigeration, changes in pH of specimen, or exposure to oxygen which can lower the survival of a variety of organisms, such as *Haemophilus*, gonococci, pneumococci, meningococci, *Salmonella*, *Shigella*, *Bordetella*, *Vibrio cholerae*, *Helicobacter pylori* and most anaerobic organisms [3, 12].

Transport of samples from the time of collection to their receipt in the laboratory should be kept relatively short and preferably not exceed two hours [3, 12]. If the two-hour time frame cannot be met, the use of a transport medium is recommended. One such transport medium is the Stuart medium which is essentially a solution of phosphate buffer containing chloride salts of sodium, potassium, calcium and magnesium. The medium also contains sodium thioglycollate and agar. The former enhances the recovery of anaerobic bacteria as a reducing agent, while the latter (agar) renders the medium semi-solid and protects against oxygenation and spillage during specimen transport.

Several transport systems are commercially available for the transport of anaerobic specimens. Most of these systems contain an indicator that monitors the maintenance of an anaerobic environment. Once the transport container is received by the laboratory, the specimen should be promptly inoculated onto an appropriate medium.

7.6.3 Respiratory specimens

Sputum

The challenge in collecting sputum samples lies in preventing contamination from oral secretions (saliva) and in obtaining truly deep-cough samples. One way to minimize contamination from oral secretions is to gargle with water prior to collection of the sputum sample. Commercially available mouth washes should not be used for gargling since they may contain antibacterial substances. Early morning sputum specimens are preferred, because they provide a concentrated sample of bacteria accumulated from overnight secretions. Either a wide-mouthed container with a tight-fitting screw-cap lid or commercially available sputum collection devices should be used for sputum collection. The container should be placed close to the mouth in such a way that all of the sputum is captured within the container without introducing any outside contamination. Sputum samples should be processed as soon as possible upon receipt in the laboratory, since delays can result in loss of organisms [13].

Criteria must be set for rejection of sputum samples based on the number of epithelial cells. The limit for acceptance of sputum specimens is less than 10 squamous epithelial cells per 10x objective field (low power) [5]. Although sputum rejection criteria historically included the number of neutrophils or inflammatory cells, this practice is in general less commonly performed due to the increasing population of immunocompromised patients who cannot mount an inflammatory response. Respiratory tract specimens submitted for the detection of *Mycobacterium tuberculosis* should not be screened for adequacy for gross contamination.

Bronchoalveolar lavage fluid and endotracheal aspirates

Bronchoalveolar lavage (BAL) specimens are collected by injecting up to 50 mL of physiologic saline with a syringe through a fiberoptic bronchoscope that has been threaded into the peripheral bronchiolar region. The aspirated saline is subjected to smear preparation and culture. Bronchial washings are similarly performed with less fluid, targeting the bronchus of the suspected affected region. Bronchial biopsies may be helpful if aseptically collected without contamination from respiratory flora.

Endotracheal aspiration may be performed if the patient is unable to provide a sputum sample. However, such specimens are frequently contaminated with many colonizing bacteria, and the utility of endotracheal aspirate culture is less than that of BAL fluid or bronchial washes. If performed, the trachea is locally anesthetized and a 16-gauge polyethylene catheter is threaded into the lower trachea using a 14-gauge needle. Subsequently, secretions are aspirated with a syringe. In order to be acceptable for culture, the endotracheal aspirate must show less than 10 squamous epithelial cells per 10x field and bacteria must be seen in at least 1 of 20 oil immersion fields [5].

Nasopharyngeal specimens

Nasopharyngeal aspirates or swabs suspected of containing *Bordetella pertussis* should be inoculated to Bordet-Gengou agar or other appropriate media as soon as possible after receipt in the laboratory.

7.6.4 Body fluids and CSF

Direct inoculation of synovial fluid and other joint fluids into blood culture bottles likely improves recovery of pathogens, as compared to direct plating of specimens in the laboratory onto solid agar media [14]. This practice has been applied to other body fluids, including peritoneal fluid. When possible, body fluids submitted in sterile tubes should be centrifuged for concentration for Gram stain and inoculation of solid media for culture. CSF should never be refrigerated if not immediately processed, since in refrigerated temperatures certain bacteria such as *Neisseria meningitidis* may not survive the cold. Gastric aspirates collected for mycobacteria should be transported immediately to the laboratory and the specimen processed rapidly, since mycobacteria die quickly in gastric washings.

Urine

Although urine is usually believed to be only transiently colonized, contamination during urine specimen collection can occur from urethral or periurethral organisms. These organisms can grow on culture plates and lead to misleading results for the clinicians if all such organisms are reported. Thus, all freshly-collected urine samples must be refrigerated if not cultured within 30 minutes of collection.

Urine specimens collected in sterile containers and refrigerated for storage should be cultured within 24 h. Urine may also be collected in a transport tube with boric acid preservative. Mid-stream collection is recommended [15, 16]. A first-voided urine specimen is expected to be concentrated with bacteria, and, as such, is preferable. Foley catheter tips should not be accepted for culture since they tend to be contaminated with urethral microorganisms [5].

Stool

Stool for bacteriologic culture may either be submitted fresh or in a stool preservative. For bacterial pathogen analysis, stools may be submitted in Cary-Blair or other transport medium if transport delay is expected. Organisms such as *Shigella* are labile and, unless the stool has been submitted in an appropriate transport medium, it should be promptly inoculated to media [5].

Wound and miscellaneous specimens

Wound specimens for culture should be submitted as tissues. Swabs of wounds are inferior to tissue or biopsies, since little material is collected on swabs [5]. Small biopsies may be collected in sterile containers with a small amount of sterile saline to keep the tissue moist if needed. Tissues may also be collected and transported to the laboratory in specified containers with transport media which support the bacteria especially if delays in transport are anticipated. If anaerobic culture is requested, tissues should be submitted in anaerobic culture collection devices. If the specimen is liquid in nature, it may be submitted either in one of the aforementioned collection devices, or it may be submitted in a capped syringe delivered to the laboratory without the needle attached. Urethral or cervical specimens suspected of harboring *Neisseria gonorrhoeae* must be inoculated immediately onto appropriate agar plates.

7.6.5 Specimens submitted for virologic studies

Samples for virologic studies typically include CSF, urine, tissue, blood, stool, vesicular fluid from skin lesions, pharyngeal washings and swab specimens taken from the nasopharynx, throat or eyes. The time of collection varies depending on the virus and the site of potential infection [17]. In general, the ability to recover viruses decreases as the duration of illness increases. Specimens for virologic culture and direct fluorescent antibody (DFA) testing should be transported rapidly to the laboratory at 4 °C in a container to maintain the stability of viruses [12, 18]. Specimens should be refrigerated at 4–8 °C or frozen at −70 °C if storage is required for a longer duration (weeks or months). They should never be stored at −20 °C due to the freeze–thaw cycling of such freezers [17]. For samples intended for molecular testing, blood samples should not be collected in heparin since the DNA amplification enzyme (taq polymerase) is inhibited by heparin [19]. When plasma is needed for molecular testing, the liquid component should be removed from the red blood cells as quickly as possible [5].

Although swabs are convenient for the purpose of collecting throat specimens, certain swabs types are more desirable than others. For instance, calcium alginate swabs have been shown to be toxic to *Neisseria gonorrhoeae* and some viruses, inhibiting growth and possibly PCR reactions [18, 20]. Flocked swabs are a relatively recent swab type which has been proven to be superior to nonflocked swabs in viral recovery for nasopharyngeal collection [21]. Viral particles adhere to nonflocked swab material and are not released by the swab when testing is performed, thus decreasing sensitivity. When used, swabs should not be allowed to dry. Instead, they should be placed in a viral transport medium that includes antibiotics to inhibit bacterial growth. The viral transport medium is usually supplemented with protein to stabilize fragile viruses.

7.6.6 Specimens submitted for parasitic studies

The type of specimen required for the identification of parasites varies depending on the type of parasite suspected. Thus, blood is the specimen of choice for the examination of parasites like *Plasmodium*, *Trypanosoma*, *Leishmania* and microfilariae. Smears can be prepared from fresh, whole blood, anticoagulated or non-anticoagulated blood [22]. The blood smear should optimally be prepared within one hour of collection for optimal morphology. Blood smears stained with Wright or Giemsa stains are ideal for the examination of plasmodia. Thin and thick smears prepared from heparinized blood can be examined for blood parasites [12].

For stool specimens, if fresh stool is examined for the presence of protozoa, particularly amoebae, it must be received in the laboratory within 30 min of passage to detect motile trophozoites [12]. If the stool is not expected to be received within 30 min of collection, it should be submitted in an appropriate preservative for the collection of ova and parasites. Various commercial preservatives are available and are listed in Table 7.6.2. Single vial stool collection systems are commercially available. Persons who received a barium enema should delay the examination of parasites in stools for at least a week after completion of the enema procedure due to interference with parasite detection. Oil-based purgatives interfere with the motility of trophozoites. In the laboratory, concentration procedures are needed for the detection of cysts, ova and trophozoites. Both flotation and sedimentation procedures are suited for this purpose.

Other types of specimens such as urine, muscle and skin biopsies can be submitted for parasitologic analysis. Direct fluorescent antibody assays, enzyme immunoassays, and now even molecular assays can be performed directly on stool and other specimens for detection of a variety of parasites. Readers should refer to the particular manufacturer's assay requirements on specimen type or to the laboratory performing the testing in order to assess whether the particular specimen type has been validated for testing on for particular assay. Parasites such as *Schistosoma hematobium* should be examined in a 24-h urine specimen. Urine specimens need to be centrifuged, and the concentrated sediment can be examined in wet mount preparations for parasites. Sputum specimens need to be liquefied with 3 % N-acetyl-L-cysteine prior to the preparation of a saline mount for the examination of larval stages or ova of specific parasites. Also, the parasites themselves (both ectoparasites such as ticks and endoparasites) can be submitted for direct examination [23]. Arthropods (e. g. fleas and lice) should be transported either fresh or in 70 % alcohol solution. Submitters should check with the Microbiology Laboratory to assess appropriateness of various specimen types, depending upon the assays and detection methodologies used.

Table 7.6.2: Preservatives used for the collection and transport of stool specimens.

Polyvinyl alcohol (PVA)-based preservative (largely being phased out due to concerns for waste disposal of mercury): mixture of PVA, 95% ethyl alcohol, mercuric chloride, glacial acetic acid and glycerin; useful for preparation of permanent-stained smears and for use in concentration procedures. Modified PVA formulas which do not contain mercury are now available.
Formalin–saline Mixture of formaldehyde and 0.85 % sodium chloride: concentrated sediment can be used for different stains, but formalin–saline is unsuitable for the preparation of permanent-stained smears and is not optimal for all immunoassays.
Sodium acetate–acetic acid–formalin (SAF) Mixture of formaldehyde, sodium acetate and glacial acetic acid: SAF can be used for parasite concentration and preparation of permanent-stained smears as well as most immunoassays. Permanent-stained fecal smears for the visualization of intestinal protozoa can be challenging with this fixative and are better seen with the use of the iron-hematoxylin stain.

7.6.7 Specimens submitted for fungal studies

In general, transport of specimens for mycological studies can be performed at room temperature if the transport time is short. However, if a Mucorales (zygomycete) infection is suspected, rapid transport may be necessary for adequate culture recovery [24].

Tissue biopsy specimens for mycologic analysis should not be allowed to dry but should be protected with moist sterile gauze, placed in a sterile screw-capped container, and transported to the laboratory as rapidly as possible. Grinding of tissue with a tissue grinder might damage some of the moulds, especially the Mucorales (zygomycetes). Thus, tissues submitted to rule out Mucorales should be minced rather than ground or crushed. Recovery of fungi from blood is facilitated by the use of lysis-centrifugation tubes, which are particularly helpful for the recovery of *Histoplasma capsulatum* and some other fungi.

Cerebrospinal fluid specimens for fungal studies should not be refrigerated. They are best stored at room temperature prior to plating the specimen. Skin, nail and hair specimens are collected by scraping the advancing edges of the lesions with a scalpel after thoroughly disinfecting the skin with 70 % alcohol [12, 24]. Such specimens should be transported dry to the laboratory in sterile containers. While a first morning urine specimen is preferred for fungi, randomly collected samples can also be used, all of which should be collected in sterile screw-capped containers. If the urine is collected fresh, it should be refrigerated to prevent bacterial overgrowth if not plated within 2 h of collection.

Conclusion

In this chapter we have summarized some of the variables in specimen collection, storage and transport that may affect test results. We discussed a wide range of specimens, including blood, respiratory specimens, and specific specimens for the examination of parasites, viruses and fungi. Minimizing the preanalytical variables which can adversely affect test results is paramount in obtaining quality results in microbiologic studies.

References

[1] Clinical and Laboratory Standards Institute (CLSI). Standard Procedures for the Handling and Transport of Domestic Diagnostic Specimens and Etiologic Agents. Wayne, PA: former National Committee for Clinical Laboratory Standards (NCCLS), 1980.

[2] Cockerill FR, III, Wilson JW, Vetter EA, Goodman KM, Torgerson CA et al. Optimal testing parameters for blood cultures. Clin Infect Dis 2004; 38:1724–30.

[3] Thompson RB. Specimen collection, transport and processing: Bacteriology. In: Baron RJ, Jorgensen JH, Landry ML, Pfaller MA (eds), Manual of Clinical Microbiology, 9th ed. Washington, DC: ASM Press, 2007, vol. 1, pp. 291–333.

[4] Bates DW, Goldman L, Lee TH. Contaminant blood cultures and resource utilization: the true consequences of false-positive results. JAMA 1991; 265:365–9.

[5] Baron EJ, Thomson RB. Specimen collection, transport and processing: Bacteriology. In: Versalovic J, Carroll KC, Funke G, Jorgensen JH, Landry ML, Warnock DW (eds), Manual of Clinical Microbiology, 10th ed. Washington DC: ASM Press, 2010, vol. 2, pp. 228–71.

[6] Baron EJ, Weinstein MP, Dunne WM, Yagupsky P, Welch DF, et al. Cumitech 1C, Blood Cultures IV. Coordinating ed. Baron EJ, Washington, DC: ASM Press, 2005.

[7] Riedel S, Bourbeau P, Swartz B, Brecher S, Carroll KC, et al. Timing of specimen collection for blood cultures from febrile patients with bacteremia. J Clin Microbiol 2008; 46:1381–5.

[8] Bartlett RC, Ellner PD, Washington JA II. Cumitech 1. Blood Cultures. Coordinating ed, JC. Sherris. Washington, DC: American Society for Microbiology, 1974.

[9] Reller LB, Murray Pr, MacLowry JD. Cumitech 1 A. Blood Cultures II. American Society for Microbiology, Washington, DC: 1982.

[10] Washington JA II. Conventional approaches to blood culture. In: Washington JA II ed. The detection of septicemia, West Palm Beach, FL: CRC Press, 1978 chap 2, pp. 41–86.

[11] Daley P, Castriciano S, Chernesky M, Smieja M. Comparison of flocked and rayon swabs for collection of respiratory epithelial cells from uninfected volunteers and symptomatic patients. J Clin Microbiol 2006; 44:2265–7.

[12] Guder WG, Narayanan S, Wisser H, Zawta B. Diagnostic Samples: From the Patient to the Laboratory. The Impact of Preanalytical Variables on the Quality of Laboratory Results, Weinheim, Germany: Wiley-Blackwell, 4th edition, 2009.

[13] Penn RL, Silberman R. Effects of overnight refrigeration on microscopic evaluation of sputum. J Clin Microbiol 1984; 19:161–5.

[14] Mathews CJ, Kingsley G, Field M, Jones A, Weston VC et al. Management of septic arthritis: a systematic review. Ann Rheum Dis. 2007; 66:440–5.

[15] Lifshitz E, Kramer L. Outpatient urine culture: does collection technique matter? Arch Intern Med 2000; 160:2537–40.

[16] Kouri T, FogazziG, Gant V., Hallander H, Hofmann W, Guder WG. European urinalysis guidelines. Scand J Clin Lab Invest, 2000; 60 suppl 231:1–96.

[17] Miller JM. A Guide to Specimen Management in Clinical Microbiology. 2nd ed. Washington DC: ASM Press, 1999.

[18] Forman MS, Valsamakis A. Specimen collection, transport and processing: Virology. In: Versalovic J, Carroll KC, Funke G, Jorgensen JH, Landry ML, Warnock DW (eds), Manual of Clinical Microbiology, 10th ed. Washington DC: ASM Press, 2010, vol. 2, pp. 1276–88.

[19] Narayanan S. Preanalytical and analytical pitfalls in molecular biology techniques. J Clin Ligand Assay 1997; 20:200–5.

[20] Lauer BA, Masters HB. Toxic effect of calcium alginate swabs on *Neisseria gonorrhoeae*. J Clin Microbiol 1988; 26:54–6.

[21] Hernes SS, Quarsten H, Hagen E, Lyngroth AL, Pripp AH et al. Swabbing for respiratory viral infections in older patients: a comparison of rayon and nylon flocked swabs. Eur J Clin Microbiol Infect Dis 2011; 30:159–65.

[22] Shimizu RY, Grimm F, Garcia LS, Deplazes P. Specimen collection, transport and processing: Parasitology. In: Versalovic J, Carroll KC, Funke G, Jorgensen JH, Landry ML, Warnock DW (eds), Manual of Clinical Microbiology, 10th ed. Washington DC: ASM Press, 2010, vol. 2, pp. 2047–63.

[23] Deplazes P, Garcia LS, Shimizu RY. Specimen collection, transport and processing: Parasitology. In: Murray PR, Baron EJ, Jorgensen JH, Landry ML, Pfaller MA (eds), Manual of Clinical Microbiology, 9th ed. Washington DC: ASM Press, 2007, vol 2, pp. 1995–2012.

[24] Sutton DA. Specimen collection, transport and processing: Mycology. In: Murray PR, Baron EJ, Jorgensen JH, Landry ML, Pfaller MA (eds), Manual of Clinical Microbiology, 9th ed. Washington DC: ASM Press, 2007, vol 2, pp. 1728–36.

Alexander Leichtle, Peter Findeisen

7.7 Biobanking

7.7.1 Preanalytical issues in biobanking

The concept of tissue banking for the collection, storing and distribution of human biological material and clinical information, is crucial to support clinical and translational research. Accordingly, human biological material (e. g. serum, plasma, tissues, cells, nucleic acids) obtained through common diagnostic procedures, has become an important resource for biomedical research [1]. The discovery and validation of new biomarkers for diagnosis, therapeutic monitoring and prognosis of many diseases can substantially be accelerated if sufficiently large number of specimens are available. However, sample quality is an important consideration, as it directly impacts the quality of ensuing research. In this context, quality should be defined as the state of the sample enabling the unbiased detection or measurement of the analytes in the intended investigation. This means, that under certain circumstances, when very stable analytes are to be measured (e. g. in newborn metabolic screening), preanalytical treatment can be performed adequately. However, where very instable analytes (e. g. short-lived intermediates of the energy metabolism) are involved quenching steps might be necessary [2], thus rendering optimal sample preprocessing impossible even in highly specialized settings of university hospitals, for example. In complex biological samples such as serum and plasma, the preanalytical variability in sample handling causes substantial in vitro changes of proteins [3], peptides [4, 5], metabolites [6, 7], nucleic acids [8, 9] or enzyme activities [10, 11]. These modifications not only affect in vitro diagnostic tests [12] but also have a major impact on multiparametric "-omics" approaches and can override disease-related patterns [13] or even abolish meaningful data interpretation completely [14]. Accordingly, the preanalytical bias has been identified as one of the most serious threats for biomarker discovery [15].

Consequently, many studies have addressed the effect of preanalytical variables on the stability of various analytes. A web-accessible library of publications, the Biospecimen Research Database (https://brd.nci.nih.gov/BRN/brnHome.seam), is constantly updating relevant publications in the field. For example, vascular endothelial growth factor – an extensively studied biomarker implicated in diabetes, arthritis and cancer – is highly unstable in plasma specimens and should never be measured in samples with repeated freeze thaw cycles [16]. Also the phosphorylation status of mTOR (mammalian target of rapamycin) in resected tissue specimens – a highly relevant parameter for individualized therapeutic options – has shown to be affected by preanalytical variability leading to unreliable quantification with resultant misclassification of patients [17]. It is extremely important to be aware that there is no "optimal" preanalytical strategy. Any interaction with the sample may cause interferences and the most serious ones are not the already known factors, which easily can be addressed, for example by an appropriate study design, but the unknown bias, which frequently is only detected when studies cannot be replicated and which is traceable by extensive documentation and elaborate exploratory statistics [18]. Especially in the field of metabolomics, which resembles a snapshot of by nature an unstable meta-state, sample breakdown might be immediate, so the only chance to obtain comparable samples is to allow for an equal grade of degradation [6].

7.7.2 Quality assurance for biobanks – present status

The preanalytical variabilities comprise tissue resection, blood collection, transport, centrifugation, storage, and subsequent sample re-preparation. The identification of standard operating procedures (SOPs) for blood/tissue sampling, storage and collection of clinical data is an essential step in modern biobank management. Sampling and collecting procedures must guarantee both the correct pathological diagnosis as well as the biological integrity of (macro-) molecules (proteins, peptides, metabolites and nucleic acids) that represent the potential subject of investigation.

Multiple working groups have defined standard operating procedures for documentation of preanalytical sample handling. These include *'Biospecimen Reporting for Improved Study Quality'* (BRISQ) [19], *'Standard PREanalytical Code'* (SPREC) [20] and *'Standardisation and Improvement of Generic Pre-analytical Tools and Procedures for In Vitro Diagnostics'* (SPIDIA) (http://www.spidia.eu/). However, the documentation of sample collection, processing and storage is important, but may be insufficient as many factors that influence sample quality can hardly be controlled or standardized under clinical routine conditions [21]. An analysis of 125 biomarker discovery papers published in open-access journals between 2004 and 2009 found that more than half included no information about how specimens had been obtained, stored or processed [22]. Due to the neglected issue of sample quality, many recently described biomarkers could not be validated independently and consequently had to be classified as false positives [23].

The stabilization of biospecimens seems to be an alternative for preservation of conditions that should be near to the in vivo situation. However, stabilization of blood specimens with additives like protease inhibitors will critically alter the specimens' composition due to release of cytokines by cellular components [24]. Also, the intraoperative resection of tissue involves hypoxia and hypothermia that lead to biomarker variations that cannot be circumvented by the postoperative stabilization of resected biospecimens [25]. For metabolomics samples, the quenching of metabolic pathways might be possible, but the faster the metabolic processes proceed the more difficult is the quencing process [2, 26].

7.7.3 Analytical quality assurance for biobanks – future aspects

At the present time, there are only limited solutions for the assessment of a given sample regarding its quality that is related to preanalytical variabilities [27]. Consequently, it is important to evaluate tools for the assessment of sample quality. Specifically, biobank repositories are facing this problem when specimens with comparable preanalytical "history" have to be identified that can be analyzed in differential display approaches without any bias.

For the majority of possible analytes, quality control (QC)-tools are not yet available and correspond to the 'QC-gap'. Up to now only few degradation markers, namely Fibrinopeptide A [5, 28], Protein S, MMP7, MMP9 and CD40L have been identified [27]. However, the concentration of these decay markers is not only related to the preanalytical time span but also to several disease states thus enabling the monitoring of a time dependent decay process difficult. A proteomic or peptidomic approach for the systematic identification and thorough validation of further decay markers that are independent from respective disease states is still missing. Recently, an external decay marker for quality control monitoring of serum and plasma has been described [29]. Briefly, blood has inherent proteolytic activity that is related to various endoproteases, for example from protease cascades of coagulation, fibrinolysis and the complement activation [30] as well as diverse exoproteases [31]. Accordingly, proteins and peptides are continuously processed in a time dependent manner. This results in the reduction of longer- and accumulation of shorter fragments during prolonged incubation. The mass spectrometry based monitoring of the proteolytic processing of synthetic reporter peptides does function as a 'proteomic degradation clock' to estimate the preanalytical quality of blood specimens.

However, this approach is restricted to prospective QC analyses as the synthetic peptide substrate has to be added to serum- and plasma tubes prior to blood withdrawal and existing collections with native samples are not feasible. In ongoing work, we are validating a mass spectrometric pattern of endogenous decay markers for quality assessment of native serum and plasma specimens [32].

7.7.4 Recommendations

To fully assess biospecimen quality, multiple quality control markers are needed that are not readily available till now [27]. A 2011 survey of more than 700 cancer researchers revealed that 47 % had trouble finding samples of sufficient quality. Because of this, 81 % have reported limiting the scope of their work, and 60 % said they question the findings of their studies [33]. Accordingly, the analytical monitoring of sample quality is obligatory for improved biomarker discovery and validation studies. As preanalytical bias cannot be avoided completely, building up a sample repository is always a prospective task. When samples are maximally standardized and potential errors as far as possible randomized, new and more sensitive analytical methods will reveal new bias, which can be traced back to its source by unabridged documentation and errors avoided for samples to follow [18].

Therefore, it will be crucial to monitor the preanalytical status and document all applied procedures of clinical specimens. Consequently, appropriate quality control analyses should be introduced for future studies. In any case, the preanalytical stability of new biomarkers will have to be investigated systematically prior to validation studies to avoid any bias related to sample quality.

7.7.5 Conclusions

The modern biobank will operate more like a clinical laboratory with formal accreditation, standard operating procedures, and quality assurance protocols [34]. Advancements in biospecimen science are rapidly growing. The National Cancer Institute and the International Society for Biological and Environmental Repositories are periodically publishing best practices that give guidance for sampling of biospecimens and should be considered accordingly.

References

[1] Botti G, Franco R, Cantile M, Ciliberto G, Ascierto PA. Tumor biobanks in translational medicine. J Translation Med 2012; 10:204.

[2] Vuckovic D. Current trends and challenges in sample preparation for global metabolomics using liquid chromatography-mass spectrometry. Anal Bioanal Chem 2012; 403:1523–48.

[3] Aguilar-Mahecha A, Kuzyk MA, Domanski D, Borchers CH, Basik M. The effect of pre-analytical variability on the measurement of MRM-MS-based mid- to high-abundance plasma protein biomarkers and a panel of cytokines. PloS one 2012; 7:e38290.

[4] Yi J, Liu Z, Craft D, O'Mullan P et al. Intrinsic peptidase activity causes a sequential multi-step reaction (SMSR) in digestion of human plasma peptides. J Proteome Res 2008; 7:5112–18.

[5] Findeisen P, Sismanidis D, Riedl M, Costina V, Neumaier M. Preanalytical impact of sample handling on proteome profiling experiments with matrix-assisted laser desorption/ionization time-of-flight mass spectrometry. Clin Chem 2005; 51:2409–11.

[6] Yin P, Peter A, Franken H, Zhao X et al. Preanalytical aspects and sample quality assessment in metabolomics studies of human blood. Clin Chem 2013; 59:833–5.

[7] Kamlage B, Maldonado SG, Bethan B, Peter E et al. Quality markers addressing preanalytical variations of blood and plasma processing identified by broad and targeted metabolite profiling. Clin Chem 2014; 60:399–412.

[8] Palmirotta R, De Marchis ML, Ludovici G, Leone B et al. Impact of preanalytical handling and timing for peripheral blood mononuclear cells isolation and RNA studies: the experience of the Interinstitutional Multidisciplinary BioBank (BioBIM). Int J Biol Markers 2012; 27:e90–8.

[9] Palmirotta R, Ludovici G, De Marchis ML, Savonarola A et al. Preanalytical Procedures for DNA Studies: The Experience of the Interinstitutional Multidisciplinary BioBank (BioBIM). Biopreservation and Biobanking 2011; 9:35–45.

[10] Rouy D, Ernens I, Jeanty C, Wagner DR. Plasma storage at −80 degrees C does not protect matrix metalloproteinase-9 from degradation. Anal Biochem 2005; 338:294–8.
[11] Sulik A, Wojtkowska M, Oldak E. Preanalytical factors affecting the stability of matrix metalloproteinase-2 concentrations in cerebrospinal fluid. Clin Cim Acta 2008; 392:73–5.
[12] World Health Organization (WHO). (Guder WG). Use of anticoagulants in diagnostic laboratory investigations and Stability of blood, plasma and serum samples. WHO/DIL/LAB/99.1 Rev.2 2002.
[13] Karsan A, Eigl B J, Flibotte S, Gelmon K et al. Analytical and preanalytical biases in serum proteomic pattern analysis for breast cancer diagnosis. Clin Chem 2005; 51:1525–8.
[14] McLerran D, Grizzle W E, Feng Z, Bigbee W L et al. Analytical validation of serum proteomic profiling for diagnosis of prostate cancer: sources of sample bias. Clin Chem 2008; 54:44–52.
[15] Ransohoff DF, Gourlay M L. Sources of bias in specimens for research about molecular markers for cancer. J Clin Oncol 2010; 28:698–704.
[16] Kisand K, Kerna I, Kumm J, Jonsson H, Tamm A. Impact of cryopreservation on serum concentration of matrix metalloproteinases (MMP)-7, TIMP-1, vascular growth factors (VEGF) and VEGF-R2 in Biobank samples. Clin Chem Lab Med 2011; 49:229–35.
[17] Bonnas C, Specht K, Spleiss O, Froehner S et al. Effects of cold ischemia and inflammatory tumor microenvironment on detection of PI3K/AKT and MAPK pathway activation patterns in clinical cancer samples. Int J Cancer 2012; 131:1621–32.
[18] Leichtle AB, Dufour JF, Fiedler GM. Potentials and pitfalls of clinical peptidomics and metabolomics. Swiss Med Weekly 2013; 143: w13801.
[19] Moore H M, Kelly A B, Jewell S D, McShane L M et al. Biospecimen reporting for improved study quality (BRISQ). Cancer Cytopathol 2011; 119:92–101.
[20] Lehmann S, Guadagni F, Moore H, Ashton G et al., Standard Preanalytical Coding for biospecimens: Review and implementation of the sample PREanalytical code (SPREC). Biopreserv Biobank 2012; 10:366–74.
[21] Baker M, Biorepositories: Building better biobanks. Nature 2012; 486:141–6.
[22] Simeon-Dubach D, Perren A. Better provenance for biobank samples. Nature 2011; 475:454–5.
[23] Findeisen P, Neumaier M. Mass spectrometry based proteomics profiling as diagnostic tool in oncology: current status and future perspective. Clin Chem Lab Med 2009; 47:666–84.
[24] Ayache S, Panelli M, Marincola FM, Stroncek DF. Effects of storage time and exogenous protease inhibitors on plasma protein levels. Am J Clin Pathol 2006; 126:174–84.
[25] Skold K, Alm H, Scholz B. The impact of biosampling procedures on molecular data interpretation. Mol Cell Proteomics 2013; 12:1489–501.
[26] Dunn W, Ellis D. Metabolomics: Current analytical platforms and methodologies. Trends Anal Chem 2005; 24:285–94.
[27] Betsou F, Gunter E, Clements J, DeSouza Y et al. Identification of evidence-based biospecimen quality-control tools: a report of the International Society for Biological and Environmental Repositories (ISBER) Biospecimen Science Working Group. J Mol Diagn 2013; 15:3–16.
[28] Govorukhina NI, de Vries M, Reijmers TH, Horvatovich P et al. Influence of clotting time on the protein composition of serum samples based on LC-MS data. J Chromatograph 2009; 877:1281–91.
[29] Findeisen P, Thumfart J O, Costina V, Hofheinz R, Neumaier M. MS-Based monitoring of proteolytic decay of synthetic reporter peptides for quality control of plasma and serum specimens. Am J Clin Pathol 2013; 140:314–23.
[30] Betsou F, Barnes R, Burke T, Coppola D et al. Human biospecimen research: experimental protocol and quality control tools. Cancer Epidemiol, Biomark Pprevent, 2009; 18:1017–25.
[31] Wildes D, Wells JA. Sampling the N-terminal proteome of human blood. Proc Nat Acad Sci U S A 2010; 107:4561–6.

[32] Thumfart J, Abidi N, Mindt S, Costina V et al. LC/MS based monitoring of endogenous decay markers for quality assessment of serum specimens. Clin Chem Lab Med 2014; 52: Abstract P121, Ann Conf German Soc Clin Chem Lab Med Sept.24.–27. 2014.
[33] Massett HA, Atkinson NL, Weber D, Myles R et al. Assessing the need for a standardized cancer HUman Biobank (caHUB): findings from a national survey with cancer researchers. J Nation Cancer Inst Monographs 2011; 2011:8–15.
[34] Yong W H, Dry S M, Shabihkhani M. A practical approach to clinical and research biobanking. Met Mol Biol 2014; 1180:137–62.

8. Quality Assurance and Auditing of the Preanalytical Phase

Astrid Petersmann, Kathrin Schlüter, Matthias Nauck

8.1 Auditing of the Preanalytical Phase

8.1.1 Introduction

Many efforts have been made to manage and improve the preanalytical phase. Hemolysis of samples or the under filling of tubes represent objective measures that allow both, judging the sample quality as well as obtaining information about the actual processes in the preanalytical phase. Some influences on the sample cannot be measured once the sample is taken and sent to the laboratory. These preanalytical errors can lead to erroneous results and therefore have the potential of harming patients. Not all errors are severe enough to be noticed at all, others can cause severe harm like a delay in finding the correct diagnosis. We like to think of standard operating procedures (SOP) as a guarantee that processes of the preanalytical phase are carried out correctly. But do they, after all, impact routine work the way we would like them to? Do they ensure a correct preanalytical phase for the majority of samples processed? Auditing of the preanalytical phase can at least give us a glance at the answers to these questions.

Among the most frequent preanalytical errors are hemolysis and, most critical, patient misidentification [1–9]. Hemolysis in samples can be detected by measuring the HIL-index, irrespective of clinically relevant in-vivo or the artificial in-vitro hemolysis [10]. This feature is offered by many, but not all modern instruments. If increased hemolysis is detected affected assay results can be flagged, commented or not be reported. Whether elevated potassium is due to a patient's disease or preanalytical errors like incorrect storage and transport of the tubes or prolonged tourniquet time or excessive fist clenching, cannot be recognized by the laboratory [11–13]. In addition, prolonged tourniquet time can lead to hemoconcentration and can impact coagulation results [14–16]. Besides tourniquet time, refrigeration of the uncentrifuged whole blood sample may cause potassium elevations. This does not cause visual hemolysis, but because the Na/K-ATPase is inhibited in cold temperatures, potassium accumulates in the diagnostic sample [17–19]. Patient misidentification and tube mislabeling may be noticed by the laboratory if a delta-check to previous samples is possible, but if it is the first sample taken, the laboratory is unable to spot these potentially severe errors. For blood culture collection, correct and thorough disinfection of the puncture site is of paramount importance to avoid contamination, yet it is impossible for the laboratory to recognize such mistakes.

The number and professions of people who take samples may vary considerably. In some countries specialized phlebotomists are very common whereas in other countries blood sampling is done by nurses, physicians and medical students (see Chapters 2.1–2.5). Furthermore, the blood collection and consequently the preanalytical processes can be organized in completely different ways. Blood collections can take place exclusively in centralized blood collection centres in hospitals, on the hospital wards or decentralized in General Practitioner's offices.

8.1.2 Audit procedures

Auditing is a challenging task for both the observer and the person who is being audited. Auditors should undergo professional training before auditing routinely. They have to establish a positive atmosphere but need to keep a personal distance at the same time. The blood collection process has to be observed and documented without interfering. For a regular venous puncture as many as 150 items can be crucial and need to be documented without delay. With some exceptions information cannot be obtained after the samples have been collected.

It is important to inform those who will be audited about the procedure a couple of days or even weeks before the actual audit takes place. How long will the audit take, who will perform the audit and how will the results of the audit be documented and returned? Returning information about the results of the audit is important for keeping up the motivation to participate in future audits. Details about what exactly the auditors will assess of all the aspects associated with a blood collection procedure should not be disclosed as this would potentially influence the behaviour of the person observed. Also, ensuring anonymity for the persons being observed is a crucial element for acceptance.

Audits by nature bear the risk of not being one hundred percent representative simply due to the presence of an auditor. When being audited, the observed individual might feel nervous or under pressure, which might negatively impact performance. On the other hand, individuals might change their habits to better practice because they are aware of being observed. The auditor has to remain in the background and has to keep an eye on many details of the workflow simultaneously. The auditor must not influence the procedure by answering questions while auditing. Audit procedures have to be adjusted to meet local needs and as a consequence can vary considerably in time and efforts.

8.1.3 Audit report form

Audits should be based on the given standard operating procedure (SOP) for blood sampling of the particular site. Special audit report forms may be developed by each institution. Alternatively, standardized audit report forms can be used by including only those items that suit the in-house SOPs. This can be helpful for benchmarking purposes, for example among hospitals. Differences in the sampling process can for example occur in the use of safety devices which are not mandatory in some countries. Also labelling of tubes is not handled the same way in every country or even hospital.

When developing or adapting an audit report form it should be carefully considered which information is worthwhile capturing. A routine blood collection may take as little as 90 sec – how many different pieces of information can be noted down in that time? How severe are the possible consequences of noncompliance? How should the data be protocolled to ensure that the analysis of the data is easy and meaningful? For example, to be as fast as possible, multiple choice tick boxes are most efficient. In case data could not be captured, it may be necessary not only to have "yes" and "no", but also "don't know" or "not applicable" to make the data more reliable. Tourniquet time for example may not be applicable if blood is taken from an indwelling catheter. Space for open comments may be a good source of further information.

In addition to the blood collection procedure itself, transport conditions, reception of samples at the laboratory and preanalytical processing in the laboratory are important aspects to capture. Examples are errors or missing information on the test request form, centrifugation conditions or visual inspection of samples to identify incomplete gel barriers, fibrin formations in serum samples, and incorrect filling of coagulation tubes.

A standardized audit form is available from Becton Dickinson (BD). It addresses patient and specimen identification procedures, infection control procedures, phlebotomy technique, health care worker safety, sample management and sample quality. On the basis of such a standardized audit report form, further items can be added to gain an even more detailed view on the preanalytical processes. To monitor performance over a longer period of time, it is important to keep the (key) items as constant as possible when improving the audit report form. Table 8.1.1 provides a list of considerations that may help to set up a questionnaire for blood collection.

Table 8.1.1: Questions and comments for the auditing procedure.

Aspect	Considerations
General information	Time, ward, observer
Blood collector	The questionnaire should be anonymous. To assess how many different persons have been audited the item "has the person been observed before" can be helpful. Other information might be: profession, where/how did the person receive training, time of experience in phlebotomy
Patient ID	Important questions may be: non-leading question, how many independent ID markers have been asked for, was the name checked against the order request, is the patient conscious and capable of answering?
Sample ID	Have the tubes been labeled before or after blood collection? If the latter, directly after blood collection at the patient bed, or later on? Has the information on the tubes been checked with the name on the order form/the patient ID?

Table 8.1.1: (continued)

Aspect	Considerations
Patient preparation	What was the patient position (sitting/lying down)? Time since the last meal?
Hygiene	Were gloves used? Were they changed between the patients? Has the tourniquet been disinfected or is it single use? Was the puncture site disinfected? Was the time for the disinfection sufficient? How was the cotton moved? Was the disinfectant still wet during venepuncture?
Venepuncture	What insertion site was chosen (elbow, hand, other)? What kind of needle was used (straight vs. wingset), which Gauge, which brand? Was a catheter used (peripheral, central, arterial)? Has it been freshly inserted or already indwelling? Was there an infusion ongoing? Has blood been discarded?
Tourniquet	Was the position of the tourniquet correct? What was the tourniquet time (shall the observer have a stopwatch, estimate the time, or other)?
Blood collection tubes	Which tubes were filled (brand, additives, volume?), in which order? Was it a needle and syringe collection? How was the blood then transferred to blood collection tubes?
Handling of blood collection tubes	Have the tubes been mixed (thoroughly or incomplete inversions or vigorously/how many times)?
Needle disposal	If it was a safety device, was the mechanism activated correctly and immediately? Was recapping observed? Where the sharps disposed of immediately and into a certified sharps container?
Transport	What was the mode of transport (pneumatic tube system, courier)? Were time and temperature during transport measured?
Other information	This may be clenching the fist, inserting the needle into a haematoma, bending the needle, patient with breast amputation, etc.

8.1.5 Audit results

The BD audit report form has been used in more than 48 hospitals in 13 countries [20]. About 25 trained auditors were involved, following approximately 3,600 tubes from 2,000 blood collections to analysis. Hospital sizes ranged from 300 to 1,400 beds with 50 % of the hospitals carrying between 350 and 750 beds. About 20 % of the samples were drawn by physicians, another approximate 20 % by phlebotomists, and the remaining majority of blood collections were done by nurses.

The audits clearly revealed that local SOPs and routine work differ. This does not necessarily imply unprofessional behavior. When deviating from a given SOP, the professional might have a good reason for doing so. In general, SOPs cover important and/or frequent processes. SOPs are usually not covering every possible situation, especially not in health care. Well trained professionals are necessary to provide individual patient care beyond SOPs. When a professional is not following a SOP the reasons for this should be carefully elucidated before drawing conclusions.

By identifying the patient through an open question patient misidentification can be prevented as one of the most severe sources of potential preanalytical errors [1–9]. This is an agreed standard in most countries. Still audits show that patient identification is not conducted according to the standard in every second blood collection [20]. A reason can be that the patient is already known to the professional who takes the sample. The person taking the sample may no longer feel the need to ask or even might find it impolite to ask somebody for her or his name, whom they had cared for during the past few days or even weeks. However, particularly if a health care worker believes he knows the patient and is mistaken, there is a considerable risk for misidentification. Leading questions also contribute to the high rate of incorrect patient identification. The response sometimes may not even be a spoken word but instead, gestures often are also accepted.

Labelling is a critical moment in linking a sample to an individual patient. If this is not correctly done it is a potential source of error [1–9]. International recommendations require labelling right after the sample has been drawn [21]. It is also accepted and sometimes recommended to label the tubes before taking the sample [22–25]. Thirty percent of hospitals in the BD study showed no consistent policy for the time of labelling. But even for those hospitals which require a certain sequence in labelling and sampling many deviations were observed [20]. Workflows can be organized in many different ways. For tube labelling the person taking the sample may not be the person printing or attaching the labels to the tubes. There are also instruments available for automated blood collection tube labelling [26].

In the majority of blood collections, the tourniquet is left tightened for more than a minute or is not opened as soon as blood flows into the first tube – the proof that the venepuncture was successful. There may be cases where the vein conditions are so poor that the tourniquet is needed throughout the blood collection process, and many professionals adopted the practice not to loosen the tourniquet. Clenching the fist, known to lead to elevated potassium results, is often done by the patients who are keen to cooperate and contribute to the process.

Proper immediate mixing of samples after drawing is important. Manufacturers have defined mixing procedures for blood collection tubes which contain some kind of additives. These recommendations are integrated into the written standards, but as shown in the audit, 80 % of the tubes are not mixed at all after blood collection.

Possible consequences for insufficient mixing are for example micro-clots in hematology samples that lead to flagging of results in the instruments.

In general, a short turnaround time improves processes in patient care in hospitals as well as in general practitioner's offices. Consequently, short transport times, for example realized by applying single sample pneumatic tube systems (Tempus600) which do not require packing and unpacking of samples, are very efficient tools. If transport time is reduced to only a few minutes the use of serum is not desirable. A sufficient amount of time (20–30 min) to complete clotting has to be allowed before analysis. Otherwise fibrin formation in serum may occur after separating the serum samples from the clot bearing a risk for laboratory instruments. Insufficient mixing can enhance the occurrence of this phenomenon. Fibrin can clog instrument probes and may lead to erroneous or delayed reporting of results also for other than the initially affected sample. Since it cannot be foreseen whether a sample is mixed well enough simply by the incoming blood flow standardized mixing improves the general quality of samples.

Safety issues are of increasing interest. Needle stick injuries may lead to infections with HIV or Hepatitis C. In Europe, it is estimated that 1,000,000 needle stick injuries happen in hospitals every year [27]. For half of the investigated hospitals, safety needles were mandatory by law. Surprisingly, only 20 % of the users activated the mechanism correctly and immediately, indicating that the devices should be further optimized by industry and that the necessary knowledge has not yet reached the consumer. Only 70 % of all needles were disposed correctly and slightly over 60 % of the professionals were wearing gloves when drawing blood. The gap between the written standards and routine work is well illustrated by these numbers and clearly shows that a written document, even if signed "read and understood" by the employee, is no guarantee for a correct workflow. This seems to be truly independent of the country or the applied quality management system.

8.1.6 Utilizing audit results

Audits can identify critical steps in the preanalytical process. This knowledge in turn should be used to focus training efforts. Some hospitals do perform very well in safety issues but do not so well in patient identification and vice versa. Instead of re-training the whole process it can be more efficient to concentrate on fewer but specific items. Furthermore, the results of an audit can help to initiate a change of mind – for the individuals taking blood to appreciate the connection between their sampling routines and habits and the consequences for the sample quality and ultimately the care of the patient. Even though audits are time consuming they can be recommended to improve the blood drawing process. Regular audits may build a basis for continuous improvement.

References

[1] Lippi G, Bassi A, Brocco G, Montagnana M, Salvagno GL, Guidi GC. Preanalytical error tracking, based on specimen acceptability, over a 1-year observational period in the 5 most representative sections of laboratory medicine department. Clin Chem 2006; 52:1442–3.

[2] Bonini P, Plebani M, Ceriotti F, Rubboli F. Errors in Laboratory Medicine. Clin Chem 2002; 48:691–8.

[3] Howanitz PJ. Errors in laboratory medicine: practical lessons to improve patient safety. Arch Pathol Lab Med. 2005; 129:1252–61.

[4] Plebani M, Ceriotti F, Messeri G, Ottomano C, Pansini N, Bonini P. Laboratory network of excellence: enhancing patient safety and service effectiveness. Clin Chem Lab Med 2006, Vol. 44:150–60.

[5] Alsina Kirchner MJ, Alvarez Funes V, Biosca Adzet C, Doménech Clar MV, Ibarz Escuer MI, Minchinela Girona J, et al. Quality indicators and specifications for key processes in clinical laboratories: a preliminary experience. Clin Chem Lab Med 2007; 45:672–7.

[6] Carraro P, Plebani M. Errors in a stat laboratory: Types and frequencies 10 years later. Clin Chem 2007, Vol. 53:1338–42.

[7] Valenstein PN, Raab SS, Walsh MK. Identification errors involving clinical laboratories. A College of American Pathologists Q-Probes Study of Patient and Specimen Identification Errors at 120 Institutions. Identification Errors Involving Clinical Laboratories. Arch Pathol Lab Med 2006, Vol. 130:1106–13.

[8] Carraro P, Zago T, Plebani M. Exploring the initial steps of the testing process: Frequency and nature of pre-preanalytic errors. Clin. Chem 2012, Vol. 58:638–42.

[9] Grecu DS, Vlad DC, Dumitrascu V. Quality Indicators in the preanalytical phase of testing in a stat laboratory. Lab Med, 2014; 45:74–81.

[10] Söderberg J, Jonsson PA, Wallin O, Grankvist K, Hultdin J. Haemolysis index-an estimate of preanalytical quality in primary health care. Clin Chem Lab Med 2009; 47:940–4.

[11] Don BR, Sebastian A, Cheitlin M, Christiansen M, Schambelan M. Pseudohyperkalemia caused by fist clenching during phlebotomy. N Engl J Med, 1990; 322:1290–2.

[12] Seimiya M, Yoshida T, Sawabe Y, Sogawa K, Umemura H, Matsushita K, Nomura F. Reducing the incidence of pseudohyperkalemia by avoiding making a fist during phlebotomy: a quality improvement report. Am J Kidney Dis, 2010; 56:686–92.

[13] Guder WG, Narayanan S, Wisser H, Zawta B. Diagnostic Samples: From the Patient to the Laboratory. 3rd ed Weinheim, Germany: Wiley-VCH 2009.

[14] Lippi G, Salvagno GL, Montagnana M, Guidi GC. Short-term venous stasis influences routine coagulation testing. Blood Coagul Fibrinolysis 2005; 16:453–8.

[15] Saleem S, Mani V, Chadwick MA, Creanor S, Ayling RM. A prospective study of causes of haemolysis during venepuncture: tourniquet time should be kept to a minimum. Ann Clin Biochem 2009; 46:244–6.

[16] Stankovic AK, Smith S. Elevated serum potassium values. The role of preanalytic variables. Am J Clin Path 2004, 121 (Suppl 1) 105–12.

[17] Ono T, Kitaguchi K, Takehara M, Shiiba M, Hayami K. Serum-constituents analyses: effect of duration and temperature of storage of clotted blood. Clin Chem 1981; 27:35–38.

[18] Rehak NN, Chiang BT. Storage of whole blood: effect of temperature on the measured concentration of analytes in serum. Clin Chem 1988; 34:2111–4.

[19] Oddoze C, Lombard E, and Portugal H. Stability study of 81 analytes in human whole blood, in serum and in plasma. Clin Biochem 2012; 45:464–9.

[20] Schlueter K, Nauck M, Petersmann A, Church S. Using BD Laboratory Consulting Services™ to understand the impact of the preanalytical phase on sample quality and safety, a multi country perspective. Biochem Med (Zagreb) 2013; 23:224.

[21] CLSI Standard H3-A6: Procedures for the Collection of Diagnostic Blood Specimens by Venipuncture; 6th ed. Approved Standard.Wayne PA: Clinical and Laboratory Standards Institute, 2007.

[22] Sciacovelli L, O'Kane M, Skaik YA, Caciagli P, Pellegrini C, Da Rin G et al. IFCC WG-LEPS. Quality Indicators in Laboratory Medicine: from theory to practice. Preliminary data from the IFCC Working Group Project "Laboratory Errors and Patient Safety". Clin Chem Lab Med 2011; 49:835–44.

[23] CAP. Laboratory General Checklist. Northfield, IL: College of American Pathologists, 2010.

[24] Lippi G, Caputo M, Banfi G, Buttarello M, Ceriotti F, Daves M, et al. per il Gruppo di Studio Intersocietario SIBioCSIMeL-CISMEL sulla Variabilita Extra-Analitica del Dato di Laboratorio. Recommendations for collection of venous blood. Biochim Clin 2008; 32: 569–77.

[25] Guder W, Fiedler M, da Fonseca-Wollheim F, Schmitt Y, Töpfer G, Wisser H, Zawta B. Quality of Diagnostic Samples. Recommendations of the working group on extraanalytical quality of the German United Society for Clinical Chemistry and Laboratory Medicine. 4th ed., Oxford: BD Europe, 2015.

[26] Söderberg J, Brulin C, Grankvist K, Wallin O. Preanalytical errors in primary healthcare: a questionnaire study of information search procedures, test request management and test tube labelling. Clin Chem Lab Med, 2009; 47:195–201.

[27] European Parliament. Preventing needle-stick injuries in the health sector, 11th February 2010.

Janne Cadamuro

8.2 Internal Quality Assurance for Preanalytical Phase

Laboratory testing has a major effect on clinical decisions, providing physicians, nurses and other health care providers with information that aids in the prevention, diagnosis, treatment and management of disease [1, 2]. Therefore the laboratory should strive for the highest possible quality of test results and errors throughout the laboratory process should be kept at a minimum. As you read earlier in this book, the majority of errors occur during the preanalytical phase [3, 4], however with a declining tendency over the past decades. This may be due to the constantly growing awareness and standardization of the respective processes according to standards like the DIN/EN/ISO 15189 [5] and others. Optimization of the preanalytical phase or a process in general, requires knowledge about the present situation in the respective setting. To acquire this information, you need tools identifying the most common preanalytical errors in your lab in order to set up preventive actions.

8.2.1 What to measure

Similar to the analytical phase of your laboratory process, you can try to assess the quality internally (internal quality controls) and externally (external quality assurance programs). For the latter, please see Chapter 8.3.

Internal quality controls (QCs), for monitoring the preanalytical processes, differ in many ways to those for analytics. The analytical process occurs within the laboratory, involves few, well-educated personnel, has limited, well-defined influencing variables, has QCs with predefined reference ranges and there mostly is international consensus due to extensive research. The preanalyical process however, occurs mostly outside of the laboratory, involves many parties, such as patients, clinicians, nurses or shippers, has many influencing variables, some difficult to define and has no QCs with predefined limits or consensus, since there is little literature in this field [6].

At the present time, the best way to monitor your preanalytical processes is to measure quality indicators (QIs), as recommended by the IFCC Working Group "Laboratory Errors and Patient Safety" [7, 8]. QIs are defined as data or group of data that help in objectively measuring the evolution of a process or activity[9]. In your preanalytical process they help to quantify the quality of your defined aspects by comparing them against defined criteria [6]. QIs should [6, 10, 11]:

- be patient – centered, to promote total quality and patient safety,
- have defined limits,
- permit the early detection of deviations from defined limits,
- be suitable for promoting corrective/preventive actions,

- be simple to implement, calculations should require a reasonable amount of workload,
- be reviewed periodically, to ensure their continued appropriateness, and
- be consistent with the requirements of the International Standard for medical laboratories accreditation [5].

Every preanalytical error could potentially be used as QI, however, it is economically reasonable to focus on those, which are most common and severely affects the sample, triggering sample rejections.

According to an evaluation of the Spanish preanalytical quality control program during the years 2001–2005, in which 105 laboratories participated and 4,715,132 tubes were monitored, the following preanalytical causes for sample rejections were the most commonly observed [12]:
- Clotted sample
- Hemolyzed sample
- Sample not received
- Insufficient sample (Sample type not suitable for ordered test)
- Improper collection container
- Inadequately labeled (or unlabeled) sample
- Improper transportation
- Inappropriate ratio anticoagulant/sample volume

In total, 81 % of rejections arose as a result of the first three of these variables.

Lippi et al additionally divided their findings into inpatients and outpatients, observing, that insufficient volume was a prevailing cause of unsuitable specimens for inpatient samples, whereas the prevalence of inappropriate containers was particularly high for outpatient specimens [13]. The overall amount of preanalytic errors varies between different studies, ranging from 0.74 % to 5.5 % of all samples [11, 13, 14].

It should be mentioned, that the prevalence of preanalytical errors differ, depending on the sample type, as samples with inappropriate volume are more common in citrate samples than in others [12, 14]. Additionally in samples, collected for coagulation analysis, other preanalytical errors might be more of interest, such as samples refrigerated or frozen before processing, contamination with other specimen or others [15].

In an attempt to harmonize QIs for preanalytical processes, the IFCC Working Group "Laboratory Errors and Patient Safety" has published a Model of Quality Indicators for the preanalytical phase, which focuses on identification, documentation and monitoring of the following [6]:
a) quality of request forms;
b) patient identification at the point of care; and
c) quality of biological specimens, particularly during and after transportation from collecting sites to the laboratory.

Table 8.2.1: Quality Indicators of the pre-analytical phase [6].
(order of priority: 1 = Mandatory; 2 = Important; 3 = Suggested; 4 = Valuable).

Quality indicator priority	Priority Score
a) Appropriateness of clinical request	
Percentage of "Number of requests without clinical question (outpatients)/Total number of requests (outpatients)"	2
Percentage of "Number of inappropriate requests, with respect to clinical question (outpatients)/Number of requests reporting clinical question (outpatients)"	4
Percentage of "Number of inappropriate requests, with respect to clinical question (inpatients)/Number of requests reporting clinical question (inpatients)"	4
b) Patient identification	
Percentage of "Number of requests with errors concerning patient identification/Total number of requests"	1
Percentage of "Number of requests with errors concerning patient identification, detected before release of results/Total number of requests"	1
Percentage of "Number of requests with errors concerning patient identification, detected after issuing results/Total number of requests"	1
c) Data entry of the request	
Percentage of "Number of outpatients requests with errors concerning physician identification/Total number of outpatients requests"	2
Percentage of "Number of unintelligible outpatients requests/Total number of outpatients requests"	3
Percentage of "Number of outpatients requests with errors concerning test input/Total number of outpatients requests"	1
Percentage of "Number of outpatients requests with errors concerning test input (missing)/Total number of outpatients requests"	1
Percentage of "Number of outpatients requests with errors concerning test imput (added)/Total number of outpatients requests"	1
Percentage of "Number of outpatients requests with errors concerning test input (misinterpreted)/Total number of outpatients requests"	1
Percentage of "Number of inpatients requests with errors concerning test input (missing)/Total number of inpatients requests"	1
Percentage of "Number of inpatients requests with errors concerning test imput (added)/Total number of inpatients requests"	1
Percentage of "Number of inpatients requests with errors concerning test input (misinterpreted)/Total number of inpatients requests"	1
d) Sample identification	
Percentage of "Number of improperly labeled samples/Total number of samples"	1
e) Sample collection	
Percentage of "Number of samples collected at inappropriate time/Total number of samples"	2
Percentage of "Number of samples collected with inappropriate sample type/Total number of samples"	1

Table 8.2.1: (continued)

Quality indicator priority	Priority Score
Percentage of "Number of samples collected in inappropriate container/Total number of samples"	1
Percentage of "Number of samples with insufficient sample volume/Total number of samples"	1
f) Transport of sample	
Percentage of "Number of damaged samples/Total number of samples"	1
Percentage of "Number of samples transported at inappropriate time/Total number of samples for which transport time is checked"	1
Percentage of "Number of samples transported under inappropriate temperature condition/Total number of samples for which the transport temperature is checked"	1
Percentage of "Number of improperly stored samples/Total number of samples"	1
Percentage of "Number of samples lost-not received/Total number of samples"	1
g) Suitability of sample	
Percentage of "Number of samples with inadequate sample-anticoagulant volume ratio/ Total number of samples with anticoagulant"	1
Percentage of "Number of haemolysed samples (hematology)/Total number of samples (hematology)"	1
Percentage of "Number of haemolysed samples (chemistry)/Total number of samples (chemistry)"	1
Percentage of "Number of clotted samples (hematology)/Total number of samples with anticoagulant (hematology)"	1
Percentage of "Number of clotted samples (chemistry)/Total number of samples with anticoagulant (chemistry)"	1
Percentage of "Number of clotted samples (immunology)/Total number of samples with anticoagulant (immunology)"	1
Percentage of "Number of haemolysed samples (immunology)/Total number of samples (immunology)"	1
Percentage of "Number of lipemic samples/Total number of samples"	1
Percentage of "Number of unacceptable samples (microbiology)/Total number of samples (microbiology)"	1
Percentage of "Number of contaminated blood cultures/Total number of blood cultures"	1

8.2.2 How to measure

Documentation of all kind of information has become overwhelming over the past years, since more and more laboratories tend to standardize their facility according to standards DIN/EN/ISO 9001[16], DIN/EN/ISO 17025 [17] or DIN/EN/ISO 15189 [5]. These documentations can be very time consuming, therefore it is utterly important to keep the workload measuring your preanalytical QIs at a minimum. We recommend using an error tracking system, which is implemented into your laboratory information system (LIMS). The technician receiving the sample in the laboratory then codes the error, using a predefined list of errors. In this way all errors are assigned to the specific request in your system. You can then evaluate your QIs statistically, since additional information to the specific requests, such as sample type, request type (stat or routine), requesting facility (Inpatient or Outpatients), requesting ward (for Inpatients), date and time when the sample was collected or date and time when the sample arrived in the lab, also is available in the LIMS. Optionally, general information about the sample collection system (vacuum or aspiration), the phlebotomy personnel or the material type (plastic or glass) can be used to assess your QIs to a higher detail.

Two of the most common preanalytical errors are hemolyzed and/or under-filled samples, as mentioned above. To measure these two variables we recommend using automatic systems, such as serum-indices [18] or preanalytical sample handling solutions, in which camera modules are able to quantify these variables (see also Chapter 6.2).

If unacceptable lag times in sample transportation are revealed by one of your QIs, you can try finding the source of delay either by interviewing the participating personnel, or by tracking selected samples using RFID-Chips.

8.2.3 How to avoid and/or prevent

When attempting to avoid preanalytical errors systematically, you need to standardize your processes and those of other health care operators involved in the preanalytical process, using some or all of the following possibilities [6, 12, 19–21]:

a) Every laboratory should have a basic set of standard operating procedures (SOP), defining procedures for test request, sample identification, sample collection, patient preparation, specimen acquisition, transportation, handling and storage. (see Chapter 8.1).
b) Health care operators should be trained in these processes according to the respective guidelines.
c) The request form should have instructions on completing the form.
d) Criteria for sample rejections should be defined and made accessible for intra- and extra-laboratory personnel.
e) Predefined QIs should be evaluated periodically.
f) After evaluation of your QIs, the participating parties (clinicians, nurses, technicians, shippers) should be informed and a consensus for future improvements should be achieved.
g) Health care operators should easily have access to a database of the laboratory's portfolio including preanalytical considerations.
h) Collection of specimens should be performed by trained personnel, for example phlebotomists and periodical meetings should be held by a respective coordinator (see Chapter 2.9).

i) When using a pneumatic tube system (PST) for sample transportation, it should be tested for analytes sensitive to agitation, such as platelet aggregometry. Additionally acceleration sensors can be used to test your PST for g-forces or vibration.

General recommendations:
Each laboratory should
- specify quality indicators, according to the current literature [6, 20, 21],
- implement an error tracking system,
- automatically measure hemolysis and filling level of specimen,
- periodically evaluate their quality indicators with an subsequent action in consensus with the involved personnel, to prevent future incidences, and
- maintain contact with all of the preanalytically involved parties. Adherence to guidelines and recommendations can only be warranted, if all involved parties have a fundamental understanding about the importance and impact of preanalytical errors.

References

[1] Hoffmann G E, Schenker M, Kammann M, Meyer-Lüerssen D, Wilke MH. The significance of laboratory testing for the German diagnosis-related group system-the systematic evaluation of comorbidities of relevance to case reimbursement and continued development of the DRG Watchdog software. Clin Lab 2004; 50:599–607.

[2] Wolcott J, Schwartz A, Goodman C. Laboratory Medicine: A National Status Report The Lewin Group. 2008.

[3] Bonini, P, Plebani M, Ceriotti F, Rubboli F. Errors in laboratory medicine. Clin Chem 2002; 48:691–8.

[4] Plebani, M (). Errors in clinical laboratories or errors in laboratory medicine? Clin Chem Lab Med 2006; 44:750–9.

[5] DIN/EN/ISO 15189: Medical laboratories - Requirements for quality and competence. Geneva, Switzerland: International Organization for Standardization (ISO) 2012.

[6] Plebani M, Sciacovelli L, Aita A, Chiozza ML. Harmonization of pre-analytical quality indicators. Biochemia medica 2014; 24:105–13.

[7] Sciacovelli L, Plebani M. The IFCC Working Group on laboratory errors and patient safety. Clin Chim Acta 2009; 404:79–85.

[8] Sciacovelli L, O'Kane M, Skaik YA, Caciagli P, Pellegrini C, Da Rin G et al. Quality indicators in laboratory medicine: from theory to practice. Preliminary data from the IFCC Working Group Project "Laboratory Errors and Patient Safety". Clin Chem Lab Med 2011; 49:835–44.

[9] Llopis, M A, Alvarez V, Martínez-Brú C et al. Quality Assurance in the Preanalytical Phase. Applications and Experiences of Quality Control. O. Ivanov (ed.) InTech (Rijeka, Croatia) 2011, pp. 187–204.

[10] CLSI-GP35-P. Development and Use of Quality Indicators for Process Improvement and Monitoring of Laboratory Quality; Proposed Guideline. Wayne, PA: Clinical and Laboratory Standards Institute, 2009.

[11] Llopis M A, Trujillo G, Llovet M I, Tarres E, Ibarz M, Biosca C, et al. Quality indicators and specifications for key analytical-extranalytical processes in the clinical laboratory. Five years' experience using the Six Sigma concept. Clin Chem Lab Med 2011; 49:463–70.

[12] Alsina MJ, Alvarez V, Barba N, Bullich S, Cortes M, Escoda I et al. Preanalytical quality control program - an overview of results (2001–2005 summary). Clin Chem Lab Med 2008; 46:849–54.
[13] Lippi G, Bassi A, Brocco G, Montagnana M, Salvagno GI, Guidi GC. Preanalytic error tracking in a laboratory medicine department: results of a 1-year experience. Clin Chem 2006; 52:1442–3.
[14] Salvagno GL, Lippi G, Bassi A, Poli G, Guidi GC. Prevalence and type of pre-analytical problems for inpatients samples in coagulation laboratory. J Eval Clin Pract 2008; 14:351–3.
[15] Clinical and Laboratory Standards Institute (CLSI) . Collection, Transport, and Processing of Blood Specimens for Testing Plasma-Based Coagulation Assays. Wayne, PA: Clinical and Laboratory Standards Institute, Document H21-A5, 2008.
[16] DIN/EN/ISO 9001. Quality Management Systems. Geneva, Switzerland: International Organization for Standardization (ISO) 2008.
[17] DIN/EN/ISO 17025. General requirements for the competence of testing and calibration laboratories. Geneva, Switzerland: International Organization for Standardization (ISO) 2005.
[18] Vermeer HJ, Thomassen E, de Jonge N. Automated processing of serum indices used for interference detection by the laboratory information system. Clin Chem 2005; 51:244–7.
[19] Kratz A, Salem RO, Van Cott EM. Effects of a pneumatic tube system on routine and novel hematology and coagulation parameters in healthy volunteers. Arch Pathol Lab Med 2007; 131:293–6.
[20] Lippi G, Banfi G, Buttarello M, Ceriotti F, Daves M, Dolci A et al. Recommendations for detection and management of unsuitable samples in clinical laboratories. Clin Chem Lab Med 2007; 45:728–36.
[21] Lippi G, Guidi GC. Risk management in the preanalytical phase of laboratory testing. Clin Chem Lab Med 2007; 45:720–7.

Gunn B. B. Kristensen, Kristin Moberg Aakre, Ann Helen Kristoffersen, Sverre Sandberg

8.3 External Quality Assurance for the Preanalytical Phase[1]

8.3.1 Introduction

In laboratory medicine, several studies have described the most frequent errors in the different phases of the total testing process (TTP) [1–12], and a large proportion of these errors occur in the preanalytical phase (which here also will include the pre-preanalytical phase) [2, 5, 13–17].

The first step in improving the quality of the preanalytical phase is to describe potential errors and to try to estimate which errors are most dangerous for the outcome of the patient [13, 18–22]. Existing preanalytical procedures should be compared to existing recommendations and thereafter improved to minimize the risk of errors. In addition, the frequency of errors should be recorded on a regular basis to detect improvement or deterioration over time, and further to explore if procedures should be changed.

The ISO 15189: 2012 [23] states that "External quality assessment programs should, as far as possible, provide clinically relevant challenges that mimic patient samples and have the effect of checking the entire examination process, including pre- and post-examination procedures" (item 5.6.4) [23]. Consequently, the EQA organizations should take upon them to set up EQA Schemes (EQAS) for the preanalytical phase. At present, most EQA organizations do not provide such service although increasing numbers of organizations are taking interest in and plan to launch preanalytical schemes. An important challenge, when developing EQAS for the extraanalytical phases, is the variety of locations and staff groups involved in the total testing process, of which several are outside the laboratory's direct control. Test ordering, data entry and specimen collection/handling often involve other than laboratory staff. Some of the preanalytical schemes, for example schemes on test ordering, could also involve other health care professionals. The contact and communication between the clinicians/nurses and the laboratory staff are often limited [7, 24]. This complexity of the service may explain the low response rate some EQA organizations experience to their preanalytical schemes. To our knowledge, and in contrast to what is required for the analytical phase, accreditation bodies do not ask laboratories for results from EQAS regarding the preanalytical phase. If this is changed, it will be a driver for the EQA organizations to set up such schemes, and for the participants to use them. Consequently, a joint effort between EQA organizations, accreditation bodies and clinical laboratories seems necessary to implement preanalytical EQA schemes. The

1 Based on previous publication [49], extended and actualized november 2014.

aim of this chapter is to present an overview of different types of EQA schemes for the preanalytical phase and to give examples of existing schemes.

8.3.2 How to perform a preanalytical EQA scheme?

Fortunately, EQAS for the preanalytical phase are increasingly being developed, and roughly three different methods have been used:

- **Type I:** Registration of procedures: Circulation of questionnaires asking about procedures on the handling of certain aspects of the preanalytical phase, for example how "sample stability" or tube filling are dealt with, which criteria are used for sample rejection, and how these issues are communicated to the requesting physicians.
- **Type II:** Circulation of samples simulating errors: For example distribution of real samples with matrixes or samples with contamination which might interfere with the measurement procedures. This is similar to usual analytical EQAS, since a sample is sent to the laboratories. Case histories can be sent together with the sample to elucidate how these samples are dealt with, and how the results are communicated to the physicians.
- **Type III:** Registration of errors/adverse events: The incidences of certain types of preanalytical errors (some of them could be used as quality indicators [QIs]) are registered and reported by the laboratories to the EQA organizer.

The EQA organizer should as usual provide feedback reports for the participating laboratories, enabling the laboratories to compare their results with the other participants. In addition, the feedback reports should include advice on how to minimize errors. In case of error reporting (Type III), the EQA organizer should also lead the work on harmonization of QIs for a valid comparison of error rates between different locations. Examples of the three different methods to conduct preanalytical EQAS will be given below. Ongoing EQAS are summarized in Table 8.3.1.

Table 8.3.1: Examples of ongoing pre-analytical EQAS[1].

Type I. Registration of procedures			
Method	**Preanalytical issues studied**	**Frequency**	**Performed by**
Registration of procedures via web-based multiple choice questionnaire	Clinical chemistry: Hemolysis, stability of samples (2011) Hemostasis testing: Phlebotomy, sample handling and sample acceptance (2012) Glucose: Sample handling and sample treatment (2013)	1x per year	Norwegian Clinical Chemistry EQA program (NKK) Not published

Table 8.3.1: (continued)

Type I. Registration of procedures			
Method	**Preanalytical issues studied**	**Frequency**	**Performed by**
Registration of procedures via web-based multiple-choice questionnaire	Hemostasis testing	2x per year	ECAT/INSTAND 2011 [47]
Clinical case-based European EQAS covering preanalytical, analytical and postanalytical phase. Five sets of multi-specimen samples	Porphyria: Case history based test ordering	2x per year	Norwegian Porphyria Centre (NAPOS)/The European Porphyria Network (EPNET) [48]
Registration of procedures via web-based multiple choice questionnaire	5 general pre- and postanalytical questions, 5 questions within specific disciplines (i. e. coagulation, hematology, immunology, microbiology)	2x per year	* Quality Control Center Switzerland (CSCQ)
Registration of procedures via a questionnaire	Pre- and potanalytical procedures in medical practice laboratories	4x per year	Quality Control (MQ) Switzerland
Registration of procedures via multiple choice questionnaire	Urine chemistry, clinical chemistry	4x per year	* WEQAS
Registration of procedures via web-based multiple choice questionnaire together with real life case studies	Clinical chemistry, Phlebotomy, Microbiology and Blood gas analysis	2x per year	** Labquality
Registration of procedures via web-based multiple-choice questionnaire	25 general preanalytical questions within specific disciplines (i. e. blood sampling, assessment and handling of probes)	Pilot (2014), to be continued 1x per year	*** INSTAND
Type II. Circulation of samples simulating errors			
Method	**Pre-analytical issues studied**	**Frequency**	**Performed by**
Circulate samples	Sample indicies – lipemic, icteric, hemolysis index	4x per year	* WEQAS

1 Based on previous publication [49], extended and actualized november 2014.

* Personal communication at the EQALM meeting in Bucuresti in 2013 where the participants were asked if they had any pre-analytical EQAS. Two EQA organizers replied: WEQAS (Annette Thomas, http://www.weqas.com/eqa/index.html) and CSCQ (Dagmar Kesseler, http://www.cscq.ch)

** Personal communication with Labquality (Jonna Pelanti, https://www.labquality.fi)

*** Personal communication with INSTAND (Cornelia Schürer, http://www.instandev.de)

Table 8.3.1: (continued)

Type II. Circulation of samples simulating errors			
Method	**Pre-analytical issues studied**	**Frequency**	**Performed by**
Circulate samples for extraction of RNA/DNA	Sample preparation for DNA and RNA testing	1x per year	SPIDIA-DNA, 2012 SPIDIA-RNA, 2011 European Commission (EC) [30, 31]
Combination of Type I and II using questionnaires, real samples and simulations	A large variety of preanalytical issues	2x per year	National Institute of Health, Portugal 2007 [32]
Type III. Registration of errors/adverse events			
Method	**Pre-analytical issues studied**	**Frequency**	**Performed by**
Q-Track (since 1998, 1x year, ongoing) programs, registration of error rates	Patient/sample identification, specimen handling/preparation, specimen acceptability, customer satisfaction	4x per year	College of American Pathologists (CAP) [35]
Registration of rejection of samples	Registration of the rejection rate and causes for rejecting the samples during 1 month or 100 rejections	2x per year	Committee for the Quality of the Extra-analytical phase (started within The Spanish Society of Clinical Chemistry and Molecular Pathology (SEQC) in 1998 [38]
Registration of key incidents which represent either the most frequent or most serious incident	Patient identification, incorrect patient preparation, phlebotomy, sample preparation/handling and sample acceptability	4x per year	Key Incident Monitoring and Management Systems Quality Assurance (KIMMS QA) 2009 [40].
Registration of preanalytical errors in samples received from primary health care (PHC) in medical laboratories	Patient identification, missing or incomplete sample prescriber or copy recipient, sampling time missing, sample material incorrect, insufficient or missing	1x year	Noklus and NKK 2014

Type I: Registration of procedures

Identification and registration of preanalytical procedures can be performed by circulating questionnaires, with open or multiple choice questions. Case histories to illustrate the potential consequences of preanalytical errors for the patients may be included. A clear advantage of this type of EQAS is the limited resources needed for distributing and completing the scheme. Several aspects of the preanalytical phase can be covered, and the questionnaire may potentially reach laboratories all over the world in a short time using an electronic contact and report form.

The Norwegian Clinical Chemistry EQA Program (NKK) has developed a preanalytical EQA scheme which aims to identify especially problematic preanalytical issues related to clinical chemistry analyses and hemostasis testing. This scheme is carried out by circulating web-based questionnaires (mostly multiple-choice questions) to the Norwegian laboratories once a year dealing with different aspects of the preanalytical phase. The laboratories receive a feedback report after each survey, showing their own results, the overall results and an overview of the relevant recommended procedures from guidelines and recent studies. The three surveys performed so far (2011, 2012 and 2013) (Table 8.3.1), concluded that there is a large variability in several preanalytical practices and considerable room for improvement.

In 2009, the Croatian Chamber of Medical biochemists (CCMB) performed a survey in Croatian laboratories to investigate appropriateness of procedures in the extra-analytical phases and to detect procedures most prone to errors of potentially clinical importance [25]. A multiple choice questionnaire including preanalytical conditions and criteria for sample acceptance and procedures of phlebotomy were circulated to members of CCMB. The study concluded that there is an urgent need for improving activities by appropriate recording and monitoring of the extra-analytical phases; emphasizing the importance of initiating pre- and postanalytical EQAS.

A kind of pre-preanalytical survey was run in 2013 by the Australasian Association of Clinical Biochemists (AACB) Vitamins Working Party: As a basis for improving the external quality program for vitamin B_1/B_6 sub-program, a detailed questionnaire, including sample collection and storage was sent to all participant laboratories in the program. The information gained through the survey is used to complement the quality assurance program of plasma vitamin B_1 and B_6 [26, 27].

INSTAND conducted a pilot preanalytical EQAS (Web-Based Quality Control (WQ)), explicitly addressing laboratory technicians, in November 2014. The WQ consisted of 25 single and multiple-choice questions that were distributed via a web-link to registered members. The main topics were: technique of blood sampling and handling of the samples. The WQ also contained educational comments, detailing every option of the proposed answers. This WQ will be continued once a year and INSTAND also plans a web based pilot study on quality control for practice nurses in 2015.

The Association for Medical Quality Control (MQ) Switzerland performs pre- and postanalytical EQAS in medical practices laboratories, distributing questionnaires concerning procedures in practice laboratories. For analysis report, attendees get a comparison with the answers of the other participants as well as a comment with recommendations. MQ started this survey, which is performed 4 times a year in 2015.

Labquality, the Finnish EQA organization, started preanalytical EQAS in 2014 (Table 8.3.1). Altogether, four preanalytical schemes are available for the areas of clinical chemistry, phlebotomy, microbiology and blood gas analysis. In these schemes, real life case studies are presented, and a multiple choice questionnaire concerning the preanalytical phase is filled out by the participants. Labquality aims to provide preanalytical EQAS also for units using POCT instruments.

Challenges regarding registration of procedures as preanalytical EQAS

To avoid misinterpretations, the questionnaires used in EQAS should be validated by experts in the field and in pilot studies to a few laboratories. This is especially important if the questionnaires are circulated in different countries and not translated into local languages. The questions should be designed to stimulate the participants to answer how the procedures really are performed, and not how they ideally should be performed (e. g. ask for written procedures). Providers of such schemes should also be aware that queuing of answers (multiple choice) might influence the results. Standardized follow-up of the surveys should be performed by offering advises to the participants on how to improve their procedures. The EQA organisations should be aware of issues that could decrease the response rate, such as too long and work-loaded questionnaires or questions that are not clear and concise. High quality feed-back reports will probably increase the number of laboratories acknowledging the importance and benefit of auditing the preanalytical phase, and could increase the response rate. Cooperation with accreditation bodies could also be useful.

Type II: Circulation of samples simulating errors

EQAS for the preanalytical phase can be performed using real samples with a matrix potentially interfering with the measurement procedures, as for example hemolysed, lipemic or icteric samples or samples containing drugs known to interfere. Another example is to send the wrong sample material (e. g. serum instead of plasma). This type is in some instances comparable to analytical EQAS, since a sample is sent to the laboratories for measurement of predefined analytes. As for analytical EQAS, case histories can be included to elucidate which preanalytical procedures are performed and how the results (or lack of results in case of sample rejection) are communicated to the physicians.

The feedback reports for these surveys are comparable to feedback reports for regular EQA surveys, and should include a comparison of the laboratory's results to the results of all participants in addition to an overview of existing guidelines/recommendations and recent publications.

Surveys focusing on problematic areas within the preanalytical phase may not be suitable for regular schemes, but can be conducted as specialized preanalytical EQAS performed once (or a couple of times). In this way, the EQAS organizer can choose to study the most relevant issue present. The Nordic committee for External Quality Assurance Programmes in Laboratory Medicine (NQLM) carried out four different EQA interference surveys between 1999 and 2002 [28]. The aim was to assess the effects of

hemolysed, icteric and lipemic samples on some common clinical chemistry serum analyses. During 2014, a similar EQA survey was performed in the Nordic countries to receive updated information on analytical performance on different method platforms and how the individual laboratories handle hemolysed samples.
One of the major goals of the European Commission funded project SPIDIA (Standardization and improvement of generic preanalytical tools and procedures for in-vitro diagnostics) has been to develop evidence-based guidelines for the preanalytical phase of blood samples used for molecular testing [29]. The SPIDIA project has resulted in two pan-European EQAS focusing on the preanalytical phase of DNA and RNA-based analyses [30, 31] (Table 8.3.1). Blood samples were sent to the laboratories which performed DNA/RNA extraction of the material, the extracted material was returned, and the EQA organizer assessed the quality. The results of these surveys will provide the basis for performing pan-European EQAS and further provide the basis for implementation of evidence-based guidelines for the preanalytical phase of DNA and RNA analyses [29].

The National External Quality Assessment program in Portugal has performed two preanalytical surveys per year since 2007, consisting of a mix of a type I and type II schemes (Table 8.3.1). Through this program, a large variety of preanalytical issues have been investigated. Results from questionnaires sent to the participating laboratories, about for example blood collection and sample handling, are compared with national guidelines. Blood samples and case histories or medical request simulations have also been circulated for analysis/evaluation. For each survey a joint report is delivered with the overall results and pertinent comments [32]. In total 105 laboratories were enrolled during the whole period (2007–2011). Enrollment in the program decreased to less than 1/3 in 2011 compared to the first year, and only 14 % continued participation for four or more years. The response rate was higher when samples/case histories were sent for evaluation, compared to the surveys with only questionnaires.

Challenges regarding the use of real samples in preanalytical EQAS

The requirements for samples to be used in preanalytical EQAS are similar to the requirements for samples in analytical EQAS. The samples should be similar to native patient samples containing the actual interferences, and should be homogenous and stable during the survey period. Additionally, they should reflect actual poor preanalytical conditions like incorrect blood sampling and sample errors due to using tubes containing incorrect additives (e. g. serum or EDTA when citrated plasma is required) or inappropriate specimen preparation, centrifugation, aliquoting, pipetting or sorting. Mimicking real life situations is a challenge when producing sample material in a large scale. Only a limited set of preanalytical problems can be investigated by utilizing this method (Type II), and for some areas it seems natural to perform the EQAS once or on an irregular basis. One should also take into consideration the bias introduced if the participants know they are to receive a sample with an interfering substance (participating in a preanalytical scheme), and it might therefore be better to contaminate samples in a regular EQA scheme.

Type III: Registration of errors/adverse events

EQAS for the preanalytical phase can be performed by initiating registration of defined and standardized types of errors (or adverse events). The error rate (or the rate of adverse events) during a prespecified time interval should then be reported to the EQA organization. Systematic registering of incidents may point out bottle necks and indicate which areas are most error prone [33]. An accredited laboratory must have a system for registration of incidents and are required to establish QIs as a measure of the laboratory's performance (ISO 15189) [23]. Utilizing this system in an EQA program (e. g. the EQA provider suggest QIs to the laboratories; the laboratories thereafter report their QI data to the EQA provider, making comparison between laboratories regarding QI performances possible) is perhaps the best EQA program for monitoring preanalytical (and postanalytical) errors. The QIs found most appropriate and their acceptable "error rates" should also be harmonized between EQA organizers.

The participants should receive a feedback report showing their own QI performance compared to the results of all participants and to desirable quality specifications. It should also include historical data showing the development of the performance of the laboratory's QIs.

In 1989, the College of American Pathologists (CAP) started the Q-Probe program, carried out as time-limited studies lasting 1–4 months, to define critical performance measures in laboratory medicine, describe error rates on these measures and provide suggestions to decrease the error rate [34, 35]. In 1998, CAP developed this concept further and started the ongoing Q-Track program [35] (Table 8.3.1). Data is collected according to defined methods and time frames by utilizing standardized forms. CAP provides a quarterly "Performance Management Report" package which helps the laboratories to identify issues providing the opportunity for improvement and to monitor the effect of changes. So far, information on error rates from more than 130 worldwide inter-laboratory studies has been included in CAP's databases [35]. The error rate has been determined for almost all the important steps of the total testing process, QIs have been established, and suggestions for error reduction provided. Improvement can be quantified as exemplified by one of the most frequently occurring errors; missing wristbands, where the median error rate was reduced from more than 4 % to less than 1 % during four years [36]. Similarly, participants who participated for two years demonstrated a significant declining trend in sample rejection rate [37].

The Spanish Society of Clinical Chemistry and Molecular Pathology (SEQC) has since 1998 conducted two E QAS per year, registering rejections of samples and the causes for rejections (Table 8.3.1). A retrospective analysis of 10 surveys (2001–2005) summarized and evaluated the data and concluded that "specimen lost or not received" (37.5 %), "hemolysis" (29.3 %) and "clotted samples" (14.4 %), were the main causes for sample rejection [38]. The percentage of sample rejections was highest if the samples were collected in wards outside the central laboratory. The retrospective evaluation of the program resulted in a simplified scheme as some of the included variables (e. g. type of serum tubes [gel, silicone or no separator], type of anticoagulant employed [liquid or solid], extraction procedure [with or without vacuum] and material employed [glass or plastic device]) turned out to be irrelevant. The results from an EQAS performed by Australian chemical pathology laboratories, aiming at measuring transcription (defined as any instances where the data on individual request forms were not identical to the data entered into the laboratory's computer system, for example patient's identification, patient's sex and age, patient's ward location or address, tests requested and requesting doctor's identification) and analytical errors, have been summarized [39]. This study showed a large inter-laboratory variation in total error rates (5–46 %). The maximum error rates detected were 39 % and 26 % for transcription and analytical errors, respectively. The study concluded that there is a need to establish broader quality assurance programs and performance requirements for the preanalytical phase.

Several pilot EQA-schemes were performed during 2006–2009 throughout Australia and New Zealand, and the experiences from these surveys were used to make the current "Key Incident Monitoring and Management Systems Quality Assurance" (KIMMS QA) program [40] (Table 8.3.1). In this scheme, the participating laboratories are asked to register a subset of pre- and postanalytical (PAPA) incidents, which represents either the most frequent or the most serious incidents, or which represents incidents with the greatest opportunity for inter-laboratory benchmarking and improvement. The data from the participants is pooled to form a national frequency distribution of PAPA incidents, and each participating laboratory can compare their own results with this distribution. For 70 % of all reported incidents, the root causes require interaction between the laboratory staff and other health care workers. In the last KIMMS EQA survey, the overall PAPA incident rate was 1.22 % and the most frequent incident was inadequate patient or sample identification (0.28 %) [40].

As a part of a project to reduce errors in laboratory testing, the IFCC Working Group on Laboratory errors and patient safety (WG-LEPS) aimed to develop a series of QIs, specially designed for clinical laboratories [41, 42]. The aim of the project is to provide a common framework and to establish a set of harmonized QIs which should cover all steps of the TTP. The group concluded that a model of QIs managed as an EQAS can serve as a tool to monitor and control the pre-, intra- and postanalytical activities. The QIs suggested by WG-LEPS were evaluated in a study performed during 2009 and 2011. The results showed that QIs in the *analytical phase* improved much more than the corresponding for the *extra-analytical phases*, probably because improvement in the extra-analytical phases may be more complex requiring close cooperation between the laboratory staff and the health care workers outside the laboratory [43]. This finding indicates that measurement of errors alone will not reduce the error rates, but corrective actions which include cooperation, teamwork and firm follow-up on achievements are necessary.

In a recent study performed in a group of clinical laboratories in Catalonia (Spain), the results obtained for the QIs of key processes were analyzed over a five years period (2004–2008) [44]. The objective was to determine the QI evolution, identify processes requiring corrective measures, and obtain robust performance specifications. Different indicators were evaluated according to efficiency and safety. The average yearly inter-laboratory median value for the different indicators over the 5-year period was used as the desirable specification for the indicators. The values obtained were transformed to the Six Sigma scale and processes with sigma values ≥4, were considered well controlled [45]. The medians for most QIs were considered stable during the study period, however, if the sigma value was less than 4, the QIs were considered robust but subject to improvement. Examples of preanalytic processes outside the laboratory that could be improved were "Total incidences in test requests" (sigma 3.4) and "Patient data missing" (sigma 3.4). Within the laboratory "Undetected requests with incorrect patient name" obtained the lowest sigma value (2.9). The method used in this study could also be performed as a regular EQA scheme for the preanalytical phase.

The Norwegian Quality Improvement of Primary Health Care Laboratories (Noklus) and NKK have initiated a preanalytical EQA program of type III, where medical laboratories are asked to register preanalytical errors in samples received from primary health care (PHC). The laboratories are asked to register four different preanalytical errors in these samples during one month, together with some basic information about the laboratory, for example accreditation status and use of electronic request forms. This program is initiated to evaluate if intense educational effort to reduce preanalytical errors in GP offices, nursing homes and other primary care centres is effective. The program started in 2014 and is planned to continue with yearly registrations. Ninety-four of 97 invited hospital laboratories participated in 2014. The main objective is to establish an ongoing preanalytical EQA program of type III.

Challenges regarding the use of QIs managed as an EQAS

The EQA organization should develop QIs and distribute registration systems to the laboratories, and this may be a quite labor intensive process. Harmonization is of uttermost importance, since variation in the methods used to register the errors might influence the results and make comparison between laboratories difficult. Harmonization of QIs may be difficult on an international basis if laboratory practices differ between regions, but harmonization within local areas should be possible. A challenge for the participating laboratories is the extra resources (time and personnel) needed to register their own errors. It is extremely important to focus on improving the system and not focus on each individual, since this might lead to under-reporting of errors. Most errors in laboratory medicine are classified as blameless and represent the accumulation of a number of contributing events, indicating the need to modify systems as opposed to disciplining individuals [33, 46].

8.3.3 Conclusion

In this chapter, three different types of preanalytical EQAS are described, having somewhat different focus and different challenges regarding implementation. A combination of the three is probably necessary to be able to detect and monitor the wide range of errors occurring in the preanalytical phase. It might be a good idea to start with preanalytical EQA schemes nationally since they are easier to adapt to local conditions and results could be discussed on local users' meetings. Based on the results of these schemes, one could in cooperation with other countries develop international schemes aiming at harmonizing preanalytical guidelines and QIs. Preanalytical EQA schemes of type II will in many instances be size-restricted by the supply of EQA sample material, while type I and III might be more suitable for international cooperation through international EQA organizations (e. g. EQALM). Development of preanalytical EQA schemes and publication of the results should be encouraged since information of such schemes and their ability to improve preanalytical routines in the laboratories are scarce in the literature.

Acknowledgment: We thank Cornelia Schürer from INSTAND for useful contribution to this chapter.

References

[1] Blumenthal D. The errors of our way. Clin Chem 1997; 43:1305.
[2] Bonini P, Plebani M, Ceriotti F, Rubboli F. Errors in laboratory medicine. Clin Chem 2002; 48:691–8.
[3] Lapworth R, Teal TK. Laboratory blunders revisited. Ann Clin Biochem 1994; 31:78–84.
[4] Nutting PA, Main DS, Fischer PM, Stull TM, Pontious M, Seifert M, Jr. et al. Toward optimal laboratory use. Problems in laboratory testing in primary care. J Am Med Assoc 1996; 275:635–9.
[5] O'Kane M. The reporting, classification and grading of quality failures in the medical laboratory. Clin Chim Acta 2009; 404:28–31.
[6] O'Kane MJ, McManus P, McGowan N, Lynch PL. Quality error rates in point-of-care testing. Clin Chem 2011; 57:1267–71.
[7] Plebani M. Exploring the iceberg of errors in laboratory medicine. Clin Chim Acta 2009; 404:16–23.
[8] Plebani M. The detection and prevention of errors in laboratory medicine. Ann Clin Biochem 2010; 47:101–10.
[9] Plebani M, Carraro P. Mistakes in a stat laboratory: types and frequency. Clin Chem 1997; 43:1348–51.
[10] Rakha EA, Clark D, Chohan BS, El-Sayed M, Sen S, Bakowski L et al. Efficacy of an incident-reporting system in cellular pathology: a practical experience. J Clin Pathol 2012; 65:643–8.
[11] Stahl M, Lund ED, Brandslund I. Reasons for a laboratory's inability to report results for requested analytical tests. Clin Chem 1998; 44:2195–7.
[12] Witte DL, VanNess SA, Angstadt DS, Pennell BJ. Errors, mistakes, blunders, outliers, or unacceptable results: how many? Clin Chem 1997; 43:1352–6.

[13] Astion ML, Shojania KG, Hamill TR, Kim S, Ng VL. Classifying laboratory incident reports to identify problems that jeopardize patient safety. Am J Clin Pathol 2003; 120:18–26.
[14] Kalra J. Medical errors: impact on clinical laboratories and other critical areas. Clinical biochemistry 2004; 37:1052–62.
[15] Kalra J. Medical errors: an introduction to concepts. Clin Biochem 2004; 37:1043–51.
[16] Kalra J. Medical errors: overcoming the challenges. Clin Biochem 2004; 37:1063–71.
[17] Rattan A, Lippi G. Frequency and type of preanalytical errors in a laboratory medicine department in India. Clin Chem Lab Med 2008; 46:1657–9.
[18] Clinical and Laboratory Standards Institute (CLSI). Development and Use of Quality Indicators for Process Improvement and Monitoring of Laboratory Quality; Proposed Guideline. CLSI document GP35-P. Clinical and Laboratory Standards Institute, West Valley Road, Wayne, PA. 2009.
[19] Fraser CG, Kallner A, Kenny D, Petersen PH. Introduction: strategies to set global quality specifications in laboratory medicine. Scand J Clin Lab Invest 1999; 59:477–8.
[20] Gandhi TK, Kachalia A, Thomas EJ, Puopolo AL, Yoon C, Brennan TA et al. Missed and delayed diagnoses in the ambulatory setting: a study of closed malpractice claims. Ann Intern Med 2006; 145:488–96.
[21] Hickner J, Graham DG, Elder NC, Brandt E, Emsermann CB, Dovey S et al. Testing process errors and their harms and consequences reported from family medicine practices: a study of the American Academy of Family Physicians National Research Network. Qual Saf Health Care 2008; 17:194–200.
[22] Wahls T. Diagnostic errors and abnormal diagnostic tests lost to follow-up: a source of needless waste and delay to treatment. J Ambul Care Manage 2007; 30:338–43.
[23] International Organization for Standardization. ISO 15189: medical laboratories: particular requirements for quality and competence. Geneva, Switzerland: International Organization for Standardization, 2012.
[24] Plebani M, Sciacovelli L, Aita A, Chiozza L. Harmonization of pre-analytical quality indicators. Biochemia Medica 2014; 24:105–13.
[25] Bilic-Zulle L, Simundic AM, Smolcic VS, Nicolac N, Honovic L. Self reported routines and procedures for the extra-analytical phase of laboratory practice in Croatia - cross-sectional survey study. Biomed Med 2010; 20:64–74.
[26] Greaves R, Jolly L, Wollard G, Hoad K. Serum vitamin A and E analysis: comparison of methods between laboratories enrolled in an external quality assurance programme. Ann Clin Biochem 2010; 47:78–80.
[27] Hoad KE, Johnson LA, Woollard GA et al. Vitamin B1 and B6 method harmonization: Comparison of performance between laboratories enrolled in the RCPA Quality Assurance Program. Clin Biochem 2013; 46:772–6.
[28] The Nordic committee for External Quality Assurance Programmes in Laboratory Medicine (NQLM). Nordic interference study 2000 (Lipid) and 2002 (Hemoglobin and Biliubin). Available at: http://www.nkk-ekv.com/84240411. Acessed: November 4, 2013.
[29] Gunther K, Malentacchi F, Verderio P, Pizzamiglio S, Ciniselli CM, Tichopad A et al. Implementation of a proficiency testing for the assessment of the preanalytical phase of blood samples used for RNA based analysis. Clin Chim Acta 2012; 413:779–86.
[30] Malentacchi F, Pazzagli M, Simi L, Orlando C, Wyrich R, Hartmann CC, et al. SPIDIA-DNA: An External Quality Assessment for the pre-analytical phase of blood samples used for DNA-based analyses. Clin Chim Acta 2013; 424:274–86.
[31] Pazzagli M, Malentacchi F, Simi L, Orlando C, Wyrich R, Gunther K et al. SPIDIA-RNA: first external quality assessment for the pre-analytical phase of blood samples used for RNA based analyses. Methods 2013; 59:20–31.
[32] Correia FA, Gomes MA. Portuguese EQAS vision versus experience on pre-analytical phase (2007–2011) Instituto Nacional de Saúde Doutor Dr. Ricardo Jorge, IP 2011.

[33] Reason J. Safety in the operating theatre - Part 2: human error and organisational failure. Qual Saf Health Care 2005; 14:56–60.
[34] Lawson NS, Howanitz PJ. The College of American Pathologists, 1946–1996. Quality Assurance Service. Arch Pathol Lab Med 1997; 121:1000–8.
[35] Howanitz PJ. Errors in laboratory medicine: practical lessons to improve patient safety. Arch Pathol Lab Med 2005; 129:1252–61.
[36] Howanitz PJ, Renner SW, Walsh MK. Continuous wristband monitoring over 2 years decreases identification errors: a College of American Pathologists Q-Tracks Study. Arch Pathol Lab Med 2002; 126:809–15.
[37] Zarbo RJ, Jones BA, Friedberg RC, Valenstein PN, Renner SW, Schifman RB et al. Q-tracks: a College of American Pathologists program of continuous laboratory monitoring and longitudinal tracking. Arch Pathol Lab Med 2002; 126:1036–44.
[38] Alsina MJ, Alvarez V, Barba N, Bullich S, Cortes M, Escoda I et al. Preanalytical quality control program - an overview of results (2001–2005 summary). Clin Chem Lab Med 2008; 46:849–54.
[39] Khoury M, Burnett L, Mackay MA. Error rates in Australian chemical pathology laboratories. Med J Aust 1996; 165:128–30.
[40] Quality Assurance Scientific and Education Committee (QASEC) of the Royal College of Pathologists of Australasia (RCPA). The Key Incident Monitoring & Management Systems (KIMMS). Available at: http://dataentry.rcpaqap.com.au/kimms/ Accessed November 4, 2013.
[41] Sciacovelli L, O'Kane M, Skaik YA, Caciagli P, Pellegrini C, Da Rin G et al. Quality Indicators in Laboratory Medicine: from theory to practice. Preliminary data from the IFCC Working Group Project "Laboratory Errors and Patient Safety". Clin Chem Lab Med 2011; 49:835–44.
[42] Sciacovelli L, Plebani M. The IFCC Working Group on laboratory errors and patient safety. Clin Chim Acta 2009; 404:79–85.
[43] Sciacovelli L, Sonntag O, Padoan A, Zambon CF, Carraro P, Plebani M. Monitoring quality indicators in laboratory medicine does not automatically result in quality improvement. Clin Chem Lab Med 2012; 50:463–9.
[44] Llopis MA, Trujillo G, Llovet MI, Tarres E, Ibarz M, Biosca C et al. Quality indicators and specifications for key analytical-extranalytical processes in the clinical laboratory. Five years' experience using the Six Sigma concept. Clin Chem Lab Med 2011; 49:463–70.
[45] Westgard JO, Westgard SA. The quality of laboratory testing today: an assessment of sigma metrics for analytic quality using performance data from proficiency testing surveys and the CLIA criteria for acceptable performance. Am J Clin Pathol 2006; 125:343–54.
[46] Thomas EJ, Sherwood GD, Helmreich RL. Lessons from aviation: teamwork to improve patient safety. Nurs Econ 2003; 21:241–3.
[47] External quality Control of diagnostic Assays and Tests (ECAT). Available at: http://www.ecat.nl/. Acessed: November 4, 2013 and http://www.instandev.de Assessed January 2015.
[48] Aarsand AK, Villanger JH, Stole E, Deybach JC, Marsden J, To-Figueras J et al. European specialist porphyria laboratories: diagnostic strategies, analytical quality, clinical interpretation, and reporting as assessed by an external quality assurance program. Clin Chem 2011; 57:1514–23.
[49] Kristensen GBB, Aakre KM, Kristoffersen AH, Sandberg S. How to conduct External Assessment Schemes for the pre-analytical phase? Biochemia Medica 2014; 24:114–22.

9 Annex

Samples and Stability of Analytes

The following table was developed over the past 20 years and will appear together with the recommendations on preanalytics in its 7th edition in German and in its 4th edition in English in 2015 (1, 2). With the collaboration of 40 corresponding members and industry representatives, the information on all measurable analytes regarding the use of anticoagulants and stability in blood, plasma/serum at different temperatures were defined according to the criteria described in Chapter 6.1. Information provided by industry representatives for the different providers of reagents are denoted with Greek symbols. If no company is named, all available representatives agreed that this material could be used.

Literature for all data given is found in the respective publications (1, 2).

Key for tables

⊕ Recommended sample
\+ Can be used without changes of result
(+) Can be used with limitations (see comments, in case of citrated plasma this indicates the need to consider dilution by citrate (3).
– Not recommended
Decreased (↘) or increased (↗) results may be obtained in comparison to recommended samples.
Blank field means no data was found in literature.
Greek letters refer to the information provided by diagnostic companies.

Stability and half-life times
min = minute(s)
h = hour(s)
d = day(s)
w = weak(s)
m = month(s)
y = year(s)

Information provided by Diagnostic Companies

α: **ORTHO-Clinical Diagnostics**; Vitros Systems
β: **Abbott**; Axsym, Architect
γ: **Roche Diagnostics**; Roche/Hitachi, Elecsys®,Modular
γγ: **Roche Diagnostics**; Cobas INTEGRA®
ε: **Siemens Healthcare Diagnostics**; Dimension®, BN Systems, Stratus CS
μ: **Siemens Healthcare Diagnostics**; ADVIA Centaur/ACS 180
κ: **Siemens Healthcare Diagnostics**; Immulite Modular
λ: **Bio-Rad**

δ: **Beckman-Coulter**; AU-Systems, UniCel DxC and Dxl, Immage 800, Access

σ: **Siemens Healthcare Diagnostics**; Enzygnost

1. Guder WG, da Fonseca-Wollheim F, Heil W, Schmitt Y, Töpfer G, Wisser H, Zawta B. Die Qualität Diagnostischer Proben. 7th ed. Heidelberg: BD – Diagnostics, 2012.
2. Guder WG, Fiedler M, da Fonseca-Wollheim F, Schmitt Y, Töpfer G, Wisser H, Zawta B. The Quality of Diagnostic Samples. 4th completely revised ed. Oxford: BD-Europe 2015.
3. Lammers M. Dilution of citrated plasma. Eur J Clin Chem Clin Biochem 1996; 34:369.

9.1 Blood

Analytes	Samples: Serum	Heparinate Plasma	EDTA Plasma	Citrated Plasma	Whole blood: Hep	Whole blood: EDTA	Whole blood: Citrate	Stability: Biological half-life	Stability in blood at room temperature	Stability in serum/plasma −20 °C	4–8 °C	20–25 °C	Stabilizer	Remarks/Comments
Acetaminophen														see Paracetamol
Acetylsalicylate	+	+ β	+ β	(+) β				15–30 min						
α_1-Acid glycoprotein (orosomucoid)	+	+ γ, δ, ε	+ γ, γγ, ε, −δ	(+)					11 d 4 w (2–8 °C)	1 y	5 m	5 m		
Adenovirus antibodies	+		(+)											Complement fixation test, ELISA IgG, IgM
Alanine aminotransferase (ALAT, ALT)	+	+	+	(+)				47 h	4 d ↘	7 d	7 d	3 d		
Albumin colorimetric	+	+*	(+) ↘, + δ	(+)				3 w	2–6 d 14 d (2–6 °C)	4 m	5 m	2.5 m		*Bichromatic assay recommended for colorimetric assay
nephelometric	+	+ ε	+ ε					3 w	6 d	3 m	1 w	4 h		
Aldosterone	+	+	⊕					min	1 d ↘	4 d	4 d	4 d	EDTA	
Alkaline phosphatase,														
– total	+↗	⊕	–	(+)				3–7 d	4 d ↘	2 m	7 d	7 d		EDTA binds essential cofactor zinc
– bone isoenzyme	+	+	–	(+)↘				9–18 h	4 d	1 m (1 m)	7 d	7 d		
Aluminium	–	–	–	–					7 d	1 y	2 w	1 w		Special tube needed
Amikacin	+	+	+ β	(+) β				30 min–3 h		2 w	7 d	2 h		
Amiodarone	+	+	+					4 h–25 d		1 w	1 w	1 d		HPLC
Amitriptyline	+	+	+					17–40 h				1 d		HPLC

Analytes	Samples: Serum	Heparinate Plasma	EDTA Plasma	Citrated Plasma	Whole blood: Hep	Whole blood: EDTA	Whole blood: Citrate	Biological half-life	Stability: Stability in blood at room temperature	Stability in serum/plasma: −20 °C	Stability in serum/plasma: 4–8 °C	Stability in serum/plasma: 20–25 °C	Stabilizer	Remarks/Comments
Ammonia (NH_4^+)	–↗	(+) ↗	⊕	–	+			min	15 min in EDTA↗	3 w	3 h	15 min	Serin 5 mmol/L + borate 2 mmol/L	Do not use ammonium heparin. Contamination by sweat ammonia.
Amphetamines	+	+	+											
Amylase														
– pancreatic	+	+	+	(+)				9–18 h	4 d ↘	1 y	1 m	7 d		*Possible decrease of the activity by Mg and Ca binding at > 25 °C
– total	+	+	+	(+)*				9–18 h	4 d ↘	1 y	1 m	7 d		
Amyloid A (SAA)	+	+								3 m ε −25 °C	8 d ε	3 d ε		
Androstendione	+								1 d ↘	1 y	4 d	1 d		
Angiotensin converting enzyme (ACE)	+	+	–	–						1 y	7 d	1d		
Anticonvulsive drugs	+													See carbamazepine, ethosuccimide, phenobarbital, phenytoine, valproic acid
Antimitochondrial antibodies (AMA)	+									1 m	7 d	1d		
Anti-Muellerian hormone	+ δ	+ δ								>2 m	6 d	1 d		2 freezing thawing cycles possible
Antineutrophil cytoplasmic antibodies (ANCA)	+									1 m	7 d	1d		

Antinuclear antibodies (ANA)	+								1 m	7 d	1d	
Antiphospholipid antibodies	+								1 m	2–3 d	1 d	
Antistaphylolysine	+	+	+						6 m	2 d	2 d	
Antistreptodornase B	+								3 m	8 d		
Antistreptokinase	+											
Antistreptolysine	+	+ β, γ, δ, −γγ	+ β, γ, γγ, δ						6 m	8 d	2 d	
Antithrombin												
– functional	–	–	–	⊕		+*	30 h	8 h	1 m	2 w	2 d	*Test by Pharmacia-Upjohn
– immunochemical			+ δ, ε	(+) δ, ε			40–135 h	2 d**	1 y	8 d		**After centrifugation
α_1- Antitrypsin	+	+	+ β, γ, γγ	(+) β, −γ, γγ				11 d 7 w (2–6 °C)	3 m	5 m	3 m	EDTA and citrate ↘
APC resistance												
– functional screening test	–	–	–	⊕				30 min	6 m (−70 °C)	3 h	3h	Centrifuge within 30 min
– genotyping factor V Leiden					⊕	⊕		1 w				
Apolipoproteins AI, AII, B	+↗	+	⊕	(+)					3 m	8 d	1 d	
Apolipoprotein CIII	+↗	+	⊕	(+)					1 m	1 m	1 m	
Apolipoprotein E	+		+					1 d	3 m	8 d		
ApoE-genotyping					⊕			1 w (4–8 °C)	3 m	1 w		Stability of $ApoE_2$ > $ApoE_4$ > $ApoE_3$
Aspartate aminotransferase (ASAT, AST)	+↗	⊕	+, −α, δ↘	(+)			12–14 h	7 d ↘	3 m	7 d	4 d	

Analytes	Samples: Serum	Heparinate Plasma	EDTA Plasma	Citrated Plasma	Whole blood: Hep	Whole blood: EDTA	Whole blood: Citrate	Stability: Biological half-life	Stability in blood at room temperature	Stability in serum/plasma: −20 °C	4–8 °C	20–25 °C	Stabilizer	Remarks/Comments
Aspergillus														
– antigen detection	+													
– antibody	+													
Atrial natriuretic peptide (ANP)			+*					8.8 min	Unstable		7 h	1 h	*Aprotinin, phospor-amidon	Centrifuge at 4 °C
– propeptide (proANP)			+					1 h	6 h	4 w	3 d	6 h		
Barbiturates	+	+						50–120 h	2 d	6 m	6 m	6 m		See also phenobarbital
Bartonella spp. antibodies	+													
Batroxobin time	–	–	–	⊕						1 m	4 h	4 h		Avoid heparinate contamination ↗
Benzodiazepine	+	+						25–50 h	<1 d		5m↘	5m↘		See also diazepam, flunitrazepam, nitrazepam
Bicarbonate	+	+	–		⊕			min	Unstable ↘ (30 min – 2 h at +4 °C)	1 m	7 d	1 d *	Keep tube closed	*1 h after opening the tube, see also blood gases
Bilirubin														
– conjugated	+	+	+	(+)				h	Unstable	6 m	7 d	2 d		Darkness required
– total (also in newborns)	+	+	+	(+)				17 d	↘,↗	6 m	7 d	1 d		when stored > 8 h
Biotin					⊕								deep frozen	UV-light sensitive

Blood cell surface markers (immunocytometry)					+	+		CD4 1d in heparinised blood					See also lymphocyte subtypes
Blood gases (CO_2, O_2, pH)					⊕		min	< 15 min ↘ pO_2 < 30 min pH, pCO_2 < 60 min on ice		2 h *		*In hep-arinised blood and closed tubes	Use closed gas tight tubes or capillaries
Bordetella pertussis antibodies	+												
Borrelia burgdorferi anti-bodies (Lyme disease)	+	+ σ	+ σ	+ σ									ELISA, Western blot
Brain natriuretic peptide (BNP)	+	+ μ	⊕			⊕	13–20 min	4–5 h	5 d –8 m	1 d	4 h	EDTA	
NT- pro BNP	+	+	+				2 h	3 d	1 y	5 d	3 d		
Brucella antibodies (Brucellosis)	+												
C_1-esterase inhibitor													
– functional assay	+		+	(+) ε					1 m	2 d	6 h		Stabilise plasma by freezing
– immunochemical	+			+ ε					1 y	8 d			
CA 125 (Cancer antigen 125)	+	+ α, γ, μ	+ α, γ, μ	(+) γ			5–10 d	2 d ↘	3 m	5 d	3 d		
CA 15–3, (Cancer antigen 15–3)	+	+ α, γ, – μ	+ α, β, γ, – μ	(+), – γ			5–7 d	7 d	3 m	7 d	7 d		
CA 19–9 (Carbohydrate antigen 19–9)	+	+	+ γ, μ	(+) –γ			4–9 d	7 d ↘	3 m	30 d	7 d		
CA 72–4 (Cancer antigen 72–4)	+	+ γ	+ γ	(+) –γ			3–7 d	3 d ↘	3 m	30 d	7 d		
Cadmium (Cd)	–		⊕	–			10–35 y	1 d in trace element tube					Special tube (Cd released from red stopper)

Analytes	Samples: Serum	Heparinate Plasma	EDTA Plasma	Citrated Plasma	Whole blood: Hep	Whole blood: EDTA	Whole blood: Citrate	Stability: Biological half-life	Stability in blood at room temperature	Stability in serum/plasma: −20 °C	4–8 °C	20–25 °C	Stabilizer	Remarks/Comments
Calcitonin	+	+	⊕					min	4 h stabilised*	1 y	1 d	4 h	*Aprotinin 400 KIU/mL	
Calcium														
– total	+	+	–↘	–↘	+			h	2 d↘	8 m	3 w	7 d	*Use calcium-titrated heparin	pH-dependent
– ionized (free)	–	(+)	–↘	–↘	⊕*			min	15 min↗ 1 d*		2 h	3 d**		**Stable in gel tubes for 25 h & 72 h after centrifugation in closed tube
Campylobacter jejuni/fetus antibodies	+													
candida albicans														
– antibodies	+													blood culture bottle
– antigen detection	+													
Carbamazepine	+	+ α↗,	+ β, γ	(+) α↗, β, δ, – γ				10–25 h	2 d	1 m	7 d	5 d		10 % higher results in plasma (α), unstable in gel separator tubes, but stable in SST II tubes up to 48h and Sarstedt plasma tubes
Carbohydrate deficient transferrin (CDT)	+	(+)						5–10 d	3 d	3 m	2 w	1 d		Method-dependent

Carcino-embryonic antigen (CEA)	+	+	+ α↘, β, γ, μ	+ γ			2–4 d	7 d	6 m	7 d	2 d		EDTA reduces by 13 % (α)
Cardiolipin antibody	+								1 m	2–3 d	1d		
Catecholamines (epinephrine, norepinephrine)	–	⊕	(+)	–			3–5 min	1 h if not stabilised	1 m 6 m stabilised	2 d	1 d	Glutathione 1.2 g/L +EGTA	EGTA plasma to be separated within 15 min and frozen at –20 °C
Ceruloplasmin	+	+	+, –γγ				4 d		1 y	2 w	8 d		
Chlamydia antibodies (C. trachomatis, C. pneumoniae)	+		(+)							7 d	5 d		DNA-PCR possible after 3–4 d at room temperature
Chloramphenicol	+	+ β	+	(+)			2–5 h						
Chloride	+	+	–	–	+		1 h	1 d ↘	y	4 w	7 d		
Cholesterol	+	+	+, – α, γ, δ	(+)				2–7 d ↗	3 m	7 d	7 d		
Cholesterol, HDL	+	+	+ β, γ, δ, – – α					2 d ↗	3 m	7 d	7 d		3 %lower cholesterol observed in EDTA plasma due to osmotic dilution effect
Cholesterol, LDL	+	+ β,γ,δ	+β, – γ, δ –					1 d ↘	3 m	7 d	7 d		
Cholinesterase, including dibucain number	+	+	+, – γ, δ				10 d*	7 d ↘	1 y	1 y	1 y		*Shorter in heavily diseased patients
Ciclosporin	–	–	–	–		⊕	10–27 h	13 d	3 m*	3 w*	3 w*	EDTA	*Stored in haemolysate
Circulating immunocomplexes (CIC)	+							4 h	1 y	8 h	4 h		
Clostridium tetani toxine antibodies	+												

Analytes	Samples							Stability						
	Serum	Heparinate Plasma	EDTA Plasma	Citrated Plasma	Whole blood			Biological half-life	Stability in blood at room temperature	Stability in serum/plasma			Stabilizer	Remarks/Comments
					Hep	EDTA	Citrate			−20 °C	4–8 °C	20–25 °C		
Clozapin	+	+	+						2 h	4 d	4 h			
Coagulation factors														
Factor II	–	–	–	⊕				41–72 h	1 d	1 m		6 h		
Factor V	–	–	–	⊕				12–15 h	4 h	1 m	2 d	6 h		Centrifuge at 4 °C
Factor VII	–	–	–	⊕				2–5 h	1 d	1 m	Unstable	6 h		
Factor VIII	–	–	–	⊕				8–12 h		2 w	4 h	3 h		
Factor VIII R: Ag	–	–	–	⊕				6–12 h		6 m	7 d*	7 d*	*Sodium azide	Five freezing-thawing cycles are possible
Factor VIII R: Co				⊕				6 h		6 m	2 w*	2 d	*Sodium azide	
Factor IX	–	–	–	⊕				18–30 h	1 d	1 m		6 h		
Factor IX: Ag	–	–	–	⊕					1 d					
Factor X	–	–	–	⊕				20–42 h	1 d	1 m		6 h		
Factor XI	–	–	–	⊕				3–4 d	1 d		Unstable	6 h		
Factor XII	–	–	–	⊕				50–70 h	1 d		Unstable	6 h		
Factor XIII	–	–	–	⊕				4–5 h		1 m		4 h		
Cocaine	+	+	–						< 10 min	4 d	30 d	<30 min	Fluoride, pH 5	Cocaine is converted in vitro into its metabolites.
Benzoyl ecgonin									5 d		5 d	5 d		
Ecgonin methyl ester									10 d		10 d	10 d		

Cold agglutinins											Keep whole blood at 37 °C (water bath)
Complement C3	+	+	+, −γγ	(+)	min	1 d 2 d (C3c) (2–6 °C)	8 d	8 d	4 d	Nafomo- stat	Dependent on antibody, during storage C3c ↗ C3 ↘
Complement C4	+	+	+	(+)	12 h–1 d	1 d 2 d (2–6 °C)	3m	8 d	2 d	Nafomo- stat	During storage C4↘,C4c↗
Copper	+	+	−	−		7 d	y	2 w	2 w		Special tube to avoid contamination
Corticotropin (ACTH)		+	⊕		min	1–4 h ↘	6 w	3 h 2 d*	1 h 1 d*	Aprotinin 400–2000 KIU/mL Mercapto-ethanol 2μL/mL	Prevent binding to glass tubes by using plastic for storage *in EDTA-plasma
Corticotropin releasing hormone	+↘	+	⊕					1 d	11–18 h		
Cortisol	+	+ α, μ	+ α, γ, μ	+ γ	1 h	7 d	3 m	7 d	7 d		11 % less in EDTA (α)
Corynebacterium diphtheriae toxine antibodies	+										
Coxiella burnetii-antibodies (Q-fever)	+										
Coxsackie virus antibodies	+										
C peptide	+	+	⊕		30 min	6 h	2 m	5 d	5 h	EDTA	Fluoride,oxalate also possible(β)
C-reactive protein (CRP)	+	(+)* + α, γ, γγ, δ, ε	+ α, δ, ε, (+), + γ −γ, γγ		2–4 h	3 w (2–6 °C), 1 d	3 y	2 m	11 d		*Patient-dependent lower results

Analytes	Samples: Serum	Samples: Heparinate Plasma	Samples: EDTA Plasma	Samples: Citrated Plasma	Whole blood: Hep	Whole blood: EDTA	Whole blood: Citrate	Biological half-life	Stability: Stability in blood at room temperature	Stability in serum/plasma: −20 °C	Stability in serum/plasma: 4–8 °C	Stability in serum/plasma: 20–25 °C	Stabilizer	Remarks/Comments
Creatinine	+	+	+	(+)				3 min	2 d ↗	3 m	7 d	7 d		
Creatine kinase (CK)	+	+	+ β, δ −γ	(+)				18 h	7 d ↘	1 m	1 m	4 h 1d*	Darkness	CK-BB not stable, *in Vacutainer SST II serum and PST II plasma tubes
Creatine kinase MB														
– enzyme activity	+	+, −α	+ γ, δ,	(+) δ				12 h	7 d ↘	1 y	7 d	2 d	SH	
– molecular mass	+	+ β, γ, δ, − μ	+ β, δ, − γ, μ	− γ				12 h	7 d ↘	4 w	7 d	2 d	reagent	
C-terminal crosslinks (CTX) (ß-Cross-Laps™)	+	+	⊕						8 h 7 d (Cross-Labs)	3 m	7 d	8h 2 d*	pH 8.0, *EDTA	Stability pH-dependent
Cyclosporin														see ciclosporin
Cyclic citrullinated peptide antibodies (CCP antibodies)	+									1 y	7 d	1 d		
Cytokeratine fragment 21–1 (CYFRA 21–1)	+	+ γ	+ γ	(+) γ				2–5 h	7 d	6 m	1 m	7 d		
Cystatin C	+	+	+					min	3 d	2 y	1 w	7 d		More stable in EDTA
Cytokines			⊕						2 h (heparinised blood)		2 d			see also Tumor necrosis factor (TNF)
– IFN−α, IFN−γ, −1α		+↗												
– IL−6	−↘	+							1 h (EDTA)					
– IL−1β, sIL−2R, sIL−6R											12 h ↘			

Cytomegalovirus														
– antigen detection (pp65)						⊕								
– DNA amplification						⊕								
– CMV antibodies	+	+ β, σ	+β,σ	(+) β, σ										
D-Dimer	(+)	+	–	⊕				6–8 h	1 w	6 m	4 d	8 h		
Dehydroepiandosterone sulfate (DHEA-S)	+	+β, γ	+β, γ	+ γ				7–9 h	2 d ↘	y	2 w	1 d		
Dengue virus antibodies	+													
Diazepam	+	+	+					25–50 h			5 m	5 m		
Differential leucocyte count	–	–	–	–		⊕	+	2 h–3 y	2 h–7 d*				Dry blood smear stable	K_3-or K_2-EDTA: Stability temperature- and instrument-dependent *Prepare blood smear within 3 h after sampling. Do not store EDTA blood in refrigerator
– Band neutrophiles									2–12 h					
– Segmented neutrophiles								6–7 h	3–12 h					
– Eosinophiles									12 h–6 d					
– Basophiles									2 h–2 d					
– Monocytes									2–12 h					
– Lymphocytes								1.5–3 y	3 h–7 d					
Digitoxin	+	+	+					6–8 d	1–2 d	6 m	3 m	2 w		
Digoxin	+	+	+	(+) β				1–2 d	1–2 d	6 m	3 m	2 w		
Disopyramide	+	+	+	(+)				4–9 h		5m	2 w			
DNA analysis by polymerase chain reaction amplification (PCR)	(+)	–*, +	+		–*	⊕	+		1 w					*Heparin inhibits Taq polymerase and restriction enzymes LiCl 1.8 mol/L eliminates this error
Dopamine		+	+					3–5 min		1 m	2 d	1 d		
Doxepin	+	+	+	(+)					2 h		4 d			

	Samples								Stability					
Analytes	Serum	Heparinate Plasma	EDTA Plasma	Citrated Plasma	Whole blood Hep	Whole blood EDTA	Whole blood Citrate	Biological half-life	Stability in blood at room temperature	Stability in serum/plasma −20 °C	Stability in serum/plasma 4–8 °C	Stability in serum/plasma 20–25 °C	Stabilizer	Remarks/Comments
Echinococcus spp. antibodies	+													
ECHO virus antibodies	+													
Elastase						+								see pancreatic elastase
Electrophoresis, protein-; see also Lipoprotein electrophoresis	⊕	(+)*								3 w	3–7 d	1 d		Fibrinogen to be considered when using heparinate plasma, may be eliminated by fibrin precipitation
Endomysium antibodies	⊕									m–ys	7 d	1 d		
Entamoeba histolytica antibodies	+													
Enterovirus antibodies	+													
Epstein Barr virus														
– heterophilic antibodies (Paul Bunnel test)	+		(+)											
– anti-EBNA, -VCA, -EA);	+	+ σ	+ σ	+ σ										IgG, IgM, IgA; ELISA, Western Blot
Erythrocyte count					(+)	⊕	(+)	2 m	4 d 7 d (4–8 °C)					

Erythrocyte sedimentation rate (ESR)					⊕		2 h	–	–	–		1 part citrate, 4 parts blood
Erythropoietin	+	+	+			4–11 h	6–24 h	5 m		2 w		Shipped frozen
Estradiol (E_2)	+	(+) γ, μ, +α, β	(+) γ, μ, + α, β	(+) γ			1 d	1 y	3 d	1 d		
Estriol (E_3)	+	+						1 y	2 d	1 d		
Ethanol	+	⊕	+ β, γ, γγ, δ	(+) β, δ	+*	2–6 h	2 w ↗ ↘**	6 m	6 m	2 w	EDTA/ Heparin	*10 g/L NaF recommended to stabilise **Evaporation, use closed tubes
Ethosuximide	+	+	+			30–60 h		5 m	4 w			
Fatty acids, non esterified (NEFA)	+	(+) ↗*	(+) ↘			2 min	30 min↗*	1 m	12 h	30 min		*Activation of lipase by heparin or clot activator. Freeze serum/ plasma immediately
Ferritin	+	+, –δ	(+)*β, γ, γγ, δ	(+) –γ, γγ			1 d	1–2 y	7 d	7 d		*Method-dependent
$α_1$-Fetoprotein (AFP)	+	+	+	(+)		2–3 d	7 d	3 m	7 d	3 d		
Fibrin(ogen) degradation products (FDP)	(+)*	–	–	(+)**			unstable↗↗	1 m	1 d	3 h	10 U thrombin and 150 KIU aprotinin / mL blood	*Special tube **Aprotinin or soybean trypsin inhibitor
Fibrin monomers	–	–	–	⊕		<1h	1 d	3 m	1 d	2 h		
Fibrinogen												Stability method-dependent
– Clauss	–	–	–	⊕		4–5 d	1 w	1 m	1–7 d	1–7 d		
– immunochemical	–	+	–	⊕		4–5 d	1 w	1 m	7 d	7 d		

Analytes	Samples							Stability						
	Serum	Heparinate Plasma	EDTA Plasma	Citrated Plasma	Whole blood Hep	Whole blood EDTA	Whole blood Citrate	Biological half-life	Stability in blood at room temperature	Stability in serum/plasma −20 °C	Stability in serum/plasma 4–8 °C	Stability in serum/plasma 20–25 °C	Stabilizer	Remarks/Comments
Fibrinopeptide A	–	–	–	⊕				3 min			2 h			
Flunitrazepam	+								< 1 d*					*Store protected from light
Folate	+	+, –μ	+β, –μ	(+) β				min	30 min ↘, 5 d (2–8 °C)	8 w	1 d	30 min	Ascorbate 2g/L	Haemolysate, prepared by 0.5 mL blood + 4.5 mL ascorbic acid (2 g/L). Na-heparin interferes with Axsym-Test (β).
– in erythrocytes					+ μ	+β, δ								
Follitropin (FSH)	+	+	+	(+) –γ				min	7 d↘	1 y	2 w	2 w		
Francisella tularensis-antibodies (tularemia)	+													
Free light chains (κ, λ) of immunoglobulins	+	+ δ, ε, –γ	+ δ, ε	(+) γ				2–6 h		κ: 6 m λ: 2 m	κ:1 m λ: 1 m	κ: 7 d λ: 7 d		
Fructosamine	+	+	+					12 d	12 h↗	2 m	2 w	3 d		
Galactose 1-p-uridyl-transferase (galactosemia sceening)						+*								*In newborns drop of blood on filter paper, analysed in erythrocytes
Gastrin	+	⊕*	+	(+)					2 h			1 w*	*With aprotinin 2000 KIU/mL	Freeze serum as soon as possible

Gastrin releasing peptide (GRP), pro GRP	+	⊕	⊕ β				2 min 1 d	1 h serum, 3 h plasma	7d	1d	3–8 h*		*Plasma 8 h, serum 3 h
Gentamicin	+	+	+	(+) β			0.5–3 h (<30 y of age) 1.5–15 h (>30 y of age)	4 h	4 w	4 w	4 h 2 d*		*Vacutainer SST II tube
Glucagon	+	+	⊕					Unstable		1.5 d	30 h	Aprotinin 500–2000 KIU/mL	Stabilise
Glucose				– ↘								Fluoride, mono-iodo-acetate, mannose, acidity	*Stabilised haemolysate and plasma **EDTA, citrate, fluoride or acid tube
– capillary	– ↘	– ↘	–↘	– ↘	(+)	⊕	min	10 min↘,	1 d*	7 d*	2 d*		
– venous	– ↘	– ↘	– ↘ +**				min	10 min↘, 2 h**	1 d*	7 d*	2 d*		
Glutamate decarboxylase autoantibodies (GADA)	+		(+)										Add 25 mmol/L CaCl to EDTA plasma, centrifuge 10 min at 10 000 *g*
Glutamate dehydrogenase	+	+	+				18 h		4 w	7 d	7 d		
Glutamate oxaloacetic transaminase (GOT)													See aspartate aminotransferase
Glutamate pyruvate transaminase (GPT)													See alanine aminotransferase
γ-Glutamyl transferase (γ-GT)	+	+	(+) ↘, + α, δ	(+) ↘, –γγ			3–4 d	1 d ↘	y	7 d	7 d		

	Samples								Stability					
Analytes	Serum	Heparinate Plasma	EDTA Plasma	Citrated Plasma	Whole blood Hep	Whole blood EDTA	Whole blood Citrate	Biological half-life	Stability in blood at room temperature	Stability in serum/plasma −20 °C	Stability in serum/plasma 4–8 °C	Stability in serum/plasma 20–25 °C	Stabilizer	Remarks/Comments
Glycated albumin														See fructosamine
Gold	+													
Hematocrit					+	⊕			1 d 4 d (4–8 °C)		4 d*		*EDTA-blood,	K_2-superior to K_3-EDTA
Hemoglobin A_{1c}					+	⊕	+	2 m	3 d (EDTA-blood)	6 m* 23 y at −70 °C	7 d*	3 d*		*Haemolysate
Hemoglobin F (HbF)						⊕		2 m						
Hemoglobin (whole blood)						⊕		2 m	4 d		7 d*	4 d*		*EDTA- blood
Hemoglobin (plasma)	(+) ↗	⊕	⊕	(+)										Haemolysis during clotting
Hanta virus	+													
– antibodies														
– RNA amplification					–	⊕	–							
Haptoglobin	+	+	+	(+) γ				3.5–4 d	8 d 7 w (2–6 °C)	3 m	8 m	3 m		
HbeAg	+	+β	+ β	(+) β				9 d			7 d			also possible from ACD-B, CPDA-1, CPD and Na-oxalate-tubes (β)
HbsAg	+	+ α, δ, σ	+ α, δ, σ	(+) α, σ ↘ δ						1 y	2 w	7 d		
Helicobacter pylori antibodies	+	+ σ	+ σ	(+) σ										

Heparin (anti Xa)				⊕							4 h	
Heparin associated thrombopenia; HEPA test	+					+		1 d		4 w		citrated blood and serum needed
Hepatitis antibodies – anti-HAV	+	+ β, δ, σ	+ β, δ, σ	(+) β, δ, σ			9 d		1 y	4 w	5 d	prevent repeated freezing and thawing of sample
– anti-HAV IgM	+	+ α, σ	+ α, σ	+ α, σ					1 y	4 w	5 d	
– anti-HBsAg	+	+ α, β, σ	+ β, σ	+ α, β, σ					1 y	4 w	7 d	
– anti-HBc	+	+ α, β, δ, σ	+ α, δ, σ	(+) α, β, δ, σ					1 y	4 w	7 d	
– anti-HBe	+	+ β, σ	+ β, σ	(+) β, σ					1 y	4 w	5 d	
– anti-HCV	+	+ α, β, δ	+ α, β, δ	+ α, – β, δ					1 y	4 w	7 d	
– anti-Hepatitis D	+	+β	+ β	(+) β								
– anti-Hepatitis E	+											
Hepatitis B virus DNA	+		+							6 h		
Hepatitis C virus – RNA amplification	+		+					6 h γ		3 d γ		
Hepatitis D virus – RNA amplification	+		+									
Hepatitis E – RNA amplification	+		+									
Herpes simplex 1 or 2 virus antibodies	+	+ σ	+ σ	+ σ								
HHV 6 antibodies (human herpes virus 6)	+											
HHV 6-, 7-, 8-DNA amplification					⊕							

Analytes	Samples								Stability					
	Serum	Heparinate Plasma	EDTA Plasma	Citrated Plasma	Whole blood			Biological half-life	Stability in blood at room temperature	Stability in serum/ plasma			Stabilizer	Remarks/Comments
					Hep	EDTA	Citrate			−20 °C	4–8 °C	20–25 °C		
HI virus 1 – (provirus) DNA amplification						⊕						7 d		Several freezing/ thawing cycles possible
– RNA amplification			⊕					5–14 d	7 d↘		5 d γ	1–2 d		
HI virus 1 and 2 antibodies	+	+ α, β, σ	+β, δ, σ	(+) α β, δ, σ							4 w	5 d		
HIV, viral load					+	⊕	+	5–14 d	7 d					
HLA-ABC typing					⊕									Ammonium heparinised blood
HLA- B27					+	⊕			1 d			1 d	Citrate-phosphate-dextrose (CPD)	
HLA DR typing						⊕								
Homocysteine	+ ↗	+	⊕	(+)		⊕ λ			1 h↗ 6 h (2–6 °C)	4 y	4 w	4 d	Sodium fluoride 4 g /L blood	Sample with EDTA/ acidic citrate (0,5 mol/L). Store blood at 0–4 °C. Haemolysed EDTA sample in detergent stable for 2 d. Serum > plasma
HTLV I – antibodies (T-cell leukemia)	+													

– (provirus) DNA amplification						⊕						
– RNA amplification			+									
Human chorion -gonadotropin (hCG)												
– free	+						0.5–1.5 d	24 h (2–8 °C)	4 w	2 d	8 h	
– total	+	+	+ β, γ	(+) α↗, γ			1–3 d	2 d	1 y	7 d	2 d	
3-Hydroxybutyrate					⊕			4 h	1 m	2 d	2 d	Deproteinisation of whole blood EDTA and citrate↘
IgA	+	+	+				6 d	8 d 1 m (2–6 °C)	8 m	8 m	8 m	
IgD	⊕		– ↘				5 d		6 m	7 d	7 d	
IgE	⊕	+	+	(+) γ			2.5 d	7 d	6 m	7 d	7 d	
antigen specific IgE	+											
IgG	+	+	+	–			3 w	11 d 1 m (2–6 °C)	8 m	8 m	4 m	
IgG subclasses	+	+										
IgM	+	+	+ γ, γγ, δ, ε				5 d	17 d 1 m (2–6 °C)	6 m	4 m	2 m	
Immunoglobulin (free) light chains (κ, λ)												See free light chains (κ, λ) of immunoglobulins
Influenza virus ABC antibodies	+											
Insulin	(+) ↘	+	+				5 min–6 h	15 min	6 m	6 d	1 d	
Iron (Fe)	+	+	–↘	– ↘			3 h	2 h↗	y	3 w	7 d	
Islet cell antibodies (1A-2A)	+		(+)*									*See also glutamate decarboxylase autoantibodies (GADA)

	Samples								Stability					
Analytes	**Serum**	**Heparinate Plasma**	**EDTA Plasma**	**Citrated Plasma**	**Whole blood**			**Biological half-life**	**Stability in blood at room temperature**	**Stability in serum/ plasma**			**Stabilizer**	**Remarks/Comments**
					Hep	**EDTA**	**Citrate**			**−20 °C**	**4–8 °C**	**20–25 °C**		
JC polyoma virus														
– antibodies (progressive multifocal leukoencephalopathy, PML)	+													
– DNA-amplification (PML)						⊕								
Lactate	– ↗	– ↗	– ↗	–	(+)			min	< 5 min, unstable↗↗	1 m*	3d 2 w*	8 h 6 d*	Mannose/ fluoride, monoiodo-acetate, deproteinisation	Use glycolysis inhibitor tube, if not immediately deproteinised *Deproteinised in whole blood
Lactate dehydrogenase (LDH)	(+) ↗	⊕	(+)	(+)				10–54 h LDH 5 < LDH 1.2	1 h ↗	6 w	4 d	7 d		LDH in serum dependent on platelet number
Lead (Pb)	–	–		–	(+)	⊕	(+)					7 d		Special tube
Legionella antibodies	+													
Leishmania spp. antibodies (visceral leishmaniosis)	+													
Leptin	+	+	+							2 y	2 m	3–6 d		Five freeze/thaw cycles possible

Leptospira spp. antibodies (Leptospirosis)	+												
Leucocyte count					+	⊕	+	6–7 h	7 d		1 d*		See also differential count, *EDTA-blood
Lidocaine	+	+ β	+ β					1–3 h			6 h		Separator gel↘
Lipase	+	+ δ, γγ ↘ α	– ↘	–				7–14 h		1 y	3 w	7 d	EDTA binds calcium (activator), 15 % less activated in heparin (α)
Lipoprotein(a)	+	+ γ, ε	+γ	–γ					1 d (4–8 °C)	3 m	2 w	2 d	Store serum in 60 % sacharose
Lipoprotein electrophoresis	⊕	–	–	–							2–5 d		Store at –20 °C with 15 % sucrose
Listeria monocytogenes antibodies	+												
– DNA amplification						⊕							
Lithium	+	+*, α	–, + α	–				8–24 h	1 h ↘	6 m	7 d	1 d	*Do not use Li-heparin
Lupus anticoagulant	–	–	–	⊕						6 m		4 h	Centrifuge platelet free
Lutropin (LH)	+	+	+ α, β, μ						7 d	1 y	5 d	3 d	
Lymphocytic choriomeningitis virus (LCM)													
– antibodies	+												
– RNA amplification						⊕							
Lymphocyte subtypes					+	(+)			1 d (7 d)*				*Special stabiliser recommended (Cyto-Chex)
α_2-Macroglobuline	+	+γ, ε							1 d		1 d	1 d	

Analytes	Samples: Serum	Heparinate Plasma	EDTA Plasma	Citrated Plasma	Whole blood Hep	Whole blood EDTA	Whole blood Citrate	Stability: Biological half-life	Stability in blood at room temperature	Stability in serum/plasma −20 °C	4–8 °C	20–25 °C	Stabilizer	Remarks/Comments
Magnesium (Mg)	+↗	+***	–	–↘	⊕				1 d↗**	1 y	7 d	7 d	*Mg-balanced heparin (15–50 kIU/L)	**Separate blood cells before analysis, do not use siliconised tubes ***higher results obtained in Terumo gel-tubes
– ionized	–	⊕*	–	–	⊕*				1 h	3 m	1 m	4 h		
Malaria														Microscopic examination of whole blood
– plasmodium antibodies	+													
– plasmodium spp.						⊕								
– trypanosoma gambiense						(+)								Blood film of capillary blood
Measles virus														
– antibodies	+													
– RNA amplification						⊕								
Mercury (Hg)					+	⊕								Special tube
Methadone	+	+												
Methotrexate	+							2–4 h		6 m	3 d			Light ↘
Microfilarias					+	+								Concentrated sample
β_2-Microglobulin	+	+ γ, δ, ε	+, δ, ε– γ (+)					40 min	1 d	6 m	1 w	3d		
Morbilli virus														
– antibodies	+	+												
– DNA amplification						⊕								

Morphine, total*	+	+					21 d 6 m (4 °C)	6 m	6 m	3 m		Light ↘ *After hydrolysis
Mumps virus antibodies	+	+ σ	+ σ	+ σ								
Mycobacterium spp. DNA amplification					⊕							
Mycoplasma pneumoniae antibodies	+											
Myeloperoxidase (MPO)	+ ↗	+ ↗	⊕						7 d	8 h		
Myoglobin	+	+	+	(+), − γ		15 min	1 h ↘	3 m	1 w	2 d		
Neisseria gonorrhoeae antibodies	+											
Netilmycin	+					2–3 h						
Neuron specific enolase (NSE)	+ ↗	⊕	+			1 d	2 h ↗	3 m 9 m (−80 °C)	3 d	2 d	Heparin	Increased in thrombocytosis, Serum > plasma
Nitrazepam	+	+ β	+ β	(+) β			1 d*	1 w	1 w			*Light ↘
Opiates	+	+					8 h	6 m	2 d	8 h		See also morphine
Osmolality	+	+						3 m	1 d	3 h		
Osteocalcin	+*	+*	⊕*			min	15 min	8 w (−30 °C) 1 y**	2 d* 4 d**	8 h 2 d**	*Aprotinin 2500 KIU/ mL + EDTA (5 mmol/L)	Three freezing/ thawing cycles are possible. **N - MID-osteocalcin in EDTA plasma
Pancreatic elastase	+		+	+				6 m	2 w			
Pancreatic polypeptide	+	+	+						6 d	2 d		
Paracetamol	+	+	+	(+) β		1–4 h	8 h	45 d	2 w	8 h		

	Samples								Stability					
Analytes	Serum	Heparinate Plasma	EDTA Plasma	Citrated Plasma	Whole blood Hep	Whole blood EDTA	Whole blood Citrate	Biological half-life	Stability in blood at room temperature	Stability in serum/plasma −20 °C	Stability in serum/plasma 4–8 °C	Stability in serum/plasma 20–25 °C	Stabilizer	Remarks/Comments
Parathyrin (PTH)	+ κ↘	+ γ, κ	⊕	(+), − γ				3–4 min	6 h (2–3 d in EDTA blood)	4 m	1 d	6 h	EDTA	15 % lower concentrations in serum compared to EDTA plasma
Partial thromboplastin time (aPTT)	–	–	–	⊕					1 d	1 m	2–8 h	2–8 h		Stability reduced in plasma of heparinised patients
Parvovirus B 19														
– antibodies (erythema infectiosum)	+													
– DNA amplification						⊕								
Phencyclidine	+													
Phenobarbital	+	+	+	(+) β, γ, δ				2–6 d	2 d	1 y	10 d	1 d 2 d*		*In Vacutainer SST II tubes
Phenytoin	+	+	+β, γ, δ	(+) β				1–8 d	2 d	5 m	1 m	2 d		Unstable in serum separator tubes, but stabile in SST II tubes and Sarstedt heparin-plasma tubes with gel separator. Biological half-life shorter in children
free	+	+γγ	−α, γγ, δ	+α, −γ										
Phosphate, inorganic	(+) ↗	⊕	−α, γγ, δ+ μ	(+) μ, −α				min	1–16 h ↗↗	1 y	7 d	3 d		Platelet-dependent in serum

Polio virus 1, 2, 3 anti-bodies	+											Neutralisation test
Potassium (K)	(+) ↗	⊕	–	–	+	min	1–16 h ↗↗	1 y	6 w	6 w		Platelet-dependent in serum > plasma, haemolysis↗
Prealbumin												see Transthyretine
Primidone	+	+	+	(+)		4–19 h	1 y	5 m	4 w			
Procainamide and N-acetyl-procainamide (NAPA)	+	+β, γ, γγ	+β, γ	(+) β		3–5 h 6–10 h		6 m	2 w			
Procalcitonin	+	+δ	+	(+)		20–26 h	1–2 d		4 d	4 h		
Procollagen type I (and its N-terminal propeptide (PINP))	+	+	+					1 y	2 d	1 d		
Pro-gastrin releasing peptide (proGRP)							7 d			7 d		see Gastrin releasing peptide
Progesterone	+	+ β, –α, μ↘	+ β, μ, – α				7 d	1 y	4d 6 d*	1 d		*in vacutainer SST II tubes
Proinsulin	+↘	+	⊕			15 min	2 d*	6 m	1 h	7 min	EDTA	*in EDTA
Prolactin	+	+ β, δ, μ	+ β, μ	–			2 d	1 y	6 d	5 d		
Propaphenone	+	+										
Propoxyphene	+	+										
Prostata specific antigen (PSA)												Three freezing thawing cycles possible
– free	+	+ γ	+ γ			2 h	2 h–7 d	1 m↗	1 d	6 h		
– total	+	+ γ, μ, – α	+ γ, μ, –κ	(+) γ		2–3 d	4–7 d	3 m↘ –2 y	30 d	7 d		

Analytes	Samples: Serum	Heparinate Plasma	EDTA Plasma	Citrated Plasma	Whole blood: Hep	Whole blood: EDTA	Whole blood: Citrate	Biological half-life	Stability: Stability in blood at room temperature	Stability in serum/plasma: −20 °C	Stability in serum/plasma: 4–8 °C	Stability in serum/plasma: 20–25 °C	Stabilizer	Remarks/Comments
Protein, total	+ ↘	⊕	+ γ, γγ, δ	(+)				Complex	1 d	1 y	4 w	6 d		Plasma results higher due to fibrinogen (Biuret method)
Protein C	–	–	–	⊕				6–8 h	1 d	3 m	7 d	7 d		Avoid freezing/ thawing cycles
Protein S	–	–	–	⊕				24–58 h	4 h	1 m	4 h	8 h		Separate cell-free plasma directly after centrifugation
Protein S100	+							2–5 h			7 d	7 d		
Prothrombin time (thromboplastin time, Quick)	–	–	–	⊕					4 h–1 w*	1 m	8 h–1 d*	4 h–1 d*		*Reagent-dependent
Pyruvate	– ↘	– ↘	–	–	+*				< 1 min*					*Only stable in deproteinized blood
Quinidin	+	+β, γγ	+ β, γ	(+) β				6–9 h		1 m	2 d	8 h		
Renin	–	–	+	–					unstable	1 y		4 h		Do not store in refrigerator (cryoactivation of prorenin possible)
Reovirus antibodies	+													
Respiratory syncytial virus (RSV) antibodies	+													
Reticulocyte count					(+)			12 h	3 d*		3 d*			*EDTA blood
maturity index						⊕			1 d*		1 d*			

Retinol binding protein (RBP)	+	+						10 h		3 m	1 w	4 h		
Rheumatoid factors and sub -fractions	+	+ γ, δ	+ γ, δ	(+) –γ					6 h	3 m	8 d	1 d		
IgA, IgG	+													
Rickettsia antibodies	+													
RNA analysis by amplification (PCR)	(+)	–*	+		–*	⊕	+		2 h, 12 h (4 °C), 4 d (EDTA) 1 m**	1 y	1 d	<1 h	RNA: 5 mmol/L Guani-dinium-isothiocy-anate **PAXge-ne™	*Heparin inhibits Taq polymerase and restriction enzymes, LiCl 1.8 mol/L eliminates this error.
Rotavirus antibodies	+													
Rubella virus														
– antibodies	+	+ β, σ	+ β, σ	(+) β, σ										
– RNA amplification						⊕								
S 100 protein														see Protein S 100
Salicylate	+	+	+	(+)				24*–30 min		6 m	2 w	7 d		*Higher at toxic concentrations
Sandfly (pappataci-) fever antibodies	+													
Selenium (Se)	–	–	–	–		+*			2 d	1 y	2 w	1 w		*Special tubes, contamination
Sirolimus						⊕		59 +/– 19 h	1d* (4–8 °C)	30 d*	7 d*	8 h*	*EDTA blood	LC-MS/MS
Sodium (Na)	+	+	–	–	+*			min	4 d↘	1 y	2 w	2 w		*Use 140 mmol/L Na-stabilized heparin 8–12 IU/mL blood

Analytes	Samples								Stability					
	Serum	Heparinate Plasma	EDTA Plasma	Citrated Plasma	Whole blood Hep	Whole blood EDTA	Whole blood Citrate	Biological half-life	Stability in blood at room temperature	Stability in serum/plasma −20 °C	Stability in serum/plasma 4–8 °C	Stability in serum/plasma 20–25 °C	Stabilizer	Remarks/Comments
Soluble transferrin receptor (sTfR)	+	+γ, γγ, ε	−ε						2–6 h	3 m	7 d	3 d		Freeze only once
Somatotropin (STH) (growth hormone)	+	+	⊕					20–50 min	1 d	3 m	8 d	3 d	EDTA	
Squamous cell carcinoma antigen (SCCA)	+	+						1.5–3 h	7 d	1 m	1 m	7 d	Closed tubes	Increase by contamination (skin)
Staphylococcal antibodies														
– anti-staphylolysin O	+	+ γ	+ γ											
Streptococcal antibodies									1 h	6 m	8 d	2 d		
– anti-DNAse B	+													
– antihyaluronidase	+	+, β, γ, δ	+, β, γ, δ											
– antistreptokinase	+													
– antistreptolysin O	+	+ β, γ, δ	+ β, γ, δ											
Tacrolismus	–	–	–	–	–	⊕		6–21 h	7 d	1 y*	2 w*	7d*		*Haemolysate LC-MS/MS
Tartrate resistant acid phosphatase (TRACP 5 b)	+	+	+						2 h	2 m		4 h		
Testosterone	+	+	+	(+) γ					7 d 1 d in women↗	1 y	7 d	1 d		

Tetrahydrocannabinol carbonic acid (THC)	+	+						~ 45 h		6 m	6 m	2m	Na azide	Unstable in plastic tubes
Theophylline	+	+	+	(+) α, β				3–12 h		3 m	7 d	1 d 2d*		*in vacutainer SST II tubes
Thrombin time	–	–	–	⊕					1–4 h↗	1 m	1 h–2 d*	1–4 h*		*Stability reagent- and heparin-dependent
Thrombocyte antibodies	+		+	+										
Thrombocyte count					(+)↘	⊕	(+)	9–10 d	4 d*, 7 d (4–8 °C)*				*in EDTA-blood	Aminoglycosides avoid pseudo-thrombocyto-penia in EDTA
Thrombocyte volume						⊕								
Thrombocyte function – using platelet function – analyzer (PFA) (ε),	–	–	–	–			⊕	9–10 d	4 h			1 h		
– using flow cytometry	–	–	–	–			⊕		2 h (7d)*					*Special stabiliser recommended
Thyreoglobulin	+							1 d	2 d	1 m	3 d–3 w	1 d		three freezing –thawing cycles possible
Thyreotropine (TSH)	+	+ β, γ, μ, – α	+ α, β, γ, – μ	(+)γ				min	7 d	3 m	7 d	1 d		Spot blood on filter paper in newborns
Thyreotropine receptor antibodies (TRAb)	+													
Thyroid antibodies	+	+									2 d			
Thyroid peroxidase antibodies (anti-TPO)														
Thyroglobulin antibodies (anti-TgAb)														

Analytes	Samples: Serum	Heparinate Plasma	EDTA Plasma	Citrated Plasma	Whole blood: Hep	Whole blood: EDTA	Whole blood: Citrate	Stability: Biological half-life	Stability in blood at room temperature	Stability in serum/plasma: −20 °C	4–8 °C	20–25 °C	Stabilizer	Remarks/Comments
Thyroxine (T_4)	⊕	+ β, γ, γγ, − α, μ	+ α, β, γ, γγ, − α, μ	(+) γ				6 m	7 d	1 m	7 d	5 d		
Thyroxine, free (fT_4)	+	+	+	(+) γ					6 h	3 m	8 d	2 d		
Thyroxine binding globulin (TBG)	+	+							7 d	1 m	5 d	5 d		
Tick borne encephalitis virus antibodies	+		(+)											
Tobramycin	+	+β, γ, δ	+ γ, δ	(+)β, γ				0.5–3 h (<30 y of age) 1.5–15 h (>30 y of age)		1 m	3 d	< 2 h 2 d *		*in Vacutainer SST II tubes Lower results obtained in heparinised plasma
Toxoplasma gondii antibodies (IgA, IgG, IgM)	+	+ β, σ	+ β, σ	+β, σ							8 d	8 d		
Transferrin	+	+	+					7–10 d	11 d 3 w (2–6 °C)	6 m–2 y	8 m	4 m		
Transthyretine (prealbumin)	+	+, − γ	+ γ					≅ 2 d		1 y	6 m	3 d		
Treponema pallidum														
– antibodies	+	+ σ	+σ	+ σ										
– DNA amplification						⊕								
Tricyclic antidepressants	+	+ β	+β	(+) β					1 w	1 y				see also Amitryptiline

Triglycerides	+	+	+, −α	(+)		3 h–3 d	7 d ↗*	y	7 d	2 d		*Decrease of triglycerides, increase of free glycerol, but only minor increase of total glycerol
Triiodothyronine (T_3)	⊕	(+) ↗ β, γ, δ, μ	+ μ	+ (+)		19 h		3 m	8 d	2 d		Serum-plasma difference method dependent
– free (fT_3)	+	+	+	(+) –				3 m	2 w	2 d		
Troponin I	+	+ β, δ, − α, μ↘	+ δ, − α, μ ↘		+	2–4 h		4 w	3 d	2 d		Half-life time increases in renal insufficiency
Troponin T	+	+ γ	+ γ			2–4 h	8 h	3 m	7 d	1 d		
Tumor necrosis factor α (TNFα)	–		⊕				1 h	7 d	7 d		EDTA	
Urea	+	+	+			min	1 d↗	1 y	7 d	7 d		Do not use NH_4-heparin
Uric acid	+	+	+ ↘	(+)		min	3–7d↗	6 m	7 d	3 d		
Valproic acid	+	+	+	(+) β		8–15 h	2 d	3 m	7 d	2d		
Vancomycin	+	+	+	(+) β		4–10 h		7 d	1 d 7d*	2 d		• In vacutainer SST-II tubes
Varicella Zoster virus												
– antibodies	+	+ σ	+ σ	+ σ								
– DNA amplification					⊕							
Vasoactive intestinale – polypeptide (VIP)	↘	↘	⊕					> 6 d	6 d	1 d	EDTA + aprotinin	
Vasopressin (ADH)	↘	+	⊕						6 d	1 d	EDTA	Freeze plasma
Vitamin B_1 (thiamine)	+	+	+		⊕		5 h*	1 y	1 d *	5 h*		light sensitive, *in EDTA blood
Vitamin B_2 (riboflavin)		+	+		⊕		1 h	1 m	1 d*	5 h*		light sensitive, *in EDTA blood

Analytes	Samples								Stability					
	Serum	Heparinate Plasma	EDTA Plasma	Citrated Plasma	Whole blood			Biological half-life	Stability in blood at room temperature	Stability in serum/ plasma			Stabilizer	Remarks/Comments
					Hep	EDTA	Citrate			−20 °C	4–8 °C	20–25 °C		
Vitamin B_6 (pyridoxal phosphate)	(+)		⊕			⊕			1 d*	30 d*	3 d*	1 d*	EDTA, darkness	light sensitive, *in EDTA blood, Plasma, Serum
Vitamin B_{12} (cobalamin)	+	+	⊕						6 h	8 w	1 d	15 min	EDTA, darkness	
Vitamin C (ascorbic acid)	+	+	+						3 h (4 °C)	3 w*	3 h		60 g/L meta-phosphate, deprotei-nised	*Only with stabilizer
Vitamin D														*Calcidiol light sensitive
1.25-dihydroxy-vitamin D (calcitriol),	+	+	+						3 d	1 y	7 d	3 d		
25-hydroxy-vitamin D (calcidiol)	+	+	+						3 d	1y	7 d	3 d		
Vitamin E (tocopherol)	+		⊕						8 h↘	1 y	1 m	8 h	EDTA	
Vitamin K (transphyllochinone)			+						unstable	3 m	unsta-ble			UV light↘
von Willebrand factor				⊕					1 w					
Yersinia enterocolitica antibodies	+													
Zinc (Zn)	–	+	–	–					30 min↗	1 y	2 w	1 w		Special tube, avoid contamination by stopper

Index

Printed and bound by CPI Group (UK) Ltd, Croydon, CR0 4YY
13/09/2026
14960858-0003